Interactive Student Edition

Reveal MATH™

Course 1 • Volume 1

McGraw Hill Education

Cover: (l to r, t to b) guvendemir/E+/Getty Images, Andrey Prokhorov/E+/Getty Images, anatols/iStock/Getty Images, iava777/iStock/Getty Images

my.mheducation.com

Send all inquiries to:
McGraw-Hill Education
STEM Learning Solutions Center
8787 Orion Place
Columbus, OH 43240

ISBN: 978-0-07-667372-8
MHID: 0-07-667372-3

Reveal Math, Course 1
Interactive Student Edition, Volume 1

Printed in the United States of America.

1 2 3 4 5 6 7 8 9 10 LMN 27 26 25 24 23 22 21 20 19 18

Contents in Brief

Reveal Math™ Makes Math Meaningful...

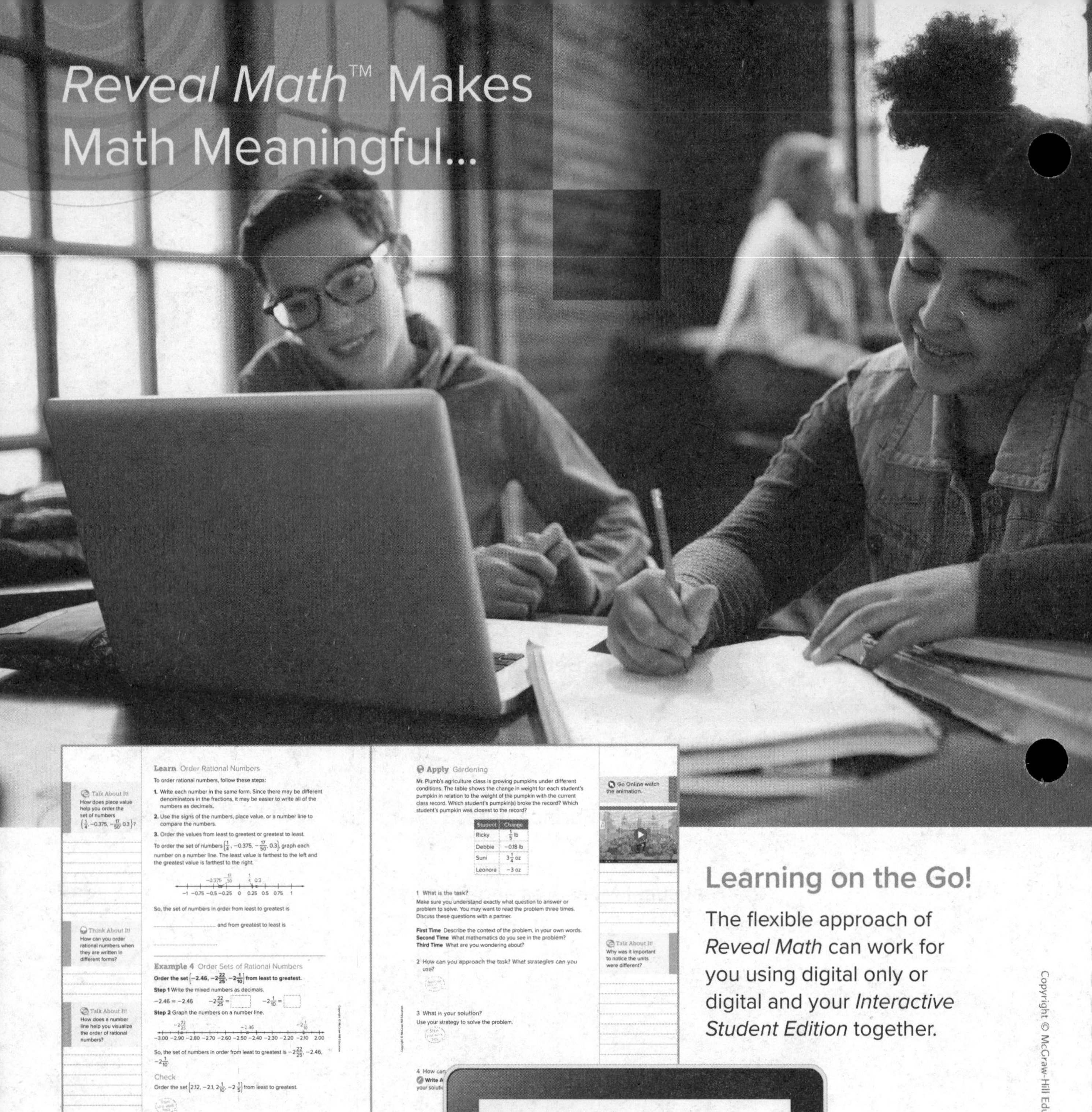

Interactive Student Edition

Learning on the Go!

The flexible approach of *Reveal Math* can work for you using digital only or digital and your *Interactive Student Edition* together.

Student Digital Center

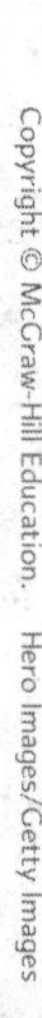

...to Reveal YOUR Full Potential!

Reveal Math™ Brings Math to Life in Every Lesson

Reveal Math is a blended print and digital program that supports access on the go. You'll find the *Interactive Student Edition* mirrors the Student Digital Center, so you can record your digital observations in class and reference your notes later, or access just the digital center, or a combination of both! The Student Digital Center provides access to the interactive lessons, interactive content, animations, videos and technology-enhanced practice questions.

Write down your username and password here

Username: ______________________

Password: ______________________

Go Online! my.mheducation.com

Web Sketchpad® Powered by The Geometer's Sketchpad®- Dynamic, exploratory, visual activities embedded at point of use within the lesson.

Animations and Videos – Learn by seeing mathematics in action.

Interactive Tools – Get involved in the content by dragging and dropping, selecting, and completing tables.

Personal Tutors – See and hear a teacher explain how to solve problems.

eTools – Math tools are available to help you solve problems and develop concepts.

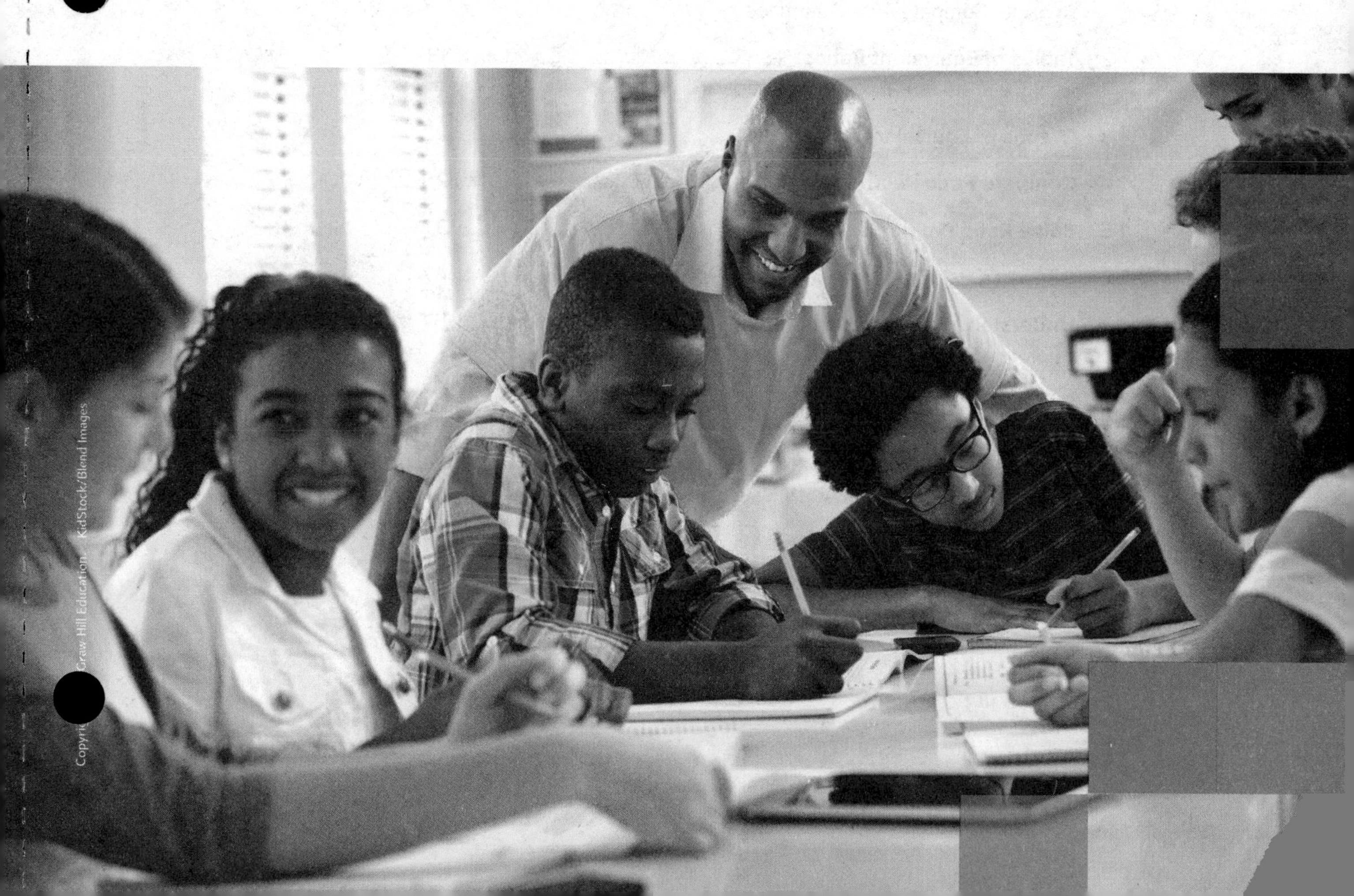

TABLE OF CONTENTS

Module 1

Ratios and Rates

Essential Question

How can you describe how two quantities are related?

Module 2

Fractions, Decimals, and Percents

Essential Question

How can you use fractions, decimals, and percents to solve everyday problems?

TABLE OF CONTENTS

Module 3

Compute with Multi-Digit Numbers and Fractions

Essential Question

How are operations with integers related to operations with whole numbers?

Module 4

Integers, Rational Numbers, and the Coordinate Plane

Essential Question

How are integers and rational numbers related to the coordinate plane?

Module 5

Numerical and Algebraic Expressions

Essential Question

How can we communicate algebraic relationships with mathematical symbols?

Module 6

Equations and Inequalities

Essential Question

How are the solutions of equations and inequalities different?

TABLE OF CONTENTS

Module 7

Relationships Between Two Variables

Essential Question

What are the ways in which a relationship between two variables can be displayed?

Module 8

Area

Essential Question

How are the areas of triangles and rectangles used to find the areas of other polygons?

TABLE OF CONTENTS

Module 9

Volume and Surface Area

Essential Question

How can you describe the size of a three-dimensional figure?

Module 10

Statistical Measures and Displays

Essential Question

Why is data collected and analyzed and how can it be displayed?

Common Core State Standards for Mathematics, Grade 6

Reveal Math, Course 1, focuses on four critical areas: (1) using concepts of ratio and rate to solve problems; (2) understanding division of fractions; (3) using expressions and equations; and (4) understanding of statistical reasoning.

MP Mathematical Practices

1. Make sense of problems and persevere in solving them.
2. Reason abstractly and quantitatively.
3. Construct viable arguments and critique the reasoning of others.
4. Model with mathematics.
5. Use appropriate tools strategically.
6. Attend to precision.
7. Look for and make use of structure.
8. Look for and express regularity in repeated reasoning.

Content Standards

Ratios and Proportional Relationships (Domain 6.RP)

- Understand ratio concepts and use ratio reasoning to solve problems.

The Number System (Domain 6.NS)

- Apply and extend previous understandings of multiplication and division to divide fractions by fractions.
- Multiply and divide multi-digit numbers and find common factors and multiples.
- Apply and extend previous understandings of numbers to the system of rational numbers.

Expressions and Equations (Domain 6.EE)

- Apply and extend previous understandings of arithmetic to algebraic expressions.
- Reason about and solve one-variable equations and inequalities.
- Represent and analyze quantitative relationships between dependent and independent variables.

Geometry (Domain 6.G)

- Solve real-world and mathematical problems involving area, surface area, and volume.

Statistics and Probability (Domain 6.SP)

- Develop understanding of statistical variability.
- Summarize and describe distributions.

Ratios and Proportional Relationships (Domain 6.RP)

Understand ratio concepts and use ratio reasoning to solve problems.

6.RP.A.1 Understand the concept of a ratio and use ratio language to describe a ratio relationship between two quantities. *For example, "The ratio of wings to beaks in the bird house at the zoo was 2:1, because for every 2 wings there was 1 beak." "For every vote candidate A received, candidate C received nearly three votes."*

6.RP.A.2 Understand the concept of a unit rate a/b associated with a ratio $a:b$ with $b \neq 0$, and use rate language in the context of a ratio relationship. *For example, "This recipe has a ratio of 3 cups of flour to 4 cups of sugar, so there is 3/4 cup of flour for each cup of sugar." "We paid $75 for 15 hamburgers, which is a rate of $5 per hamburger."*

6.RP.A.3 Use ratio and rate reasoning to solve real-world and mathematical problems, e.g., by reasoning about tables of equivalent ratios, tape diagrams, double number line diagrams, or equations.

A. Make tables of equivalent ratios relating quantities with whole-number measurements, find missing values in the tables, and plot the pairs of values on the coordinate plane. Use tables to compare ratios.

B. Solve unit rate problems including those involving unit pricing and constant speed. *For example, if it took 7 hours to mow 4 lawns, then at that rate, how many lawns could be mowed in 35 hours? At what rate were lawns being mowed?*

C. Find a percent of a quantity as a rate per 100 (e.g., 30% of a quantity means 30/100 times the quantity); solve problems involving finding the whole, given a part and the percent.

D. Use ratio reasoning to convert measurement units; manipulate and transform units appropriately when multiplying or dividing quantities.

The Number System (Domain 6.NS)

Apply and extend previous understandings of multiplication and division to divide fractions by fractions.

6.NS.A.1 Interpret and compute quotients of fractions, and solve word problems involving division of fractions by fractions, e.g., by using visual fraction models and equations to represent the problem. *For example, create a story context for $(2/3) \div (3/4)$ and use a visual fraction model to show the quotient; use the relationship between multiplication and division to explain that $(2/3) \div (3/4) = 8/9$ because 3/4 of 8/9 is 2/3. (In general, $(a/b) \div (c/d) = ad/bc$.) How much chocolate will each person get if 3 people share 1/2 lb of chocolate equally? How many 3/4-cup servings are in 2/3 of a cup of yogurt? How wide is a rectangular strip of land with length 3/4 mi and area 1/2 square mi?*

Compute fluently with multi-digit numbers and find common factors and multiples.

6.NS.B.2 Fluently divide multi-digit numbers using the standard algorithm.

6.NS.B.3 Fluently add, subtract, multiply, and divide multi-digit decimals using the standard algorithm for each operation.

Common Core State Standards for Mathematics, Grade 6

6.NS.B.4 Find the greatest common factor of two whole numbers less than or equal to 100 and the least common multiple of two whole numbers less than or equal to 12. Use the distributive property to express a sum of two whole numbers 1–100 with a common factor as a multiple of a sum of two whole numbers with no common factor. *For example, express 36 + 8 as 4(9 + 2).*

Apply and extend previous understandings of numbers to the system of rational numbers.

6.NS.C.5 Understand that positive and negative numbers are used together to describe quantities having opposite directions or values (e.g., temperature above/below zero, elevation above/below sea level, credits/debits, positive/negative electric charge); use positive and negative numbers to represent quantities in real-world contexts, explaining the meaning of 0 in each situation.

6.NS.C.6 Understand a rational number as a point on the number line. Extend number line diagrams and coordinate axes familiar from previous grades to represent points on the line and in the plane with negative number coordinates.

A. Recognize opposite signs of numbers as indicating locations on opposite sides of 0 on the number line; recognize that the opposite of the opposite of a number is the number itself, e.g., –(–3) = 3, and that 0 is its own opposite.

B. Understand signs of numbers in ordered pairs as indicating locations in quadrants of the coordinate plane; recognize that when two ordered pairs differ only by signs, the locations of the points are related by reflections across one or both axes.

C. Find and position integers and other rational numbers on a horizontal or vertical number line diagram; find and position pairs of integers and other rational numbers on a coordinate plane.

6.NS.C.7 Understand ordering and absolute value of rational numbers.

A. Interpret statements of inequality as statements about the relative position of two numbers on a number line diagram. *For example, interpret –3 > –7 as a statement that –3 is located to the right of –7 on a number line oriented from left to right.*

B. Write, interpret, and explain statements of order for rational numbers in real-world contexts. *For example, write –3°C > –7°C to express the fact that –3°C is warmer than –7°C.*

C. Understand the absolute value of a rational number as its distance from 0 on the number line; interpret absolute value as magnitude for a positive or negative quantity in a real-world situation. *For example, for an account balance of –30 dollars, write |–30| = 30 to describe the size of the debt in dollars.*

D. Distinguish comparisons of absolute value from statements about order. *For example, recognize that an account balance less than –30 dollars represents a debt greater than 30 dollars.*

6.NS.C.8 Solve real-world and mathematical problems by graphing points in all four quadrants of the coordinate plane. Include use of coordinates and absolute value to find distances between points with the same first coordinate or the same second coordinate.

Expressions and Equations (Domain 6.EE)

Apply and extend previous understandings of arithmetic to algebraic expressions.

6.EE.A.1 Write and evaluate numerical expressions involving whole-number exponents.

6.EE.A.2 Write, read, and evaluate expressions in which letters stand for numbers.

A. Write expressions that record operations with numbers and with letters standing for numbers. *For example, express the calculation "Subtract y from 5" as* $5 - y$.

B. Identify parts of an expression using mathematical terms (sum, term, product, factor, quotient, coefficient); view one or more parts of an expression as a single entity. *For example, describe the expression* $2(8 + 7)$ *as a product of two factors; view* $(8 + 7)$ *as both a single entity and a sum of two terms.*

C. Evaluate expressions at specific values of their variables. Include expressions that arise from formulas used in real-world problems. Perform arithmetic operations, including those involving whole-number exponents, in the conventional order when there are no parentheses to specify a particular order (Order of Operations). *For example, use the formulas* $V = s^3$ *and* $A = 6s^2$ *to find the volume and surface area of a cube with sides of length* $s = 1/2$.

6.EE.A.3 Apply the properties of operations to generate equivalent expressions. *For example, apply the distributive property to the expression* $3(2 + x)$ *to produce the equivalent expression* $6 + 3x$*; apply the distributive property to the expression* $24x + 18y$ *to produce the equivalent expression* $6(4x + 3y)$*; apply properties of operations to* $y + y + y$ *to produce the equivalent expression* $3y$.

6.EE.A.4 Identify when two expressions are equivalent (i.e., when the two expressions name the same number regardless of which value is substituted into them). *For example, the expressions* $y + y + y$ *and* $3y$ *are equivalent because they name the same number regardless of which number y stands for.*

Reason about and solve one-variable equations and inequalities.

6.EE.B.5 Understand solving an equation or inequality as a process of answering a question: which values from a specified set, if any, make the equation or inequality true? Use substitution to determine whether a given number in a specified set makes an equation or inequality true.

6.EE.B.6 Use variables to represent numbers and write expressions when solving a real-world or mathematical problem; understand that a variable can represent an unknown number, or, depending on the purpose at hand, any number in a specified set.

6.EE.B.7 Solve real-world and mathematical problems by writing and solving equations of the form $x + p = q$ and $px = q$ for cases in which p, q and x are all non-negative rational numbers.

6.EE.B.8 Write an inequality of the form $x > c$ or $x < c$ to represent a constraint or condition in a real-world or mathematical problem. Recognize that inequalities of the form $x > c$ or $x < c$ have infinitely many solutions; represent solutions of such inequalities on number line diagrams.

Represent and analyze quantitative relationships between dependent and independent variables.

6.EE.C.9 Use variables to represent two quantities in a real-world problem that change in relationship to one another; write an equation to express one quantity, thought of as the dependent variable, in terms of the other quantity, thought of as the independent variable. Analyze the relationship between the dependent and independent variables using graphs and tables, and relate these to the equation. *For example, in a problem involving motion at constant speed, list and graph ordered pairs of distances and times, and write the equation $d = 65t$ to represent the relationship between distance and time.*

Geometry (Domain 6.G)

Solve real-world and mathematical problems involving area, surface area, and volume.

6.G.A.1 Find the area of right triangles, other triangles, special quadrilaterals, and polygons by composing into rectangles or decomposing into triangles and other shapes; apply these techniques in the context of solving real-world and mathematical problems.

6.G.A.2 Find the volume of a right rectangular prism with fractional edge lengths by packing it with unit cubes of the appropriate unit fraction edge lengths, and show that the volume is the same as would be found by multiplying the edge lengths of the prism. Apply the formulas $V = lwh$ and $V = bh$ to find volumes of right rectangular prisms with fractional edge lengths in the context of solving real-world and mathematical problems.

6.G.A.3 Draw polygons in the coordinate plane given coordinates for the vertices; use coordinates to find the length of a side joining points with the same first coordinate or the same second coordinate. Apply these techniques in the context of solving real-world and mathematical problems.

6.G.A.4 Represent three-dimensional figures using nets made up of rectangles and triangles, and use the nets to find the surface area of these figures. Apply these techniques in the context of solving real-world and mathematical problems.

Statistics and Probability (Domain 6.SP)

Develop understanding of statistical variability.

6.SP.A.1 Recognize a statistical question as one that anticipates variability in the data related to the question and accounts for it in the answers. *For example, "How old am I?" is not a statistical question, but "How old are the students in my school?" is a statistical question because one anticipates variability in students' ages.*

6.SP.A.2 Understand that a set of data collected to answer a statistical question has a distribution which can be described by its center, spread, and overall shape.

6.SP.A.3 Recognize that a measure of center for a numerical data set summarizes all of its values with a single number, while a measure of variation describes how its values vary with a single number.

Summarize and describe distributions.

6.SP.B.4 Display numerical data in plots on a number line, including dot plots, histograms, and box plots.

6.SP.B.5 Summarize numerical data sets in relation to their context, such as by:

A. Reporting the number of observations.

B. Describing the nature of the attribute under investigation, including how it was measured and its units of measurement.

C. Giving quantitative measures of center (median and/or mean) and variability (interquartile range and/or mean absolute deviation), as well as describing any overall pattern and any striking deviations from the overall pattern with reference to the context in which the data were gathered.

D. Relating the choice of measures of center and variability to the shape of the data distribution and the context in which the data were gathered.

Module 1

Ratios and Rates

Essential Question

How can you describe how two quantities are related?

6.RP.A.1, 6.RP.A.2, 6.RP.A.3, 6.RP.A.3.A, 6.RP.A.3.B, 6.RP.A.3.D
Mathematical Practices: MP1, MP2, MP3, MP4, MP5, MP6, MP7

What Will You Learn?

Place a checkmark (✓) in each row that corresponds with how much you already know about each topic **before** starting this module.

KEY

□ — I don't know. ◇ — I've heard of it. ☆ — I know it!

	Before			After		
	□	◇	☆	□	◇	☆
writing ratios to compare quantities						
finding unit rates						
using equivalent ratios to solve ratio problems						
graphing and describing ratio relationships						
comparing ratio relationships						
using bar diagrams to solve ratio and rate problems						
using equivalent ratios to solve ratio and rate problems						
using double number lines to solve ratio and rate problems						
converting measurements						

Foldables Cut out the Foldable and tape it to the Module Review at the end of the module. You can use the Foldable throughout the module as you learn about ratios and rates.

What Vocabulary Will You Learn?

Check the box next to each vocabulary term that you may already know.

- ☐ double number line
- ☐ equivalent ratios
- ☐ part-to-part ratio
- ☐ part-to-whole ratio
- ☐ rate
- ☐ ratio
- ☐ ratio table
- ☐ scaling
- ☐ unit price
- ☐ unit rate
- ☐ unit ratio

Are You Ready?

Study the Quick Review to see if you are ready to start this module.
Then complete the Quick Check.

Quick Review

Example 1

Divide whole numbers.

Find $6\overline{)348}$.

$$\begin{array}{r} 58 \\ 6\overline{)348} \\ -30 \\ \hline 48 \\ -48 \\ \hline 0 \end{array}$$

Divide each place-value position from left to right.

Since 48 − 48 = 0, there is no remainder.

Example 2

Write fractions to express part of a whole.

Write a fraction to represent the shaded part of the bar diagram.

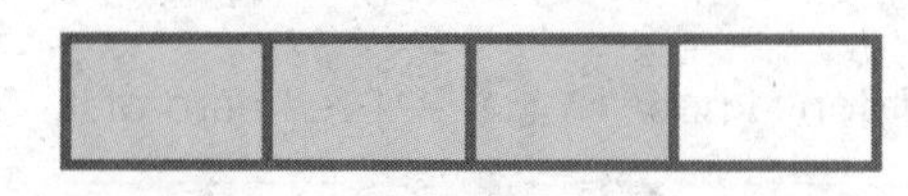

The shaded part of the bar diagram represents the fraction $\frac{3}{4}$.

Quick Check

Find each quotient.

1. $3\overline{)87}$

2. $8\overline{)584}$

3. Write a fraction to represent the shaded part of the bar diagram.

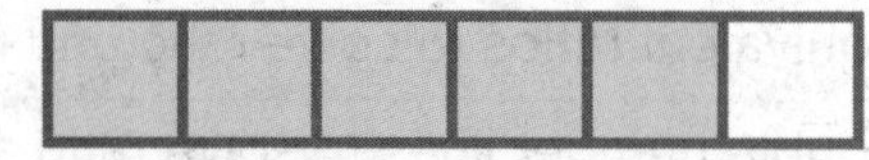

How Did You Do?

Which exercises did you answer correctly in the Quick Check? Shade those exercise numbers at the right.

Understand Ratios

I Can... show a ratio relationship between two quantities using different representations, and describe the relationship using correct mathematical language.

Today's Standards
6.RP.A.1
MP1, MP2, MP3, MP5, MP6

What Vocabulary Will You Learn?
part-to-part ratio
part-to-whole ratio
ratio

Explore Compare Two Quantities

Online Activity You will use Web Sketchpad to determine how many students and teachers should be on various buses to maintain a relationship of one teacher for every eight students.

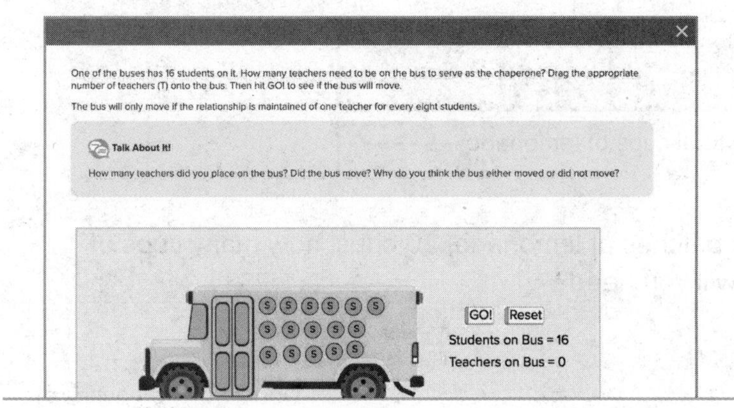

Learn Understand Ratio Relationships

The table shows the ingredients needed to make 10 cups of lemonade. How does the number of cups of lemon juice compare to the total number of cups of lemonade?

Ingredient	Number of Cups
Lemon Juice	2
Simple Syrup	1
Water	7

One way to make a comparison is to use a bar diagram. There are 8 more cups of lemonade than there are cups of lemon juice. This is an *additive comparison* because $2 + 8 = 10$.

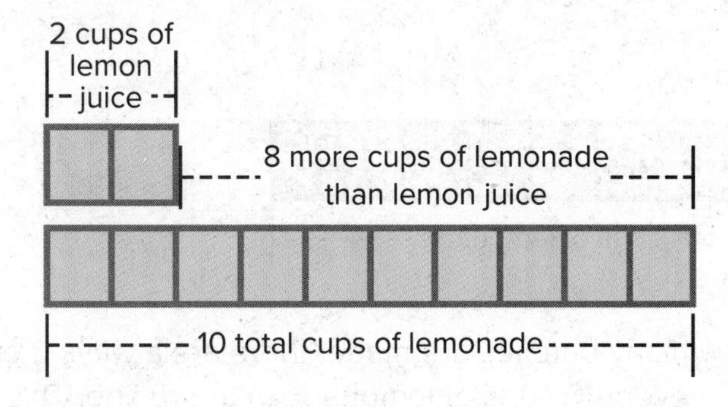

(continued on next page)

Your Notes

Another way to make a comparison is to use a ratio. A **ratio** is a comparison between two quantities, in which for every a units of one quantity, there are b units of another quantity. The phrases *for every* and *for each* are used to define and describe ratios.

The relationships of the quantities of ingredients in recipes are examples of ratio relationships. To make one batch of lemonade, 10 cups, you need 2 cups of lemon juice.

Ingredient	Number of Cups
Lemon Juice	2
Simple Syrup	1
Water	7

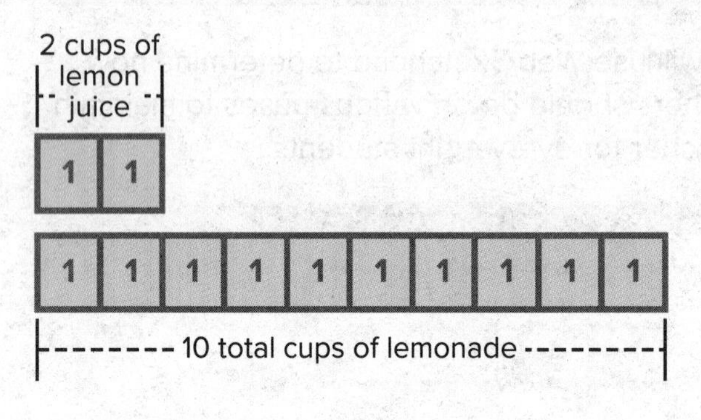

For every 2 cups of lemon juice, there are 10 total cups of lemonade. Each section represents 1 cup.

To make two batches of lemonade, 20 cups, how many cups of lemon juice will you need?

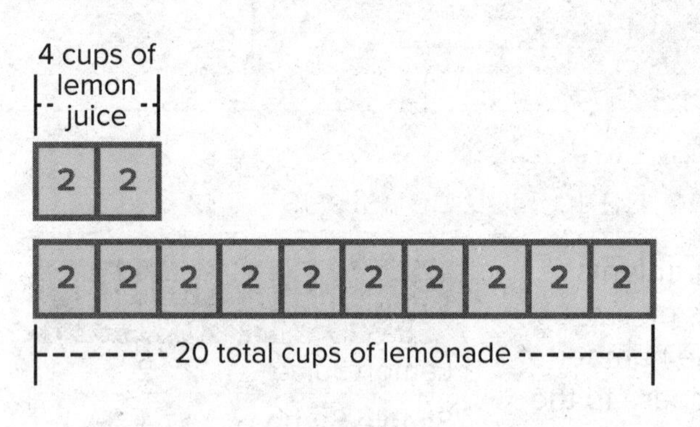

Double the quantities of lemon juice and lemonade to keep the same ratio. Each section represents 2 cups. You need 4 cups of lemon juice.

To make three batches of lemonade, 30 cups, how many cups of lemon juice will you need?

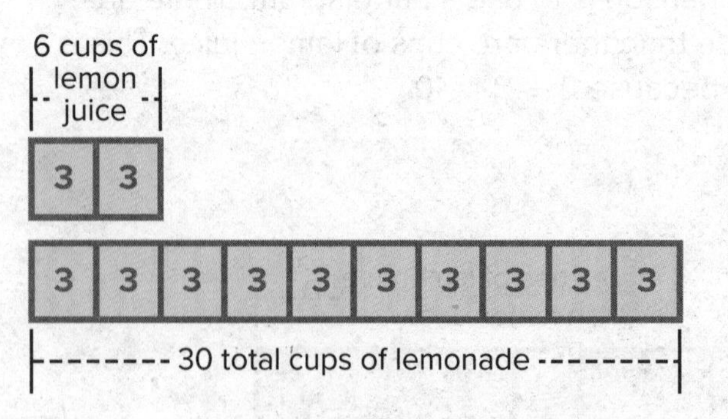

Triple the quantities of lemon juice and lemonade to keep the same ratio. Each section represents 3 cups. You need 6 cups of lemon juice.

No matter how many batches are made, there are always 2 cups of lemon juice for every 10 cups of lemonade in the recipe. This confirms the relationship is a ratio relationship.

Talk About It!

If you did not maintain the same ratio of lemon juice to total cups of lemonade when making 2 or 3 batches, what might happen to your lemonade? Justify your response.

Example 1 Understand Ratio Relationships

Pedro mixed two sample containers of blue paint with three sample containers of yellow paint to create his favorite shade of green paint. Pedro realized he did not have enough paint, so he added two more sample containers of each color.

Will the new mixture result in the same shade of green? Justify your response.

To create his favorite shade of green, Pedro used a ratio of 2 to 3. For every 2 containers of blue paint, there are 3 containers of yellow paint.

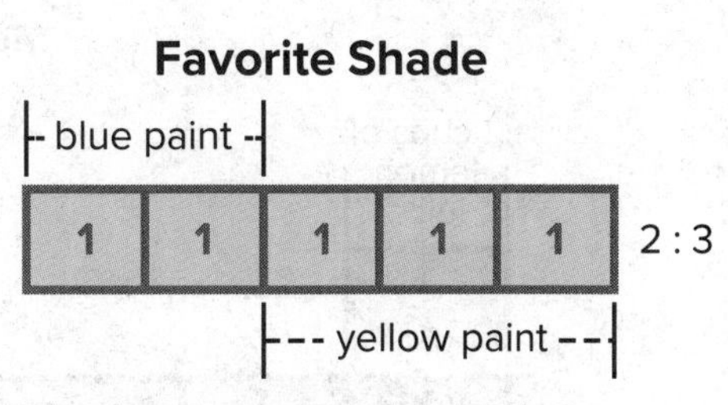

Pedro added two more containers of each color. The ratio of blue paint to yellow paint in the new mixture is 4 to 5.

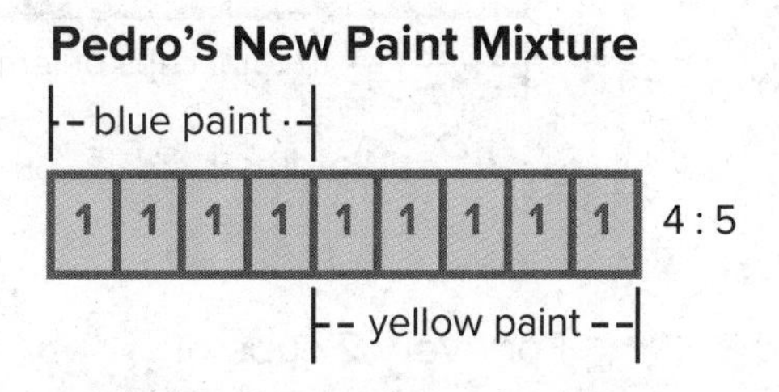

The amount of blue paint in the new mixture is twice that of Pedro's favorite shade. To maintain the same ratio, the amount of yellow paint should also be twice that of his favorite shade. Because $3 \times 2 \neq 5$, the ratio relationship was not maintained. The resulting shade of green will not have enough yellow in it to match Pedro's favorite shade.

If Pedro adds one more container of yellow paint to his new mixture, he will be able to create his favorite shade of green.

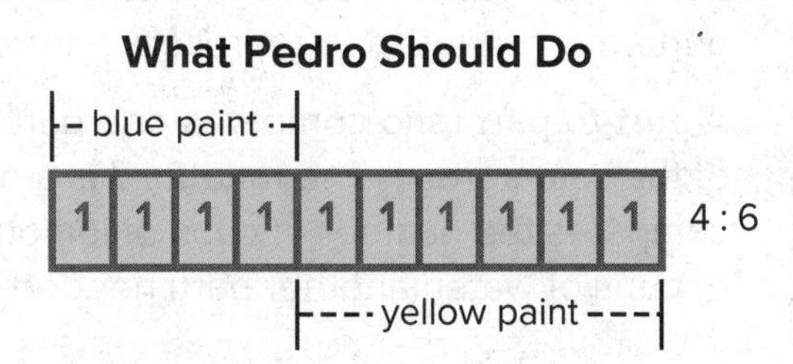

Check

A recipe for rice calls for 6 cups of water and 3 cups of uncooked rice. Trinity only has 2 cups of uncooked rice. She reasons that because she subtracted 1 cup of rice, she needs to use a total of 6 − 1, or 5 cups of water. Is her reasoning correct? Explain.

Think About It!
How will you begin solving the problem?

Talk About It!
What are some other ways that Pedro could make his mixture and still end up with his favorite shade of green?

Go Online You can complete an Extra Example online.

Learn Part-to-Whole and Part-to-Part Ratios

A **part-to-whole ratio** compares one part of a group to the whole group. The ratio 2 : 10 is a part-to-whole ratio because it compares the number of cups of lemon juice (the part) to the total number of cups of lemonade (the whole).

Ingredient	Number of Cups
Lemon Juice	2
Simple Syrup	1
Water	7

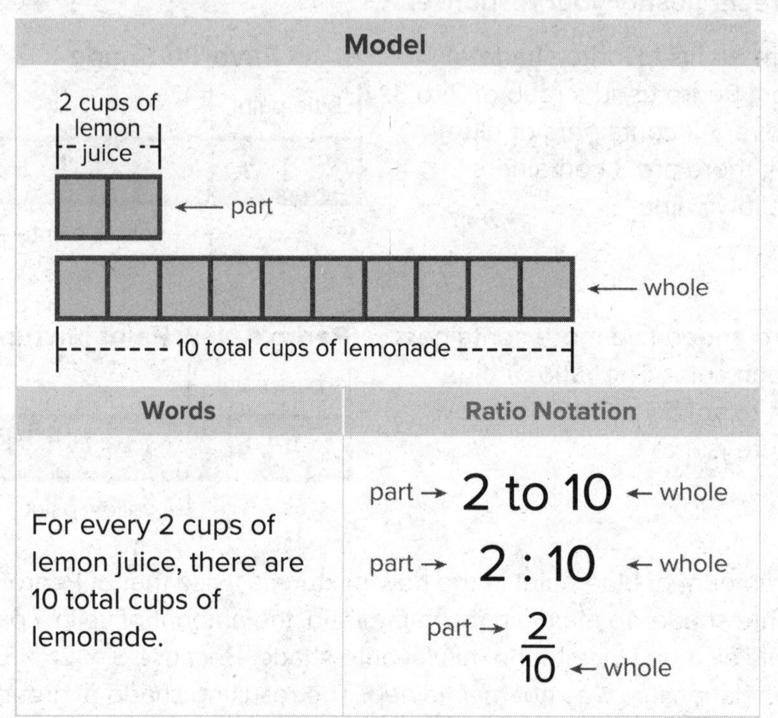

Words	Ratio Notation
For every 2 cups of lemon juice, there are 10 total cups of lemonade.	part → 2 to 10 ← whole part → 2 : 10 ← whole part → $\frac{2}{10}$ ← whole

Talk About It!
Why would fraction notation not be the best ratio notation to use to represent the ratio of the number of cups of lemon juice to the number of cups of water?

Because a fraction represents part of a whole, fraction notation is generally only used to represent part-to-whole ratios.

A **part-to-part ratio** compares one part of a group to another part of the same group. The ratio 2 : 7 is a part-to-part ratio because it compares the number of cups of lemon juice (one part) to the number of cups of water (another part) needed to make the lemonade.

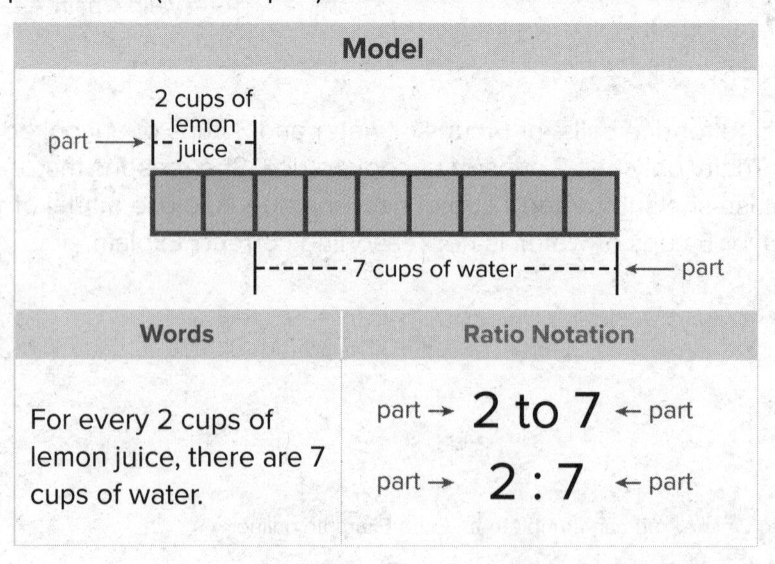

Words	Ratio Notation
For every 2 cups of lemon juice, there are 7 cups of water.	part → 2 to 7 ← part part → 2 : 7 ← part

Talk About It!
No matter how many batches of lemonade are made, will there always be 2 cups of lemon juice for every 7 cups of water? Justify your response.

Example 2 Part-to-Whole Ratios

A florist is arranging flowers in vases to sell to her customers. She has two sizes of vases available: small and large. She wants the large vase to have the same ratio of flowers as the small vase.

Small Vase	
Flower	Quantity
Carnations	6
Sunflowers	2
Tulips	4

If the large vase has a total of 36 flowers, how many are tulips?

Step 1 Use a bar diagram to represent the ratio of tulips to total flowers for the small vase.

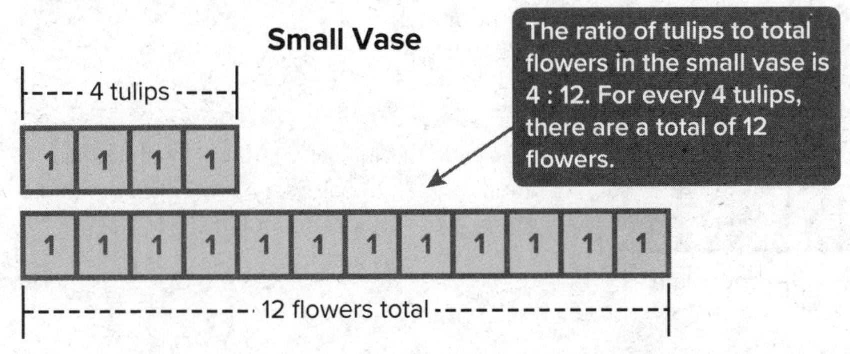

Step 2 Use the same ratio to find the number of tulips in the large vase.

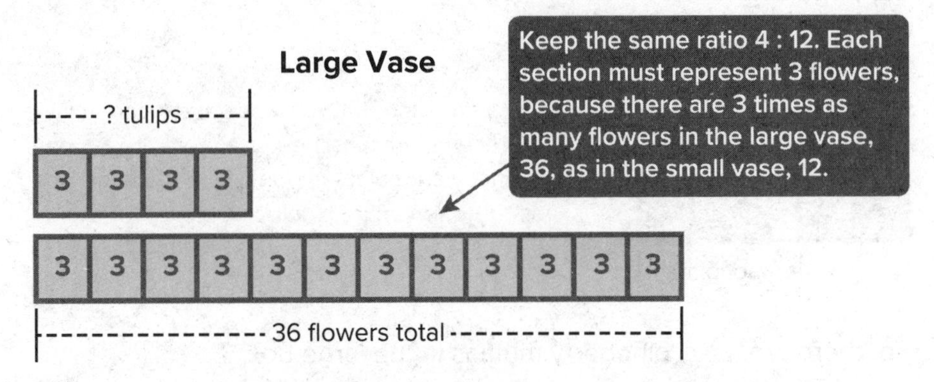

Each section in the diagram represents 3 flowers. There are four sections for tulips, so the large vase will contain 4×3, or _____ tulips.

Check

Refer to the table in Example 2. If the large vase has a total of 36 flowers, how many are carnations? Use bar diagrams and ratio reasoning to justify your response.

Go Online You can complete an Extra Example online.

Think About It!

Why is the ratio of tulips to total flowers a part-to-whole ratio?

Talk About It!

Why does each section of the bar diagram have to represent the same amount, in this case, 3 flowers?

Talk About It!

Suppose the florist wanted to place the flowers in a medium vase, using the same ratio. What quantities of tulips and total flowers might be reasonable for a medium vase? Justify your response.

Example 3 Part-to-Part Ratios

A bakery sells fresh-baked muffins, sold in small or large boxes. The manager of the bakery wants to maintain the same ratio of each type of muffin in the large box as in the small box.

Small Box	
Muffin	Quantity
Blueberry	2
Cinnamon	1
Chocolate	3

If the large box contains 9 chocolate muffins, how many blueberry muffins are in the large box?

Think About It!
Why is the ratio of blueberry muffins to chocolate muffins a part-to-part ratio?

Step 1 Use a bar diagram to represent the ratio of blueberry muffins to chocolate muffins for the small box.

Small Box

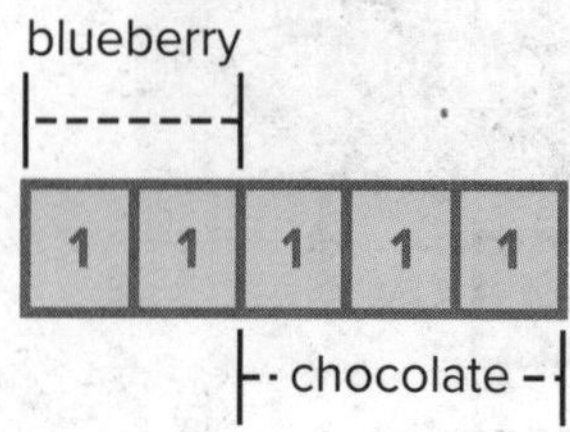

The ratio of blueberry muffins to chocolate muffins is 2 : 3. For every 2 blueberry muffins, there are 3 chocolate muffins.

Step 2 Use the same ratio to find the number of blueberry muffins in the large box.

Talk About It!
Describe a part-to-whole relationship between the blueberry muffins and the total muffins in the small box. What ratio represents that relationship?

Large Box

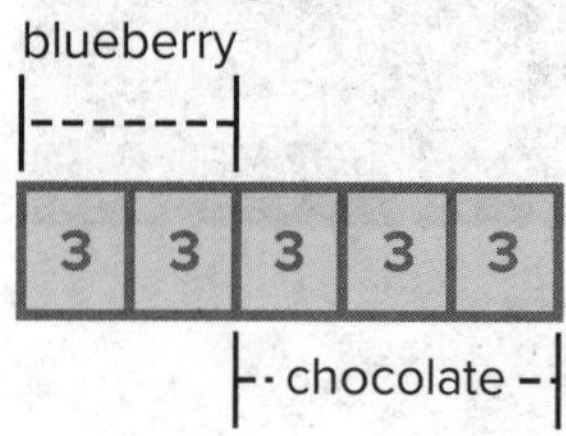

Keep the same ratio 2 : 3. Each section must represent 3 muffins, because there are 3 times as many chocolate muffins, 9, in the large box as there are in the small box, 3.

So, there are _____ blueberry muffins in the large box.

Check

Refer to the table in Example 3. If the large box contains 9 chocolate muffins, how many cinnamon muffins are in the large box? Use bar diagrams and ratio reasoning to justify your response.

Go Online You can complete an Extra Example online.

Apply Fundraising

The students at Lake Meadow Middle School will sell bags of honey granola for a fundraising event. The table shows a recipe that makes 6 cups of granola. The students will place 3 cups of granola in each bag. If forty people are expected to buy one bag of granola each, how many cups of rolled oats do they need?

Honey Granola
4 cups rolled oats
1 cup chopped almonds
$\frac{2}{3}$ cup honey
1 cup coconut oil
$\frac{1}{2}$ teaspoon salt
1 tablespoon ground cinnamon
1 teaspoon vanilla extract

1 What is the task?

Make sure you understand exactly what question to answer or problem to solve. You may want to read the problem three times. Discuss these questions with a partner.

First Time Describe the context of the problem, in your own words.
Second Time What mathematics do you see in the problem?
Third Time What are you wondering about?

2 How can you approach the task? What strategies can you use?

3 What is your solution?

Use your strategy to solve the problem.

4 How can you show your solution is reasonable?

Write About It! Write an argument that can be used to defend your solution.

Talk About It!

How can you solve this problem another way?

Check

The ingredients to make two servings of a fruit smoothie are shown in the table. Suppose you have 12 cups of frozen strawberries. If you use the entire amount, how many cups of plain yogurt do you need to maintain the same ratio? How many servings will this make?

Ingredient	Cups
Plain Yogurt	2
Fruit Juice	1
Frozen Strawberries	3

Go Online You can complete an Extra Example online.

Pause and Reflect

Create a graphic organizer that shows your understanding of ratios. Include examples of each of the following in your graphic organizer.

- bar diagrams
- words
- ratio notation
- part-to-whole ratios
- part-to-part ratios

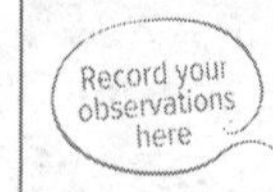

Name _______________ Period _______ Date _______

Practice

Go Online You can complete your homework online.

1. In Suri's coin purse, she has 6 dimes and 4 quarters. Martha has 5 dimes and 3 quarters. Suri thinks that the ratio of dimes to quarters in both purses is the same because they each have 2 more quarters than dimes. Is the ratio relationship the same? Justify your response. **(Example 1)**

2. In a trivia game, Levi answered 8 questions correctly out of 10 turns in the game. He then answered the next three questions correctly. He reasoned that because he added 3 to both the total questions and his correct responses, that the ratio of correct answers to total questions remained the same. Is he correct? Justify your response. **(Example 1)**

3. Riley needs to make fruit punch for the family reunion. One batch of punch has the ingredients shown. If the punch bowl holds 27 cups, how many cups of orange juice will she need to keep the ratio in a full punch bowl the same? **(Example 2)**

Item	Cups
Cranberry Juice	4
Lemon Lime Soda	1
Orange Juice	2
Pineapple Juice	2

4. A small fruit basket contains the fruits shown. A large basket has the same ratio of fruits as the small basket. If the large basket has 42 total pieces of fruit, how many are pears? **(Example 2)**

Type of Fruit	Amount
Apple	6
Orange	5
Pear	3

5. Mrs. Santiago is buying doughnuts for her office. Each box contains 6 glazed, 4 cream filled, and 2 chocolate flavored doughnuts. If there were 20 total cream filled doughnuts, how many chocolate doughnuts did she buy? **(Example 3)**

6. A small batch of trail mix contains 2 cups of raisins, 2 cups of peanuts, 1 cup of sunflower seeds, and 1 cup of chocolate coated candies. A large batch has the same ratio of ingredients as a small batch. If the large batch has 8 cups of peanuts, how many cups of sunflower seeds are in a large batch? **(Example 3)**

Test Practice

7. **Open Response** A football coach needs to divide 48 players into two groups. He wants the ratio of players in Group 1 to players in Group 2 to be 1 to 3. How many players will be in group 2?

Apply

8. To make a homemade all-purpose household cleaner, you can mix the ingredients shown in the table. Samuel has 1 cup of rubbing alcohol and will use the entire amount. He plans to store the cleaning solution in containers that each hold a maximum of 6 cups. How many containers does he need? Write an argument to defend your solution.

All-Purpose Cleaner
1 cup vinegar
$\frac{1}{2}$ cup rubbing alcohol
1 gallon water (16 cups)

9. The table shows the ingredients needed to make one batch of homemade slime. Dodi has 2 cups of liquid starch and will use the entire amount. She plans to store the slime in containers that each hold a maximum of 6 ounces. How many containers will she need? Write an argument to defend your solution. *(Hint: 2 cups = 16 ounces)*

Ingredient	Amount (oz)
Glue	4
Liquid Starch	4
Water	4

10. MP **Find the Error** The ratio of quarts of white paint to red paint is 3 : 4. A student says that to keep the ratio the same, he will need 7 quarts of white paint if he has 8 quarts of red paint, because originally there was one more quart of red paint than white paint. Find the student's mistake and correct it.

11. MP **Justify Conclusions** Rowan found that 4 out of 28 students in her class bike to school. What is the ratio of students that bike to school to the number of students that do not bike to school? Write an argument to defend your solution.

12. Create Write your own real-world problem involving part-to-whole or part-to-part ratios. Trade problems with a partner and solve each other's problem. Discuss with your partner how your knowledge of ratios helped you solve each problem.

13. The ratio of the distance around a circle, the circumference, to the distance across a circle, the diameter, is represented by the Greek letter π. If the circumference of a circle is 6.28 inches and the diameter of the same circle is 2 inches, what is the approximate value of π?

Tables of Equivalent Ratios

I Can... represent a collection of equivalent ratios and show the ratio relationship between two quantities using tables of equivalent ratios and double number lines.

Today's Standards
6.RP.A.3, 6.RP.A.3.A
MP1, MP2, MP3, MP5, MP7

What Vocabulary Will You Learn?
double number line
equivalent ratios
ratio table
scaling

Explore Equivalent Ratios

Online Activity You will use equivalent ratios to find the number of cups of flour and Greek yogurt to make 8 pizzas.

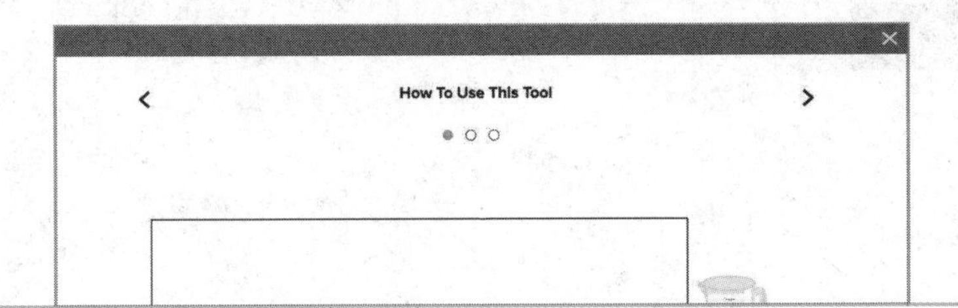

Learn Equivalent Ratios and Ratio Tables

The table shows the ingredients needed to make the dough for one pizza. You used this information in the Explore activity to find the number of cups of each ingredient needed to make 1, 2, and 3 pizzas by maintaining the ratio of 2 : 3.

Ingredient	Number of Cups
Greek Yogurt	2
Self-Rising Flour	3

The bar diagrams also show how the ratio of 2 : 3 is maintained, by using two sections that represent Greek yogurt and three sections that represent flour. The resulting ratios for 1, 2, and 3 pizzas are 2 : 3, 4 : 6, and 6 : 9, respectively. The ratios 2 : 3, 4 : 6, and 6 : 9 are **equivalent ratios** because they express the same ratio relationship between the quantities.

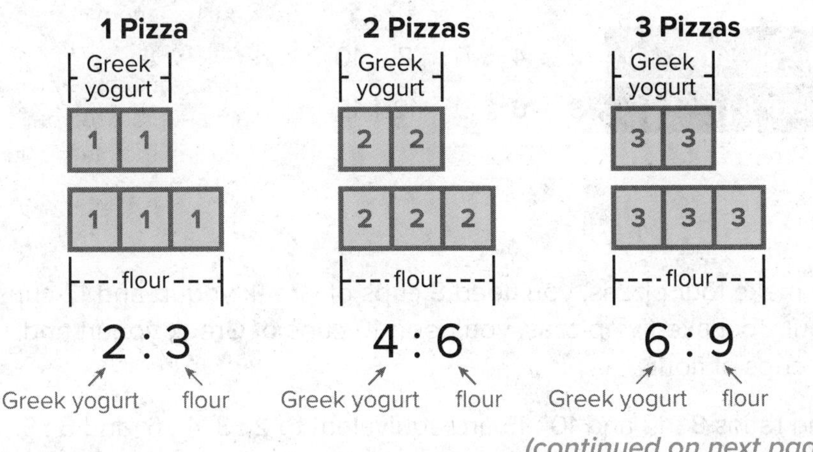

(continued on next page)

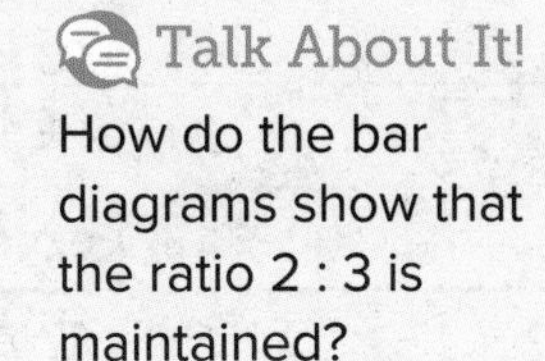

Your Notes

A table of equivalent ratios, or **ratio table**, is a collection of equivalent ratios that are organized in a table. Each column consists of a pair of quantities that have the same ratio as the pairs of quantities in the other columns.

In the ratio table shown, the ratios 2 : 3, 4 : 6, and 6 : 9 are all equivalent.

Greek Yogurt (c)	2	4	6
Flour (c)	3	6	9

Ratio tables show both an additive structure and a multiplicative structure.

+2 +2

Greek Yogurt (c)	2	4	6
Flour (c)	3	6	9

+3 +3

Add 2 to the cups of yogurt for each new column. Add 3 to the cups of flour for each new column.

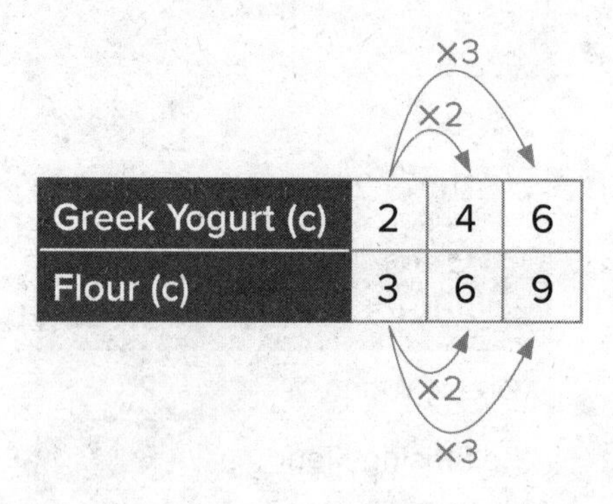

Greek Yogurt (c)	2	4	6
Flour (c)	3	6	9

Multiply each of the original quantities by the same number to obtain the values in each of the other columns.

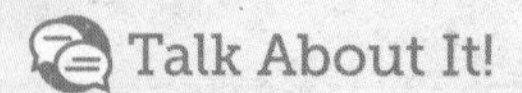

Talk About It!

Why might a ratio table be more advantageous to use than a bar diagram when finding the quantity of each ingredient needed to make 5 pizzas?

The process of multiplying each quantity in a ratio by the same number to obtain equivalent ratios is called **scaling**.

You can use scaling to extend the ratio table to find the number of cups of each ingredient needed to make additional pizzas. By doing so, you find more equivalent ratios.

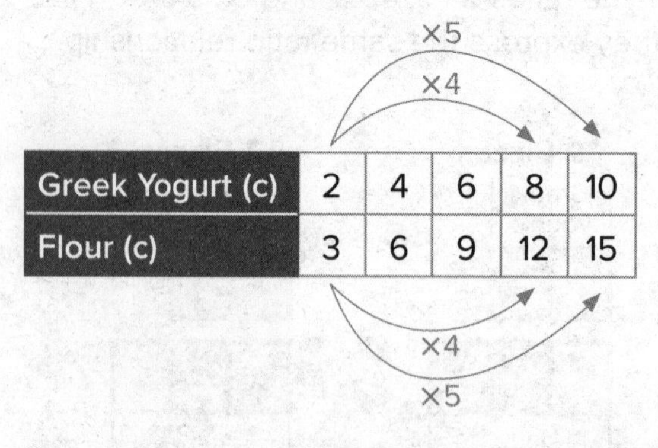

Greek Yogurt (c)	2	4	6	8	10
Flour (c)	3	6	9	12	15

Continue the pattern by multiplying each of the original quantities by the same number to obtain the values in the other columns.

To make four pizzas, you need 8 cups of Greek yogurt and 12 cups of flour. To make five pizzas, you need 10 cups of Greek yogurt and 15 cups of flour.

The ratios 8 : 12 and 10 : 15 are equivalent to 2 : 3, 4 : 6, and 6 : 9.

Example 1 Scale Forward to Find Equivalent Ratios

To make yellow icing, Amida mixes 6 drops of yellow food coloring with 2 cups of white icing.

How many drops of yellow food coloring should Amida mix with 8 cups of white icing to get the same shade of yellow?

Step 1 Create a ratio table with the given information.

For every 6 drops of yellow food coloring, there are 2 cups of icing. The unknown is the number of drops of yellow needed to mix with 8 cups of icing.

Drops of Yellow	6	?
Cups of Icing	2	8

Step 2 Scale forward to find how many drops of yellow Amida needs to mix with 8 cups of icing.

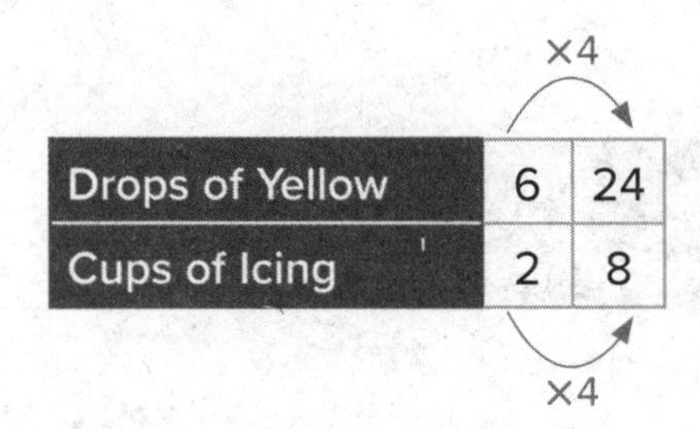

Drops of Yellow	6	24
Cups of Icing	2	8

Because 2 × ________ = 8, multiply 6 by ________ to obtain 24.

The ratios 6 : 2 and 24 : 8 are equivalent ratios.

So, Amida should mix ________ drops of yellow food coloring with 8 cups of white icing to get the same shade of yellow.

Check

There are 16 ounces in every pound. How many ounces are in 8 pounds? Construct a ratio table and an argument to support your answer.

Go Online You can complete an Extra Example online.

Think About It!

Should Amida add less than, more than, or the same number of drops, 6, of yellow food coloring to mix with the 8 cups of icing? Why?

Talk About It!

Describe how you can use a bar diagram to solve this problem. Is there a representation that is more advantageous to use in this case? Why or why not?

Example 2 Scale Backward to Find Equivalent Ratios

Think About It!

How do you know that you cannot scale forward to solve this problem?

Akeno mixes three sample containers of yellow paint with four sample containers of red paint to create his favorite shade of orange paint. His little sister Aiko wants to create the same shade of orange paint, but she only has two sample containers of red paint.

What should Aiko do to create the same shade of orange paint?

Step 1 Create a ratio table with the given information.

For every 3 containers of yellow paint, there are 4 containers of red paint. The unknown is the amount of yellow paint needed to mix with 2 containers of red paint.

Yellow Paint (containers)	?	3
Red Paint (containers)	2	4

Step 2 Scale backward to find the equivalent ratio.

÷2

Yellow Paint (containers)	1.5	3
Red Paint (containers)	2	4

÷2

Talk About It!

Suppose Aiko said that since she has half of the amount of red paint that Akeno has, she can mix that with half of the amount of yellow paint that Akeno has. Is she correct? Explain.

Because $4 \div 2 = 2$, divide 3 by ______ to obtain ______.

The ratios 1.5 to 2 and 3 to 4 are equivalent.

So, Aiko should mix ______ containers of yellow paint with 2 containers of red paint to create the same shade of orange paint.

Check

To make three loaves of banana bread, you need 9 bananas. How many bananas are needed to make one loaf of banana bread? Construct a ratio table and an argument to support your answer.

Go Online You can complete an Extra Example online.

Example 3 Scale in Both Directions

Natasha made raspberry punch for a party by mixing 9 fluid ounces of fruit punch, 3 liters of soda, and 6 scoops of raspberry ice cream. Halfway through the party, the punch bowl was empty.

If Natasha only has 6 fluid ounces of fruit punch left, how much ice cream does she need to make another batch of punch?

Step 1 Create a ratio table with the given information.

For every 9 fluid ounces of fruit punch, there are 6 scoops of raspberry ice cream. The unknown is the amount of ice cream needed to mix with 6 fluid ounces of fruit punch.

Fruit Punch (fl oz)	6	9
Ice Cream (scoops)	?	6

There is no whole number by which you can multiply 6 to obtain a product of 9.

Step 2 Scale backward to find an equivalent ratio.

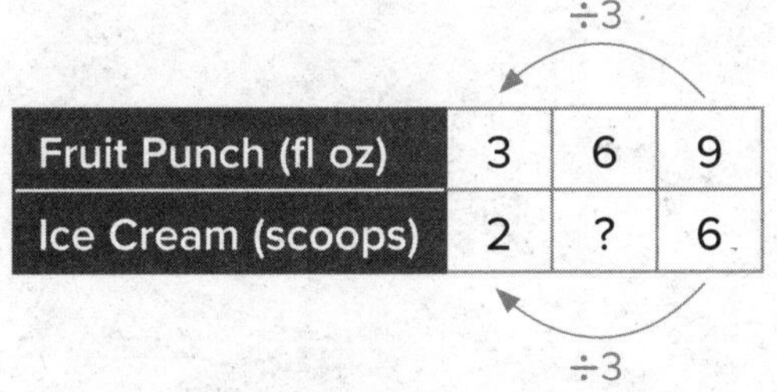

Fruit Punch (fl oz)	3	6	9
Ice Cream (scoops)	2	?	6

To scale back, you can divide both 9 and 6 by 3. This results in the equivalent ratio 3 : 2.

Step 3 Use the equivalent ratio you found to scale forward to find the desired equivalent ratio.

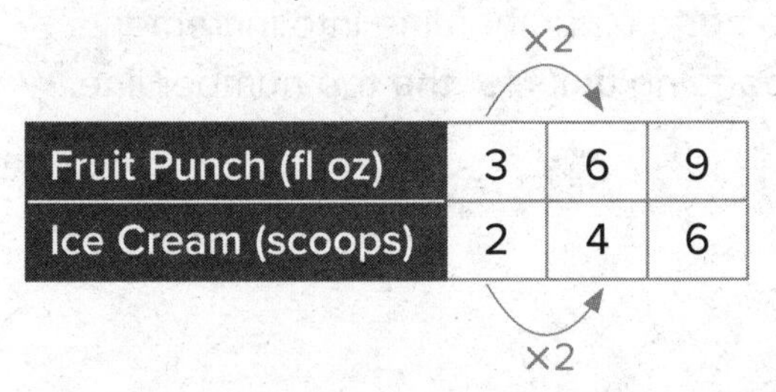

Fruit Punch (fl oz)	3	6	9
Ice Cream (scoops)	2	4	6

To scale forward, you can multiply both 3 and 2 by 2. This results in the equivalent ratio 6 : 4.

So, Natasha should mix ______ scoops of raspberry ice cream with the remaining 6 fluid ounces of fruit punch.

Check

Refer to Example 3. How many liters of soda should Natasha mix with the 6 fluid ounces of fruit punch? Construct a ratio table and an argument to support your answer.

Go Online You can complete an Extra Example online.

Think About It!

To mix with the remaining amount of fruit punch, will the number of scoops of ice cream that Natasha needs be less than, more than, or equal to 6? Explain.

Talk About It!

Why was scaling back to find the equivalent ratio 3 : 2 helpful in solving the problem?

Example 4 Use a Double Number Line to Find Equivalent Ratios

Think About It!

To make 4 biscuits, will the number of cups of flour be less than, greater than, or equal to 2? Explain.

Talk About It!

Compare and contrast using a ratio table and using a double number line to solve this problem.

The ingredients needed to make 24 biscuits are shown in the table.

Homemade Biscuits
4 c flour
8 tsp baking powder
2 tbsp sugar
1 tsp salt
1 c shortening
2 large eggs
2 c milk

If Portia wants to only make 18 biscuits, how many cups of flour does she need?

Use a double number line to solve this problem. A **double number line** consists of two number lines, in which the coordinated quantities are equivalent ratios.

Step 1 Draw a double number line.

The top number line represents the cups of flour and the bottom number line represents the number of biscuits.

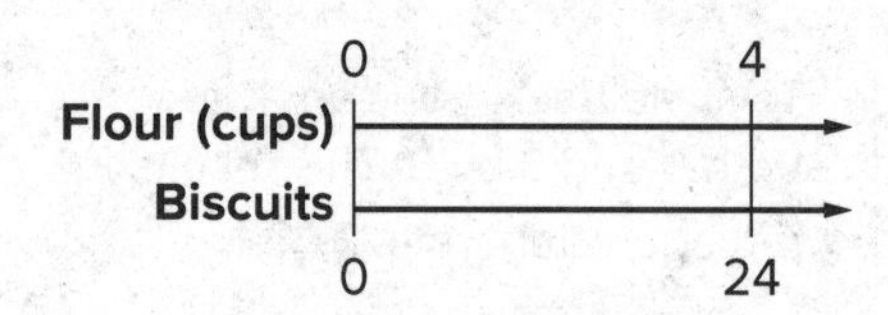

To make 24 biscuits, Portia needs 4 cups of flour.

Step 2 Find the equivalent ratio.

To scale back, you can divide both 4 and 24 by 4. This results in the equivalent ratio 1 : 6. Divide the bottom number line into increments of 6 units and label the corresponding units for the top number line.

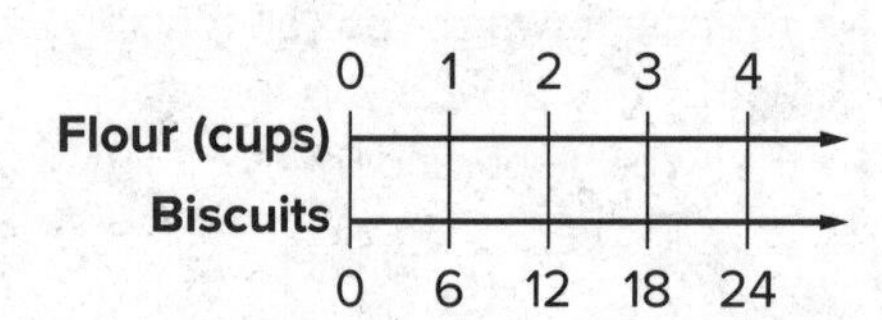

The value on the top number line that corresponds with 18 is 3.

So, to make 18 biscuits, Portia needs ________ cups of flour.

Check

Refer to Example 4. If you only wanted to make 6 biscuits, how many teaspoons of baking powder do you need? Construct a double number line and an argument to support your answer.

Go Online You can complete an Extra Example online.

Apply Packaging

A toy store sells assorted marbles, sold in small or large bags. The table shows the number of each color of marble in the small bag. The manager of the store wants to maintain the same ratio of each color of marble in the large bag as in the small bag. Each marble costs 20 cents. If the large bag contains 20 green marbles, how much does the large bag cost?

Color	Quantity
Blue	14
Red	12
Green	8
Orange	6

1 What is the task?

Make sure you understand exactly what question to answer or problem to solve. You may want to read the problem three times. Discuss these questions with a partner.

First Time Describe the context of the problem, in your own words.
Second Time What mathematics do you see in the problem?
Third Time What are you wondering about?

2 How can you approach the task? What strategies can you use?

Record your observations here

3 What is your solution?

Use your strategy to solve the problem.

Show your work here

4 How can you show your solution is reasonable?

Write About It! Write an argument that can be used to defend your solution.

Talk About It!
How many red marbles are in the large bag? Provide a mathematical argument to support your answer.

Check

The table shows the number of slices of turkey and cheese in the regular Totally Turkey Sandwich at Dave's Deli.

Totally Turkey Sandwich (Regular)	
Ingredient	**Quantity**
Turkey Slices	3
Cheese Slices	2

The ingredients are doubled in the large Totally Turkey Sandwich. On Wednesday, three times as many customers ordered the regular sandwich as the large sandwich. If 27 customers ordered the regular sandwich, how many total slices of turkey were used to make the sandwiches that day?

Go Online You can complete an Extra Example online.

Foldables It's time to update your Foldable, located in the Module Review, based on what you learned in this lesson. If you haven't already assembled your Foldable, you can find the instructions on page FL1.

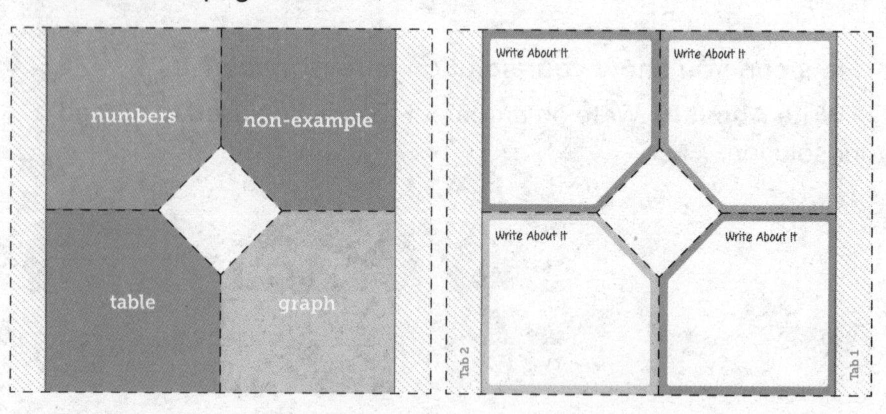

Name ______________________ Period __________ Date __________

Practice

Go Online You can complete your homework online.

Use any strategy to solve each problem.

1. Jayden's snow cone machine makes 3 snow cones from 0.5 pound of ice. How many snow cones can be made with 5 pounds of ice? **(Example 1)**

2. Nyoko is having a pizza party. Two large pizzas serve 9 people. How many large pizzas should she order to serve 36 guests at the party? **(Example 1)**

3. The world record for the most number of speed skips in 60 seconds is 332 skips. If the record holder skipped at a constant ratio of seconds to skips, how many skips did she make in 15 seconds? **(Example 2)**

4. A recipe for homemade clay calls for 6 cups of water for every 12 cups of flour. How many cups of water are needed when 4 cups of flour are used? **(Example 2)**

5. Adrian decorated 16 cupcakes in 28 minutes. If he continues at this pace, how many minutes will it take him to decorate 56 cupcakes? **(Example 3)**

6. A comic book store is having a sale. You can buy 20 comic books for $35. What is the cost of 8 comic books during the sale? **(Example 3)**

7. A certain store is selling packages of 10 pencils and 4 pens for back to school. The store manager wants to make a larger package in the same ratio. If the large package has 10 pens, how many pencils are in the large package? **(Example 4)**

Test Practice

8. Open Response Ben made trail mix for his camping trip that contained 8 ounces of peanuts, 6 ounces of raisins, and 10 ounces of chocolate candies. He wants to make a larger batch for his next camping trip with 28 ounces of peanuts. How many ounces of raisins will he need?

Apply

9. The table shows the items in a family chicken taster meal at a restaurant. The restaurant wants to create a larger meal to accommodate larger groups of people. They also want to limit the number of chicken tenders to 15. If the ratio remains the same, how many biscuits are in the larger meal?

Family Taster Meal
4 chicken sliders
6 chicken tenders
8 biscuits
1 pint of cole slaw

10. **MP Identify Structure** Generate an equivalent ratio to $\frac{\$10}{15 \text{ tickets}}$ by scaling backward. Then generate an equivalent ratio by scaling forward.

11. **MP Justify Conclusions** There are 21 goats and 35 chickens on a farm. If 5 more goats and 5 more chickens are added, is the ratio of goats to chickens the same? Write an argument to defend your solution.

12. **MP Reason Inductively** A student said you can add the same number to both terms of a ratio to find an equivalent ratio. Is the student correct? Explain why or why not.

13. **Create** Write and solve a real-world problem where you determine if two ratios are equivalent.

Graphs of Equivalent Ratios

I Can... represent a collection of equivalent ratios as ordered pairs and graph the ratio relationship on the coordinate plane.

Today's Standards
6.RP.A.3, 6.RP.A.3.A
MP1, MP2, MP3, MP5, MP7

Learn Ratios as Ordered Pairs

You previously learned how to create a ratio table and extend it by finding equivalent ratios. You can also represent a ratio relationship by creating a table of ordered pairs and graphing the ordered pairs on the coordinate plane.

To make a simple salad dressing, you can use 3 cups of olive oil for every cup of vinegar. You can then add herbs, salt, and/or pepper for seasoning. This ratio relationship is shown in the table.

Vinegar (c), *x*	Olive Oil (c), *y*
1	3
2	6
3	9
4	12
5	15

Each pair of equivalent ratios can be expressed as an ordered pair. The *x*-coordinate represents the number of cups of vinegar. The *y*-coordinate represents the number of cups of olive oil.

Talk About It!

Compare and contrast the ratio table and the graph. How do they both illustrate the same ratio relationship? How does the graph help you visualize the ratio relationship?

Recall that to graph a point, start at the origin. Move right along the *x*-axis the number of units indicated by the *x*-coordinate. From that location, move up along the *y*-axis the number of units indicated by the *y*-coordinate. Place a dot at that location.

The graph illustrates the ratio relationship of the cups of olive oil to the cups of vinegar in the salad dressing.

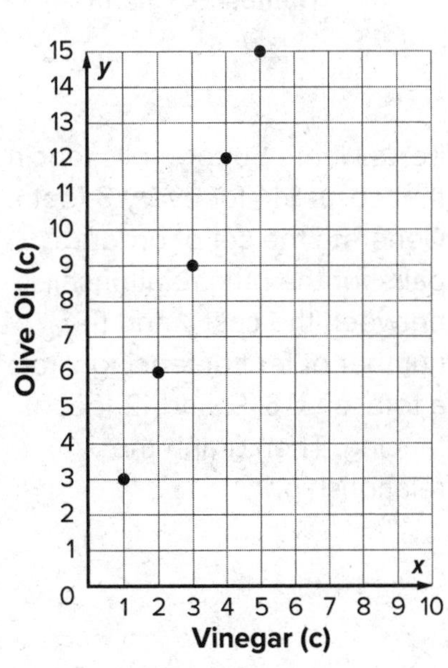

What to you noticed about the graphed points? You might notice that to travel from each point to the next point, you move up 3 units and to the right 1 unit. These are the same numbers in the ratio of 3 cups of olive oil for every 1 cup of vinegar.

Your Notes

Think About It!

What is the ratio of charms to beads? Beads to charms?

Talk About It!

What do you notice about the points on graph?

Example 1 Graph Ratio Relationships

Tamara is making charm bracelets for several friends. She uses 6 beads for every charm.

Generate the set of ordered pairs for the ratio relationship between the number of beads and the number of charms for a total of 1, 2, 3, and 4 charms. Then graph the relationship.

Part A Create a table of ordered pairs.

Let the x-coordinates represent the number of charms and the y-coordinates represent the number of beads.

Charms, x	Beads, y
1	6
2	
3	
4	

Use scaling to complete the table to write the equivalent ratios for 2, 3, and 4 charms.

Part B Graph the ordered pairs on the coordinate plane.

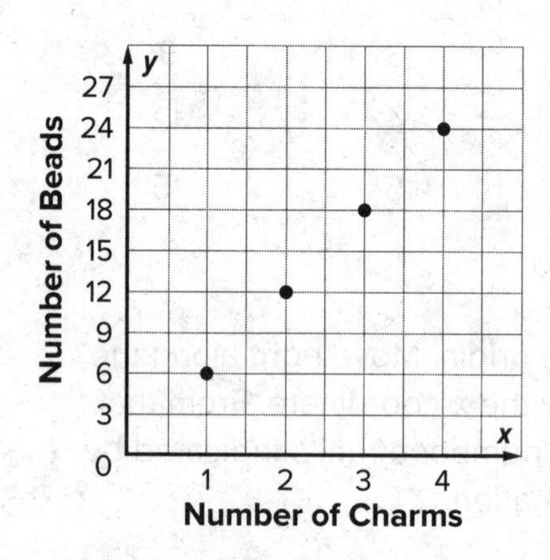

Check

Ken's Home Supply sells fencing that costs $14 for every 3 feet. Generate the set of ordered pairs for the ratio relationship between the cost y and the number of feet of fencing x for a total of 3, 6, 9, and 12 feet of fencing. Then graph the relationship.

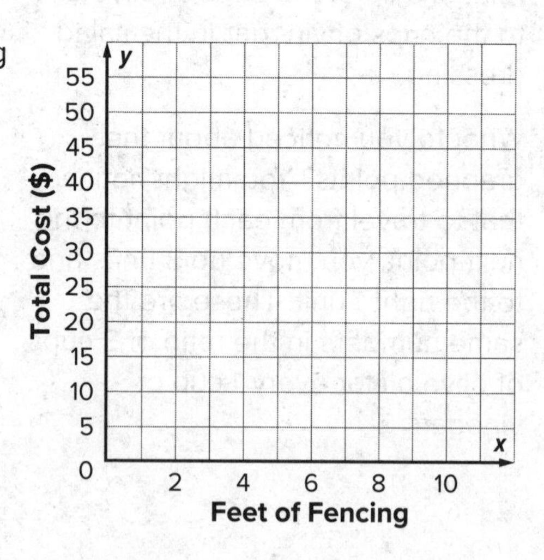

Go Online You can complete an Extra Example online.

Example 2 Graph and Interpret Ratio Relationships

To make one batch of homemade modeling clay that can be used in arts and crafts, Sequoia mixed the ingredients shown in the table.

Homemade Clay
4 cups flour
1 cups salt
2 cups water
food coloring

Graph the ratio relationship between the number of cups of water and flour for a total of 5 batches. Then describe the pattern in the relationship.

Part A Graph the ratio relationship.

Step 1 Generate a set of ordered pairs.

For every 4 cups of flour, there are 2 cups of water. Let the *x*-coordinates represent the number of cups of flour and the *y*-coordinates represent the number of cups of water.

Flour (c), *x*	Water (c), *y*
4	2
8	4
12	6
16	8
20	10

Use scaling to write the equivalent ratios for 1, 2, 3, 4, and 5 batches.

← 1 batch
← 2 batches
← 3 batches
← 4 batches
← 5 batches

Step 2 Graph the relationship.

The *x*-coordinates increase from 4 to 20, so let each grid unit along the *x*-axis on the coordinate plane represent 2 units.

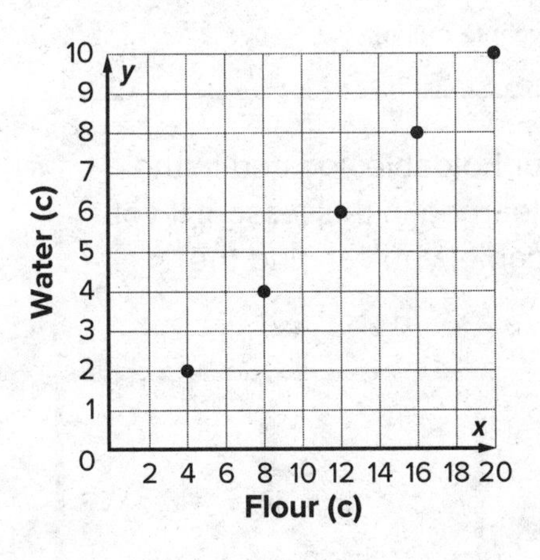

Part B Describe the pattern in the ratio relationship.

In the graph, the points appear to fall on a straight line. Each new point is 4 units to the right and 2 units up from the previous point. This confirms the relationship that for every ________ cups of flour, there are ________ cups of water.

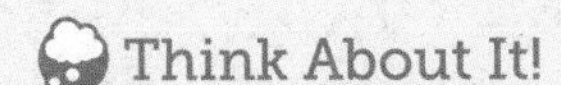

How do you know that the relationship between flour and water is a ratio relationship?

Talk About It!

Do you think that all ratio relationships will have graphs that appear to fall on a straight line? Why or why not?

Check

To make one batch of nectar to feed hummingbirds, Melanie added 4 cups of boiling water for every cup of refined white sugar.

Part A Graph the ratio relationship between cups of boiling water y and cups of refined white sugar x for a total of 1, 2, 3, 4, and 5 batches.

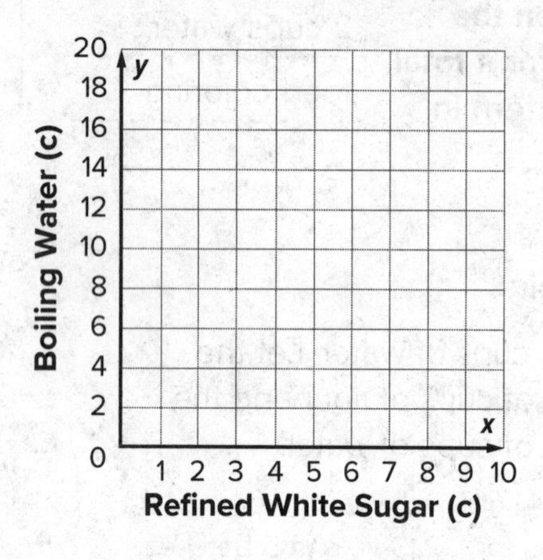

Part B Describe the pattern in the relationship.

Go Online You can complete an Extra Example online.

Foldables It's time to update your Foldable, located in the Module Review, based on what you learned in this lesson. If you haven't already assembled your Foldable, you can find the instructions on page FL1.

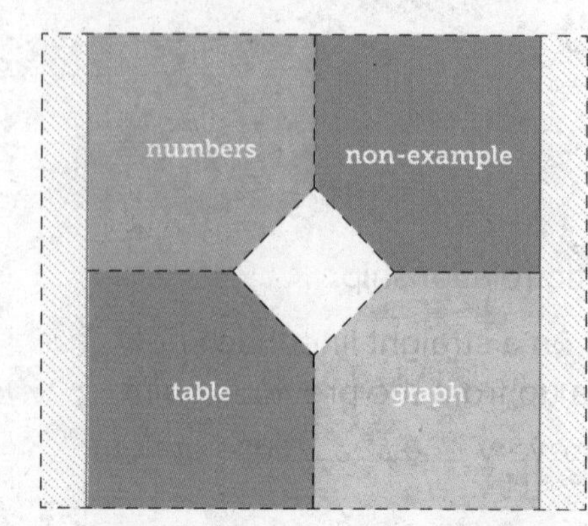

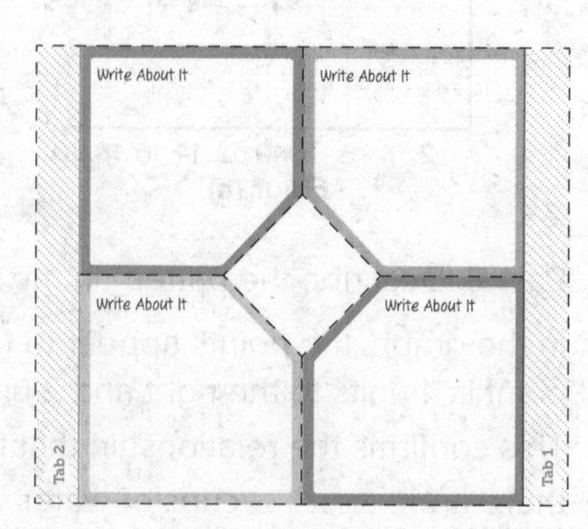

Name ________________________ Period ____________ Date ____________

Practice

Go Online You can complete your homework online.

1. Lulah is buying beach balls for her beach themed party. Each package contains 6 beach balls. Generate the set of ordered pairs for the ratio relationship between the number of beach balls y and the number of packages x for a total of 1, 2, 3, and 4 packages. Then graph the relationship on the coordinate plane and describe the pattern in the graph. **(Examples 1 and 2)**

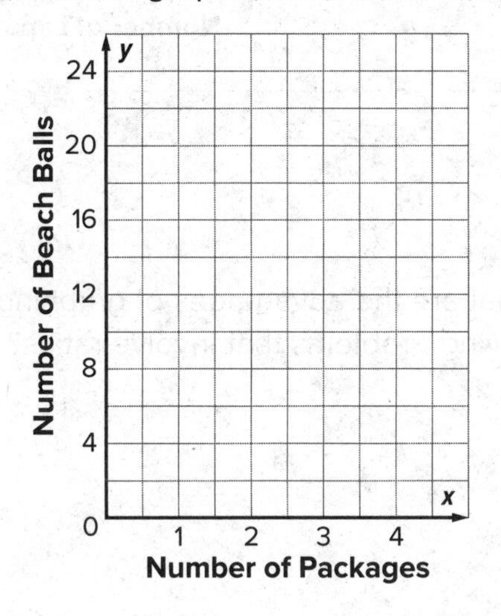

2. A sloth travels about 7 feet every minute. Generate the set of ordered pairs for the ratio relationship between the total distance traveled y and the number of minutes x for a total of 1, 2, 3, and 4 minutes. Then graph the relationship on the coordinate plane and describe the pattern in the graph. **(Examples 1 and 2)**

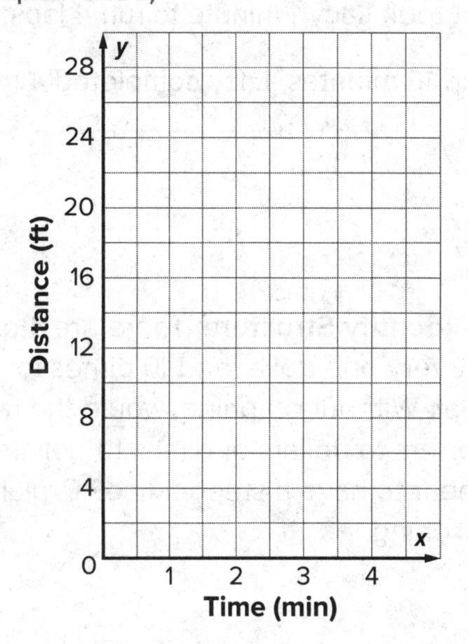

3. Two friends are making scrapbooks. The number of photos Lexi and Audrey place on each page of their scrapbooks is shown in the graph. Describe the ratio relationship for each person.

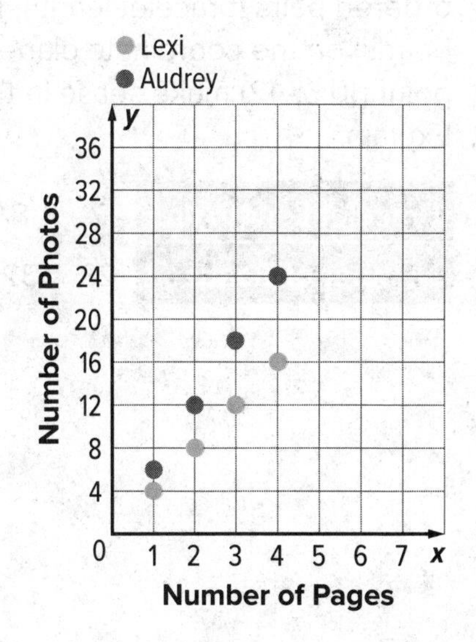

4. Multiselect Lacy is running laps around the track. The time in minutes and the number of laps ran are shown in the graph. Which of the following is true about the ratio relationship shown in the graph?

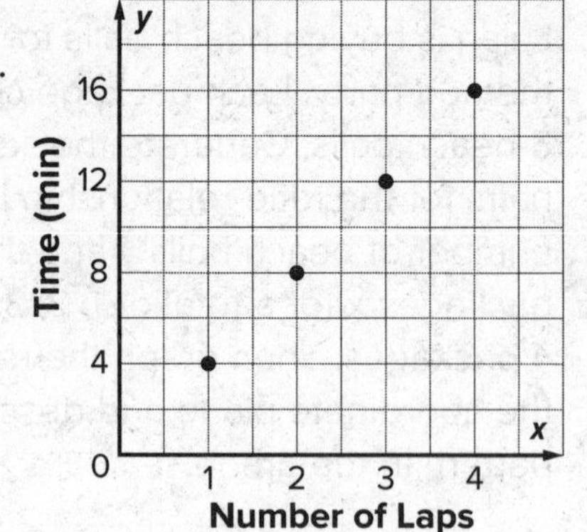

- ☐ Every 4 minutes, Lacy ran 1 lap.
- ☐ Lacy ran 8 laps in 2 minutes.
- ☐ It took Lacy 1 minute to run 4 laps.
- ☐ In 16 minutes, Lacy completed 4 laps.

5. MP **Identify Structure** There are 4 quarters for every one dollar and 10 dimes for every dollar. Without graphing, would the ratio of quarters to dollars or dimes to dollars appear to have a steeper line? Explain your reasoning.

6. What are the advantages of graphing when solving problems that involve ratios?

7. MP **Reason Abstractly** The table gives the number of beads needed to make bracelets of certain lengths. Suppose you graph the ordered pairs (bracelet length, number of beads) on the coordinate plane. Would the point (10.5, 42) make sense in this context? Explain.

Bracelet Length (in.)	7	8	9	10
Number of Beads	28	32	36	40

8. Multiple Relationships For every second, the average green sea turtle can swim 9 meters. Represent how far a green sea turtle can swim in 1, 2, 3 and 4 seconds in a table. Then graph the points on a coordinate plane.

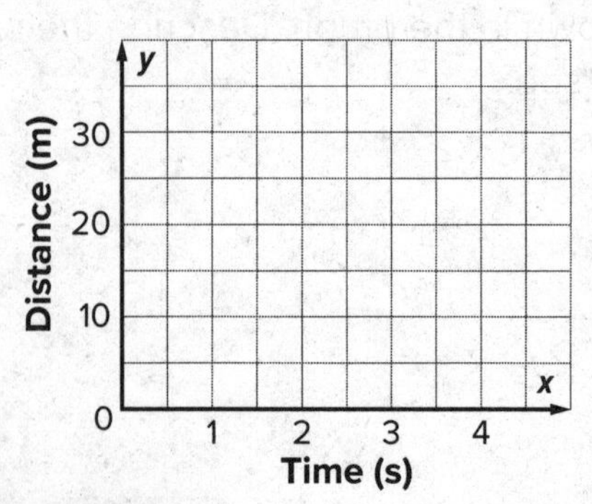

Time (s)				
Distance (m)				

Compare Ratio Relationships

I Can... compare ratio relationships that are shown using different representations.

Today's Standards
6.RP.A.3, 6.RP.A.3.A
MP1, MP2, MP3, MP5, MP7

Learn Use Graphs to Compare Ratio Relationships

Ratios for ingredients in dog food vary among companies that manufacture it. Company A advertises 25 grams of protein for every cup of dog food. The relationship between protein and cups of dog food for two other companies is shown in the table and graph.

Company B

Dog Food (c), x	Protein (g), y
2	44
3	66
4	88

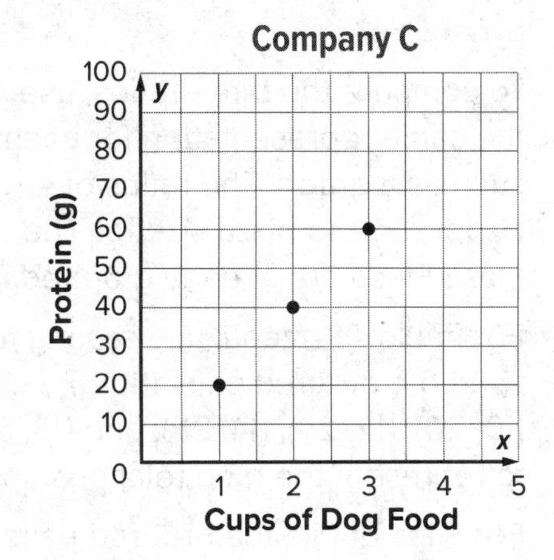

How can you compare the ratios of protein to cups of dog food for the three companies?

The ratios for each of the three companies is shown using a different representation. To compare them, you can use the same representation for each, such as a graph.

The ratios for Company C are already graphed. For Company A, you can generate equivalent ratios to find the ordered pairs (1, 25), (2, 50), and (3, 75).

For Company B, the ordered pairs are (2, 44), (3, 66), and (4, 88).

Draw a dotted line through the points to determine which relationship has the steepest graph. The graph for Company A is the steepest, and the graph for Company B is steeper than the graph for Company C. This means that Company A has the greatest ratio of protein to cups of dog food.

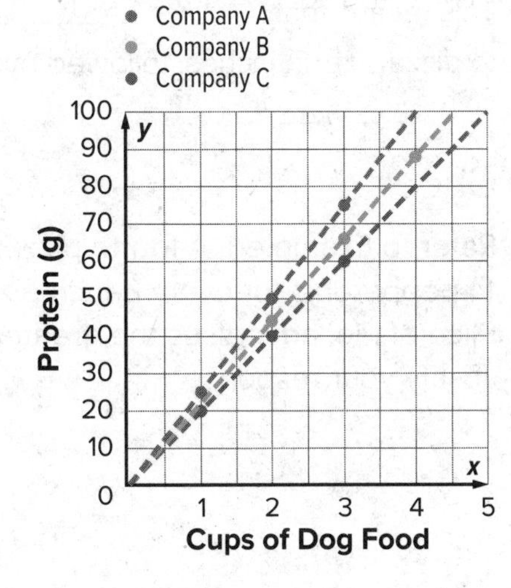

Talk About It!
If the ratio compared cups of dog food to protein, how would the graph change? Which line would be the steepest?

Your Notes

Think About It!

Just by studying the table, which pizzeria, Slice of Pie or Paulo's Pizzeria, has more pepperonis on a 12-inch pizza?

Example 1 Use Graphs to Compare Ratio Relationships

Paulo's Pizzeria advertises 24 pepperonis on every 12-inch pizza. The relationship of pepperonis to pizza size for two other pizzerias is shown in the table and graph.

Slice of Pie

Pizza Size (in.)	Pepperonis
10	15
12	18
14	21

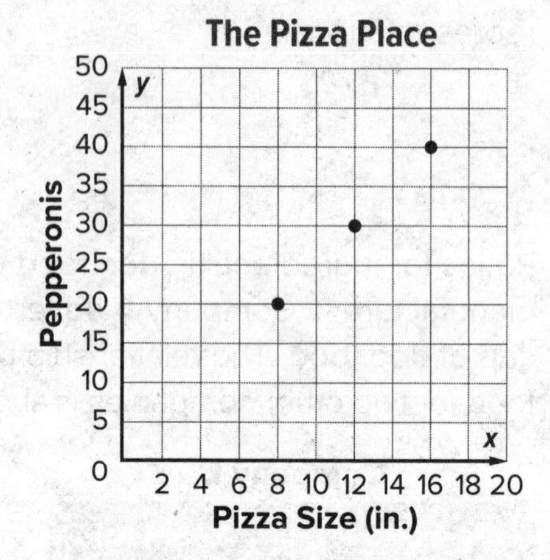

Which pizzeria advertises the greatest ratio of pepperonis to pizza size?

To compare the three ratios, use the same representation for each, such as a graph. The ratios of pepperonis to pizza size for The Pizza Place are already graphed.

For Paulo's Pizzeria, use scaling to write the ordered pairs (8, 16), (10, 20), (12, 24), (14, 28), and (16, 32) to represent the ratio relationship.

For Slice of Pie, the ordered pairs are (10, 15), (12, 18), and (14, 21).

Draw lines through the points. The graph for The Pizza Place is the steepest, and the graph for Paulo's Pizzeria is steeper than the graph for Slice of Pie.

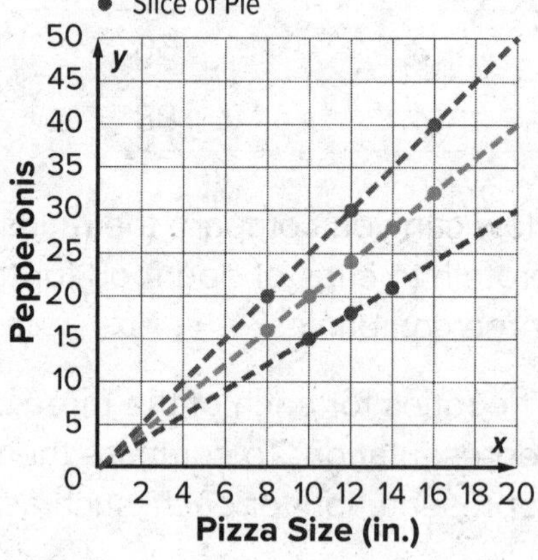

This means that ________________ has the greatest ratio of pepperonis to pizza size, in inches, followed by ________________, and then ________________.

Talk About It!

How many more pepperonis would be on an 18-inch pizza from The Pizza Place than on an 18-inch pizza from Paulo's Pizzeria? Justify your response.

Check

Refer to Example 1. A fourth pizzeria, Pizza Café, advertises 14 pepperonis for every 8-inch pizza. Which pizzeria, Pizza Café or Slice of Pie, advertises the greater ratio of pepperonis to pizza size? Justify your response.

Go Online You can complete an Extra Example online.

Learn Use Tables to Compare Ratio Relationships

Another way to compare ratio relationships is to use tables. For example, a comparison of three smoothie recipes shows that Recipe A has a blueberry to strawberry ratio of 8 to 2, Recipe B has a ratio of 5 to 1, and Recipe C has a ratio of 10 to 3. You can use tables of equivalent ratios to determine which recipe has the greatest ratio of blueberries to strawberries.

Recipe A

Blueberries	8	16	24
Strawberries	2	4	6

Recipe B

Blueberries	5	10	15	20	25	30
Strawberries	1	2	3	4	5	6

Recipe C

Blueberries	10	20	30
Strawberries	3	6	9

Use scaling to write equivalent ratios for each recipe. You can compare the ratios when one of the quantities in each relationship is the same.

Recipe B has a ratio of 30 blueberries for every 6 strawberries, followed by Recipe C, 20 to 6. Recipe A has the least ratio 24 to 6. So, Recipe B has the greatest ratio of blueberries to strawberries.

Talk About It!

If the ratio relationships were graphed with blueberries on the *y*-axis and strawberries on the *x*-axis, the line for which recipe would have the steepest line? Explain.

Example 2 Use Tables to Compare Ratio Relationships

Roman is considering different recipes for bird seed to fill his hanging bird feeder. Measured in ounces, Recipe A has a sunflower seed to peanut ratio of 2 to 3, Recipe B has a ratio of 3 to 4, and Recipe C has a ratio of 5 to 6.

Which recipe has the greatest ratio of ounces of sunflower seeds to ounces of peanuts?

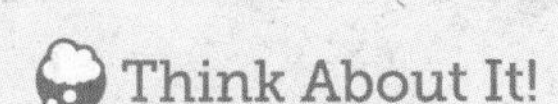

Think About It!

Which quantity will you make equivalent in each ratio in order to compare the other quantity?

Step 1 Create a ratio table for each recipe. Find equivalent ratios to compare the relationships.

Recipe A

Sunflower Seeds (oz)	2			
Peanuts (oz)	3			

Recipe B

Sunflower Seeds (oz)	3			
Peanuts (oz)	4			

Recipe C

Sunflower Seeds (oz)	5			
Peanuts (oz)	6			

(continued on next page)

Step 2 Determine the recipe with the greatest ratio of sunflower seeds to peanuts.

Recipe A: 8 : 12 **Recipe B:** 9 : 12 **Recipe C:** 10 : 12

Since 10 is greater than 9 and 8, the recipe with the greatest ratio of sunflower seeds to peanuts is Recipe ______.

Talk About It!

Compare and contrast using graphs and using tables to compare ratio relationships.

Check

When working on homework, Bailey spends 15 minutes reading for every 20 minutes spent on math, Aisha spends 12 minutes reading for every 15 minutes of math, and Tyler spends 7 minutes reading for every 10 minutes of math. Which person has the greatest ratio of minutes spent on reading to minutes spent on math?

Bailey

Reading (min)						
Math (min)						

Aisha

Reading (min)						
Math (min)						

Tyler

Reading (min)						
Math (min)						

Go Online You can complete an Extra Example online.

Pause and Reflect

Have you ever wondered when you might use the concepts you learn in math class? What are some everyday scenarios in which you might use what you learned today?

Apply Mixing Paint

Three friends are each mixing containers of red and blue paint, according to the ratios shown, to create their favorite shades of purple paint. Each container is the same size. If each person uses 6 quarts of red paint, whose paint mixture will have the most blue?

	Marcus	Cassidy	Hiram
Red (qt)	2	3	2
Blue (qt)	3	4	2

1 What is the task?

Make sure you understand exactly what question to answer or problem to solve. You may want to read the problem three times. Discuss these questions with a partner.

First Time Describe the context of the problem, in your own words.
Second Time What mathematics do you see in the problem?
Third Time What are you wondering about?

2 How can you approach the task? What strategies can you use?

3 What is your solution?

Use your strategy to solve the problem. A coordinate grid is provided should you choose to use it.

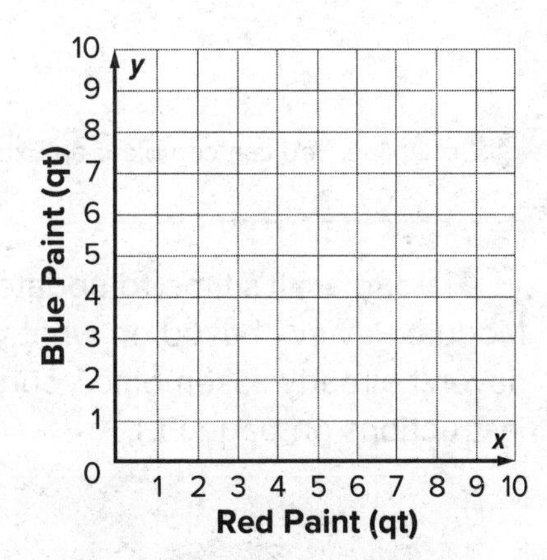

4 How can you show your solution is reasonable?

Write About It! Write an argument that can be used to defend your solution.

Talk About It!
Will the friend who has the most blue in his or her paint mixture always have the most blue, no matter how many quarts of red paint are used? Why or why not?

Check

Three cereal brands advertise the average number of berries for every cup of whole-grain cereal flakes as shown in the table. Each box is the same size. Which company advertises the greatest ratio of berries for every cup of flakes?

	Brand A	Brand B	Brand C
Cups of Flakes	1	2	3
Berries	5	6	12

A coordinate grid is provided should you choose to use it.

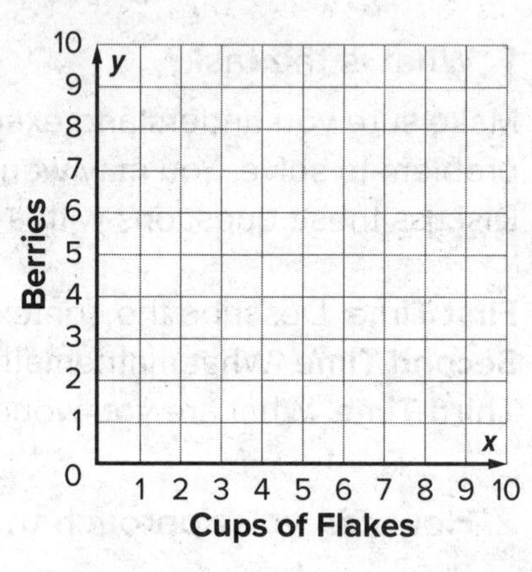

Go Online You can complete an Extra Example online.

Foldables It's time to update your Foldable, located in the Module Review, based on what you learned in this lesson. If you haven't already assembled your Foldable, you can find the instructions on page FL1.

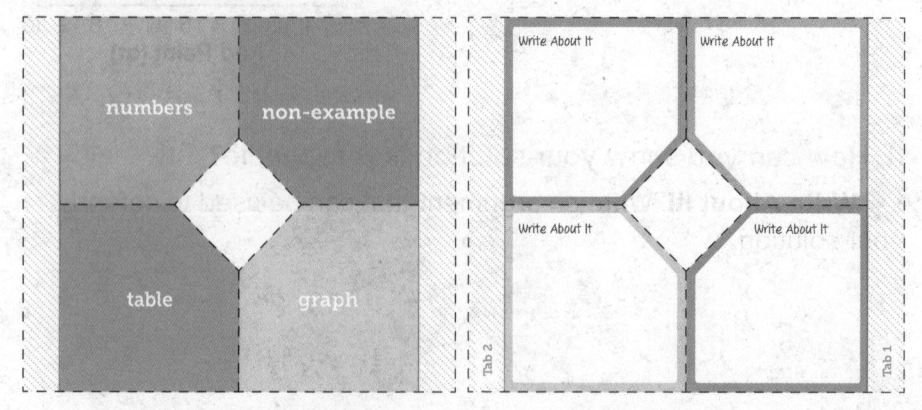

Name ______________________ Period __________ Date __________

Practice

Go Online You can complete your Homework online.

1. Cereal Brand A advertises that they have 60 raisins in their 24-ounce box of cereal. The advertised ratio of raisins to ounces for two other cereal brands are shown in the table and graph. Which brand advertises the greatest ratio of raisins to ounces of cereal? Justify your response. **(Example 1)**

Brand B

Ounces of Cereal	6	12	20	24
Raisins	18	36	60	72

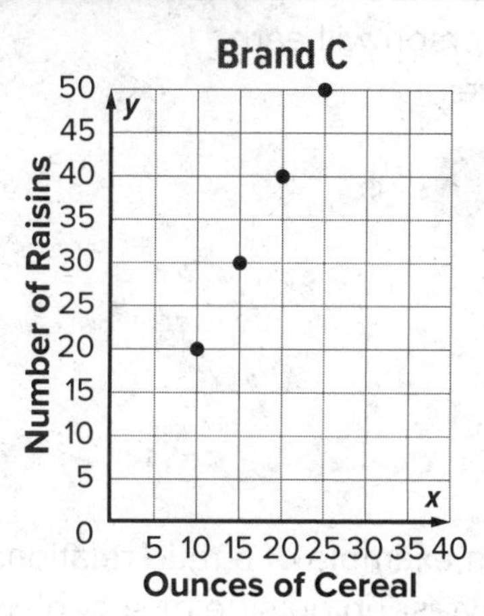

2. At the gym, Alex spends 24 minutes doing resistance training for every 30 minutes spent doing cardio exercises, Carisa spends 15 minutes on resistance for every 20 minutes on cardio, and Manuel spends 14 minutes on resistance for every 15 minutes on cardio. Which person has the greatest ratio of minutes spent on resistance to minutes spent on cardio? **(Example 2)**

Alex

Resistance (min)			
Cardio (min)			

Carisa

Resistance (min)			
Cardio (min)			

Manuel

Resistance (min)			
Cardio (min)			

Test Practice

3. **Open Response** Mrs. Quinto is comparing the Calories in different types of bread. The wheat bread has 150 Calories for every 2 slices. The Calories in two other types of bread are shown in the table and graph. Which bread has the greatest ratio of Calories to slices?

White Bread

Slices	Calories
2	160
4	320
6	480

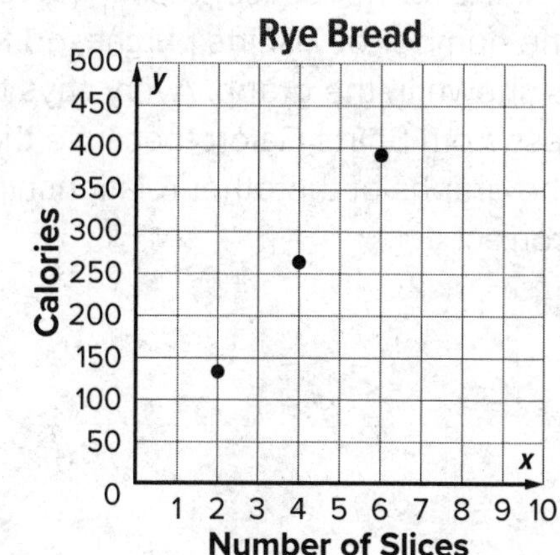

Apply

4. Mrs. Gonzalez wants to hire a catering company for her daughter's quinceañera. The ratios of the cost per person for a child and an adult for two different companies are shown in the table. Mrs. Gonzalez is planning on 25 adults and 12 children adding the party. How much less will it cost for her to hire Planning Pros than Party Time?

	Party Time	Planning Pros
Cost per Adult ($)	$10.50	$9
Cost per Child ($)	$6	$7.50

5. Charlie, Beth, and Miguel all babysit kids in their neighborhood. The table shows the number of hours and the amount each of them earned last night. If each person babysits for 5 hours next weekend, which person will earn the most money? Use a coordinate plane if needed to solve.

	Charlie	Beth	Miguel
Number of Hours	3	4.5	4
Total Earned ($)	28.50	42	40

6. MP **Construct an Argument** Ratio relationships can be described with words or they can be displayed using bar diagrams, tables, and graphs. Which display is more advantageous to use when comparing ratio relationships? Explain your reasoning.

7. Give an example of a ratio relationship that you have seen outside of school. How was the ratio relationship displayed, and why was the relationship displayed that way?

8. MP **Find the Error** Avery wants to order new practice T-shirts for her soccer team. The ratio of the total cost to the number of T-shirts purchased for three different stores is shown in the graph. Avery says that the shirts will cost less from Shirts Galore because the graph is steeper than the graphs of the other relationships. Find her mistake and correct it.

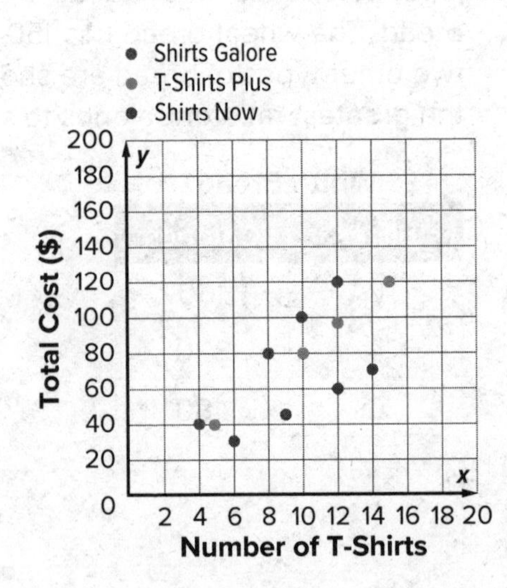

Solve Ratio Problems

I Can... solve real-world problems involving ratio relationships by using bar diagrams, double number lines, and equivalent ratios.

Today's Standards
6.RP.A.3, *Also addresses 6.RP.A.1*
MP1, MP2, MP3, MP5, MP7

Learn Use Bar Diagrams to Solve Ratio Problems

Suppose three out of five randomly selected students at a certain school play sports. There are 525 students at the school. You can create a bar diagram to predict how many of the students play sports.

Step 1 Draw a bar.

Three out of five students play sports, so divide the bar into 5 equal sections.

Step 2 Shade and label the diagram.

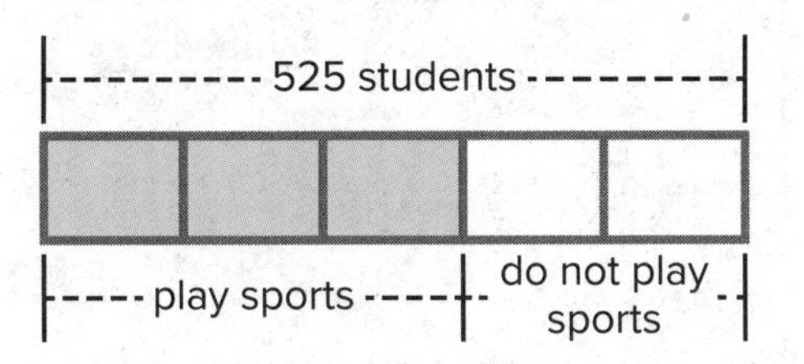

Shade three sections to represent the three out of five students who play sports. Label each group and the total number of students at the school, 525.

Step 3 Find the value of each section.

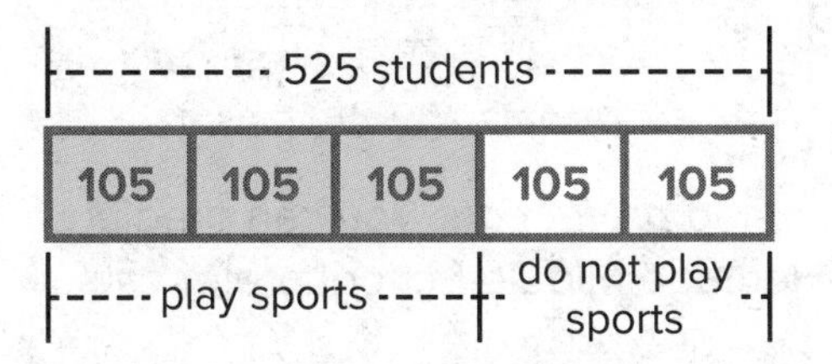

Divide the total number of students by 5 to determine the value of each section. Because $525 \div 5 = 105$, each section represents 105 students.

Because there are three sections labeled *play sports*, you can predict that 3×105, or 315 students at the school play sports.

Pause and Reflect

How does the bar diagram illustrate what you have previously learned in this module about part-to-whole and part-to-part ratios?

Record your observations here

Your Notes

Think About It!

How do you know that the number of students at Heritage Middle School who prefer cats can be expected to be greater than 375?

Example 1 Use Bar Diagrams to Solve Ratio Problems

Two out of three randomly selected students in Mrs. Mason's class at Heritage Middle School prefer cats as a household pet than any other pet.

If there are 750 students at Heritage Middle School, how many students can be expected to prefer cats as a household pet?

You can use a bar diagram to solve the problem.

Step 1 Draw a bar.

Two out of three students prefer cats, so divide the bar into three equal sections.

Step 2 Shade and label the diagram.

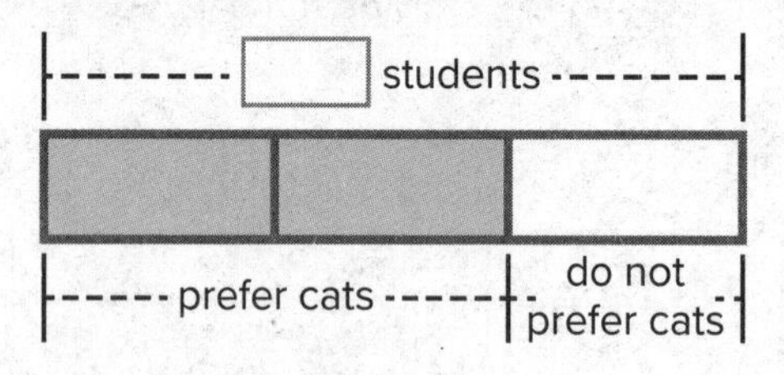

Shade two sections to represent the two out of three students who prefer cats. Label each group and the total number of students at the school, 750.

Step 3 Find the value of each section.

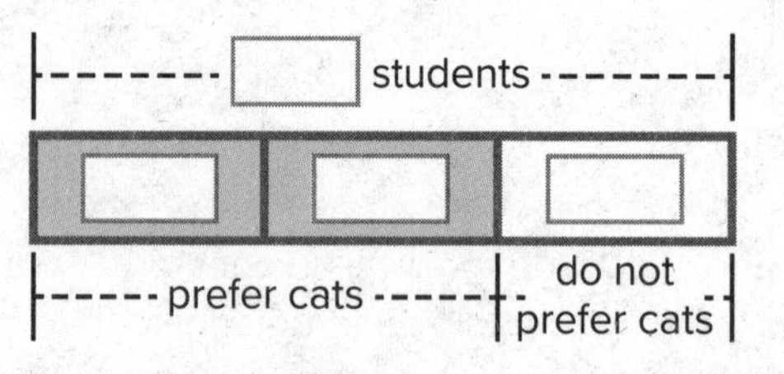

Divide the total number of students by 3 to determine the value of each section. Because $750 \div 3 = 250$, each section represents 250 students.

Because there are two sections labeled *prefer cats*, you can predict that 250×2, or 500 students at Heritage Middle School prefer cats as a household pet.

Talk About It!

How can you use ratio reasoning to check your solution?

Check

A survey of randomly selected students found that out of every ten students, three said they get their news from their cell phone. If there are 750 students at Heritage Middle School, how many students can be expected to get their news from their cell phone? Draw a bar diagram to support your solution.

Go Online You can complete an Extra Example online.

Example 2 Use Bar Diagrams to Solve Ratio Problems

During their family vacation, Marcus took 18 photos on his cell phone. The ratio of the number of photos Marcus took to the number of photos his sister Maribel took is 3 to 4.

How many photos did Maribel take?

You can use a bar diagram to solve the problem.

Step 1 Draw a bar.

The ratio of the number of photos Marcus took to the number Maribel took is 3 : 4, so divide the bar into four equal sections.

Step 2 Shade and label the diagram.

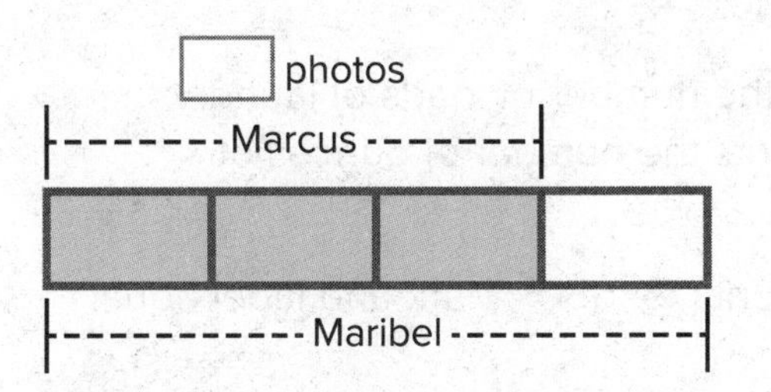

Shade three sections to represent the ratio 3 : 4 and add labels for Marcus and Maribel. Because Marcus took 18 photos, label the three shaded sections as 18 photos.

Step 3 Find the value of each section.

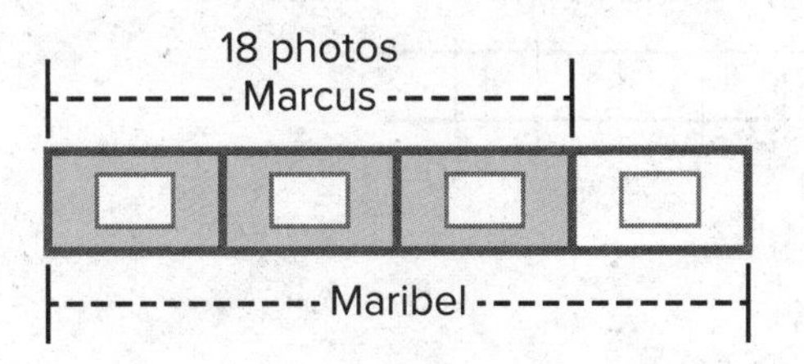

Divide the total number of photos Marcus took by 3 to determine the value of each section. Because $18 \div 3 = 6$, each section represents 6 photos.

There are four sections that represent the number of photos Maribel took. Multiply 6 by 4. So, Maribel took a total of 6×4, or 24 photos on their vacation.

Check

Refer to Example 2. Suppose Marcus took a total of 36 photos. If the ratio of the number of photos Marcus took to the number of photos Maribel took remained 3 : 4, how many photos did Maribel take?

Go Online You can complete an Extra Example online.

Think About It!
Is the number of photos Maribel took less than, greater than, or equal to 18? How do you know?

Talk About It!
How does the bar diagram indicate how many more photos Maribel took than Marcus?

Learn Use Double Number Lines and Equivalent Ratios to Solve Ratio Problems

The manager of a small hotel determines that it takes 30 loads of laundry to clean the towels and sheets of the hotel's rooms each day. A large bottle of laundry detergent contains 150 ounces and the label indicates that the contents of the bottle can clean 75 loads. How many ounces of detergent are needed to clean the hotel's towels and sheets each day?

You can represent this ratio relationship and solve the problem by using double number lines and equivalent ratios.

Talk About It!

Why might a bar diagram not be the best representation to help solve this problem?

Method 1 Use a double number line.

Step 1 Draw a double number line.

The top number line represents the number of loads of laundry. The bottom number line represents the number of ounces of detergent needed.

Mark the ratio of loads to detergent (75 : 150). Mark and label equal increments to show 30 loads.

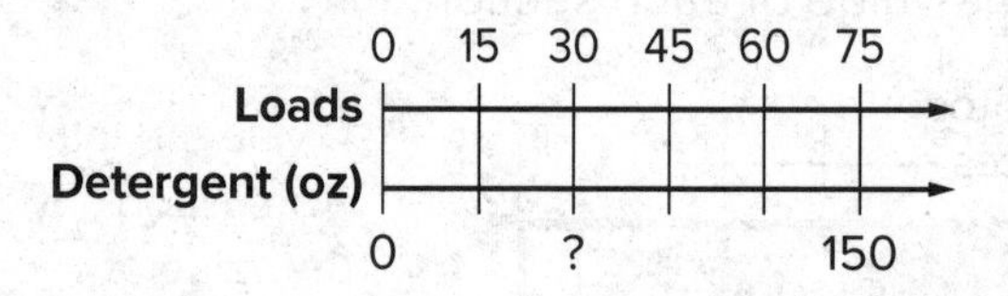

Step 2 Find the equivalent ratio.

There are 5 equal sections. Because $150 \div 5 = 30$, label equal increments of 30 on the bottom number line.

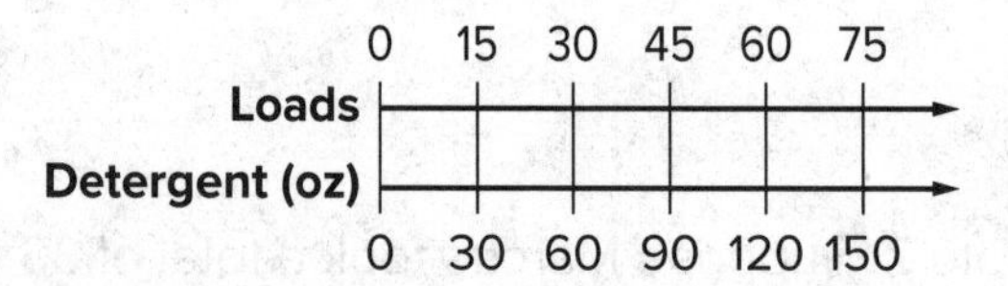

The value on the bottom number line that corresponds with 30 loads is 60 ounces of detergent.

So, 60 ounces of detergent are needed each day.

(continued on next page)

Method 2 Use equivalent ratios.

Write and solve an equation stating that two ratios are equivalent. Let d represent the unknown number of ounces of detergent needed to clean 30 loads of laundry.

loads of laundry → $\frac{30}{d} = \frac{75}{150}$ ← loads of laundry
ounces of detergent → ← ounces of detergent

$$\frac{30}{d} = \frac{75}{150} \quad (\div 2.5)$$

Because $75 \div 2.5 = 30$, divide 150 by 2.5 to find the value of d.

$$\frac{30}{60} = \frac{75}{150}$$

$150 \div 2.5 = 60$; So, $d = 60$.

So, using either method, 60 ounces of detergent are needed to clean the hotel's towels and sheets each day.

Talk About It!

Compare and contrast using a double number line and equivalent ratios. Which method might be more advantageous to use if the numbers are large?

Example 3 Use Double Number Lines and Equivalent Ratios to Solve Ratio Problems

The manager of a grocery store determines that an average of 480 jars of peanut butter are sold each week. Two cases of peanut butter contain 96 jars.

How many cases of peanut butter should the manager order each week?

Method 1 Use a double number line.

Step 1 Draw the double number line.

The top number line represents the number of cases of peanut butter. The bottom number line represents the number of jars of peanut butter.

Mark the ratio of cases to jars (2 : 96). Mark and label equal increments to show 480 jars.

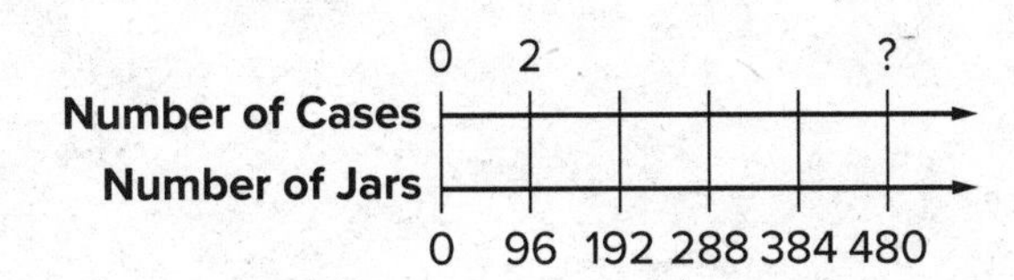

Think About It!

Can you solve this problem mentally without using any diagrams? Explain.

(continued on next page)

Step 2 Find the equivalent ratio.

There are 5 equal sections. Label equal increments of 2 on the top number line.

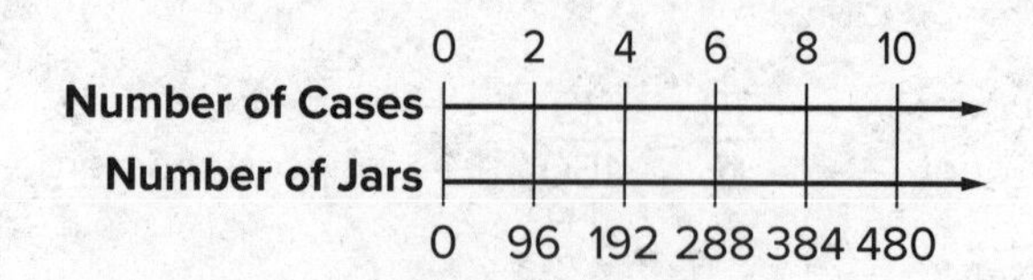

The value on the top number line that corresponds with 480 jars is 10 cases. So, 10 cases should be ordered each week.

Method 2 Use equivalent ratios.

Write and solve an equation stating that two ratios are equivalent. Let c represent the unknown number of cases the manager should order each week.

number of cases → $\frac{2}{96} = \frac{c}{480}$ ← number of cases
number of jars → ← number of jars

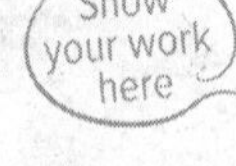

$$\frac{2}{96} = \frac{c}{480}$$ (× 5 on both numerator and denominator)

Because $96 \times 5 = 480$, multiply 2 by 5 to find the value of c.

$$\frac{2}{96} = \frac{10}{480}$$

$2 \times 5 = 10$; So, $c = 10$.

So, using either method, the manager should order ______ cases of peanut butter each week.

Check

The manager of a bakery determines that an average of 112 loaves of cheese bread are sold each week. For every 2 loaves of cheese bread that are sold, about 3 loaves of whole wheat bread are sold. About how many loaves of whole wheat bread are sold each week?

Show your work here

Talk About It!

How can you use scaling and a table of equivalent ratios to solve this problem?

 Go Online You can complete an Extra Example online.

Apply Inventory

The manager of an office supply store decides to hold a *Buy 2, Get 1 Free* sale on all reams of paper. A *ream* of paper holds 500 sheets of paper. The sale is held for one week and a total of 154 reams of paper were sold (not including the ones given away for free). If each ream of paper cost the store $4.50, how much money did the store lose by giving away the free reams of paper?

1 What is the task?

Make sure you understand exactly what question to answer or problem to solve. You may want to read the problem three times. Discuss these questions with a partner.

First Time Describe the context of the problem, in your own words.
Second Time What mathematics do you see in the problem?
Third Time What are you wondering about?

2 How can you approach the task? What strategies can you use?

3 What is your solution?

Use your strategy to solve the problem.

Talk About It!
Why do you think stores offer sales, such as *Buy 2, Get 1 Free?*

4 How can you show your solution is reasonable?

Write About It! Write an argument that can be used to defend your solution.

Check

The manager of a clothing store decides to hold a *Buy 1, Get 2 Free* sale on all pairs of socks. The sale is held for one week and a total of 182 pairs of socks were sold (not including the ones given away for free). If each pair of socks cost the store $2.50, how much money did the store lose by giving away the free socks?

Go Online You can complete an Extra Example online.

Pause and Reflect

What are the advantages of using a bar diagram to solve ratio problems? When might it be more advantageous to use double number lines or equivalent ratios?

Name _______________ Period _______ Date _______

Practice

Go Online You can complete your homework online.

Use any strategy to solve each problem. (Examples 1–3)

1. A survey showed that 4 out of 5 students own a bicycle. Based on this result, how many of the 800 students in a school own a bicycle?

2. A survey of Mr. Thorne's class shows that 5 out of 8 students will buy lunch today. Based on this result, how many of the 720 students in the school will buy today?

3. The ratio of the number of baskets made by Tony to the number of baskets made by Colin is 2 to 3. Tony made 10 baskets. How many baskets did Colin make?

4. In the school choir, there is 1 boy for every 4 girls. There are a total of 11 boys. How many girls are in the choir?

5. Liberty Middle School has 600 students. In Anna's class, 3 out of 8 students walk to school. How many students at the school can be expected to walk to school?

6. Pine Hill Middle School has 300 students. In Zoey's class, 2 out of 5 students belong to a club. How many students at the school would you expect belong to a club?

7. In a survey, the ratio of students who prefer popcorn to potato chips is 3 to 4. If the number of students surveyed who prefer popcorn is 360, how many preferred potato chips?

Test Practice

8. **Open Response** In a neighborhood, the ratio of houses with swing sets to houses without swing sets is 3 to 5. If the number of houses with swing sets is 270, how many houses do not have swing sets?

Apply

9. The manager of an art supply store decides to hold a *Buy 2, Get 1 Free* sale on tubes of watercolor paints. The sale is held for one week and a total of 280 tubes of paint were sold (not including the ones given away for free). If each tube of watercolor paint cost the store $7.25, how much money did the store lose by giving away the free tubes of paint?

10. The manager of a garden store decides to hold a *Buy 3, Get 1 Free* sale on vegetable plants. The sale is held for one week and a total of 636 vegetable plants were sold (not including the ones given away for free). If each plant cost the store $2.90, how much money did the store lose by giving away the free plants?

11. MP **Construct an Argument** Determine if the following statement is *true* or *false*. Construct an argument to defend your response.

In equivalent ratios, if the first quantity of the first ratio is greater than the second quantity of the first ratio, then the first quantity of the second ratio is less than the second quantity of the second ratio.

12. Compare and contrast the use of bar diagrams and equivalent ratios to solve ratio problems.

13. MP **Persevere with Problems** Suppose 20 out of 140 people said they play tennis and 1 out of every 9 of those players have a tennis coach. Using these same ratios, in a group of 504 people, predict how many you would expect to have a tennis coach. Explain how you made the prediction.

14. Use Math Tools Write and solve a real-world ratio problem that can be solved by using a bar diagram.

Convert Customary Measurement Units

I Can... use ratio reasoning to convert between customary units of measurement.

Today's Standards
6.RP.A.3, 6.RP.A.3.D,
Also addresses 6.RP.A.1
MP1, MP2, MP3, MP5, MP7

What Vocabulary Will You Learn?
unit ratio

Learn Unit Ratios and Measurement Conversions

The table shows the Customary measurement conversions of length, weight, and capacity.

Customary Conversions

Type of Measure	Larger Unit	→	Smaller Unit
Length	1 foot (ft)	=	12 inches (in.)
	1 yard (yd)	=	3 feet
	1 mile (mi)	=	5,280 feet
Weight	1 pound (lb)	=	16 ounces (oz)
	1 ton (T)	=	2,000 pounds
Capacity	1 cup (c)	=	8 fluid ounces (fl oz)
	1 pint (pt)	=	2 cups
	1 quart (qt)	=	2 pints
	1 gallon (gal)	=	4 quarts

Each relationship listed in the table is a ratio relationship. Because there are 12 inches for every 1 foot, the relationship between number of inches and number of feet is a ratio relationship. The ratio of inches to feet is 12 : 1 or 12 to 1.

A **unit ratio** is a ratio in which the first quantity is compared to 1 unit of the second quantity. Each of the conversions can be written as unit ratios. Some examples of unit ratios are shown.

inches to feet	12 : 1
feet to yards	3 : 1
feet to miles	5,280 : 1

What unit ratio can you use to represent the relationship between ounces and pounds? ________

What unit ratio can you use to represent the relationship between pints and quarts? ________

What unit ratio can you use to represent the relationship between feet and miles? ________

Talk About It!
What are some other unit ratios that you can describe from the conversions listed in the table?

Your Notes

Think About It!
Do you think the number of fluid ounces will be less than, greater than, or equal to 6? Why?

Talk About It!
Explain why the number of fluid ounces, 96, is greater than the number of pints, 6.

Learn Convert Larger Units to Smaller Units

You can use reasoning about ratios to convert a measurement from a larger unit to a smaller unit. The numerical value of the measurement is greater when a smaller unit is used. To see why, consider the following problem. Suppose you have 6 pints and you want to know how many fluid ounces are in 6 pints.

Method 1 Use a bar diagram.

Step 1 Draw a bar to represent 6 pints.

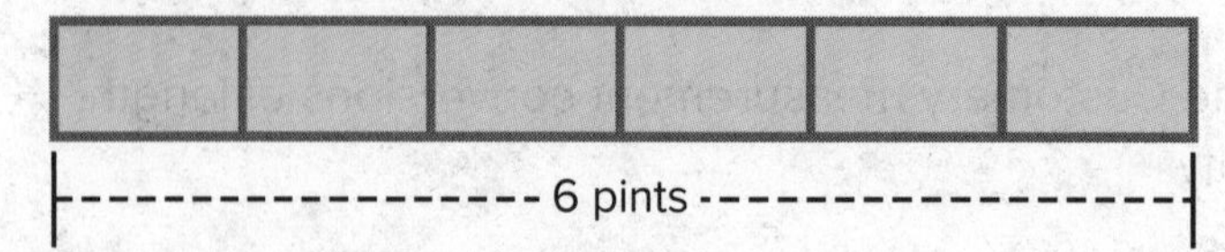

Divide the bar into six equal sections. Each section represents 1 pint.

Step 2 Find the number of cups.

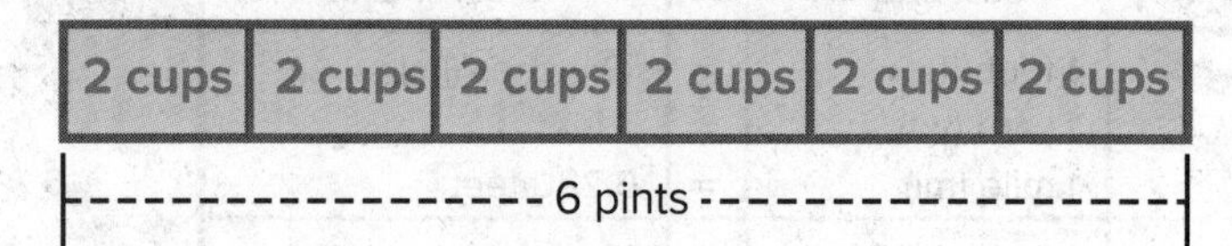

Label each section as 2 cups, because there are 2 cups for every 1 pint.

Step 3 Find the number of fluid ounces.

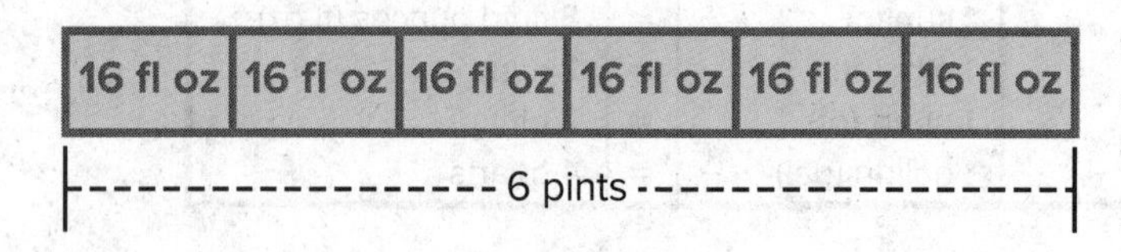

For every 1 cup, there are 8 fluid ounces. This means that for every 2 cups, there are 16 fluid ounces.

Multiply 6 by 16 to find the number of fluid ounces that are in 6 pints. Because $6 \times 16 = 96$, there are 96 fluid ounces in 6 pints.

Method 2 Use unit ratios and equivalent ratios.

Step 1 Convert 6 pints to cups.

There are 2 cups in every 1 pint. The unit ratio of cups to pints is 2 : 1. Let c represent the unknown number of cups that are in 6 pints.

cups → $\frac{2}{1} = \frac{c}{6}$ ← cups
pints → ← pints

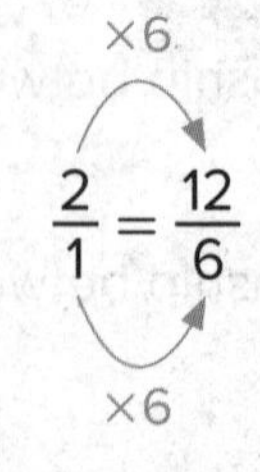

$\frac{2}{1} = \frac{12}{6}$

Because $1 \times 6 = 6$, multiply 2 by 6 to find the value of c. There are 12 cups.

(continued on next page)

Step 2 Convert 12 cups to fluid ounces.

There are 8 fluid ounces in every 1 cup. The unit ratio of fluid ounces to cups is 8 : 1. Let *f* represent the unknown number of fluid ounces.

fluid ounces → $\frac{8}{1} = \frac{f}{12}$ ← fluid ounces
cups → ← cups

$$\frac{8}{1} \overset{\times 12}{=} \frac{96}{12}$$

Because $1 \times 12 = 12$, multiply 8 by 12 to find the value of *f*. There are 96 fluid ounces.

Using either method, there are ______ fluid ounces in 6 pints.

Talk About It!
Compare the use of the bar diagram to using equivalent ratios. Which method is more advantageous to use to visualize the relationship?

Example 1 Convert Larger Units to Smaller Units

Marco needs to mix $\frac{1}{2}$ gallon of fertilizer with some soil before planting his tulip bulbs.

How many cups of fertilizer should Marco use?

Method 1 Use a bar diagram.

Step 1 Draw a bar to represent 1 gallon.

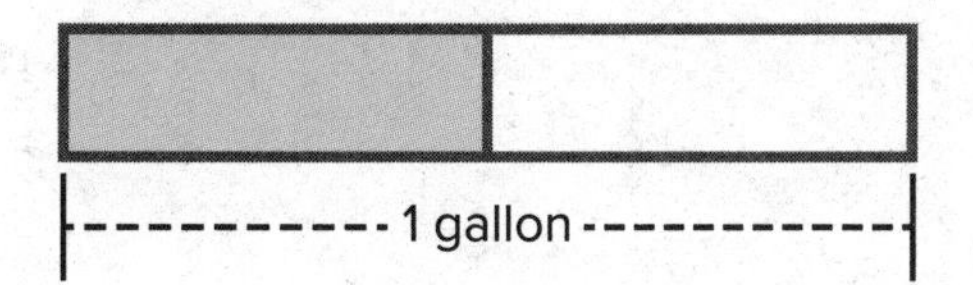

Divide the bar into two equal sections. Shade one section to represent $\frac{1}{2}$ gallon.

Step 2 Find the number of quarts.

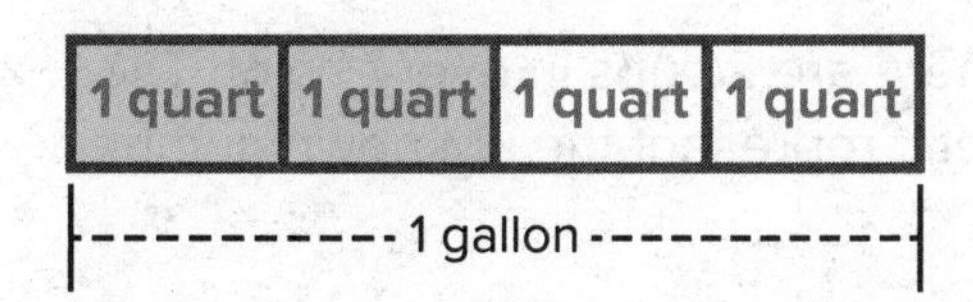

There are 4 quarts in 1 gallon so there are 2 quarts in a $\frac{1}{2}$ gallon. Divide each half into two sections. Label each section as 1 quart.

Step 3 Find the number of pints.

2 pints	2 pints	2 pints	2 pints

1 gallon

For every 1 quart, there are 2 pints. Label each section as 2 pints.

Talk About It!
Suppose Marco needed to find the number of cups that are in $\frac{1}{3}$ gallon. Why might a bar diagram not be the most advantageous method to use in this case?

(continued on next page)

Step 4 Find the number of cups. For every 1 pint, there are 2 cups. This means that for every 2 pints, there are 4 cups.

4 cups	4 cups	4 cups	4 cups

1 gallon

There are two shaded sections that each represent 4 cups. So there are 2×4 or 8 cups in $\frac{1}{2}$ gallon.

Method 2 Use unit ratios and equivalent ratios.

Step 1 Convert $\frac{1}{2}$ gallon to quarts. There are 4 quarts in every 1 gallon. The unit ratio of quarts to gallons is 4 : 1. Let q represent the unknown number of quarts.

$$\frac{4}{1} = \frac{q}{\frac{1}{2}}$$ (quarts over gallons)

$$\frac{4}{1} \overset{\div 2}{=} \frac{2}{\frac{1}{2}}$$

Because $1 \div 2 = \frac{1}{2}$, divide 4 by 2 to find the value of q. There are 2 quarts.

Step 2 Convert 2 quarts to pints. There are 2 pints in every 1 quart. The unit ratio of pints to quarts is 2 : 1. Let p represent the unknown number of pints.

$$\frac{2}{1} = \frac{p}{2}$$ (pints over quarts)

$$\frac{2}{1} \overset{\times 2}{=} \frac{4}{2}$$

Because $1 \times 2 = 2$, multiply 2 by 2 to find the value of p. There are 4 pints.

Step 3 Convert 4 pints to cups. There are 2 cups in every 1 pint. The unit ratio of cups to pints is 2 : 1. Let c represent the unknown number of cups.

$$\frac{2}{1} = \frac{c}{4}$$ (cups over pints)

$$\frac{2}{1} \overset{\times 4}{=} \frac{8}{4}$$

Because $1 \times 4 = 4$, multiply 2 by 4 to find the value of c. There are 8 cups.

So, Marco should use ____ cups of fertilizer.

(continued on next page)

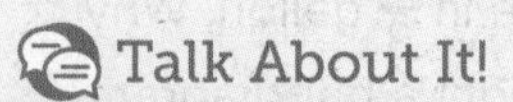
Talk About It!
Explain why it makes sense that the number of cups of fertilizer that are in $\frac{1}{2}$ gallon is greater than $\frac{1}{2}$.

Check

How many ounces are in $\frac{1}{4}$ pound?

 **Go Online** You can complete an Extra Example online.

Learn Convert Smaller Units to Larger Units

You can use reasoning about ratios to convert a measurement from a smaller unit to a larger unit. The numerical value of the measurement is less when a larger unit is used. To see why, consider the following problem. Suppose you want to convert 24 inches to yards.

Method 1 Use a bar diagram.

Step 1 Find the number of feet.

Draw a bar with 24 equal sections to represent 24 inches. For every 12 inches, there is 1 foot. Mark equal increments of 12 inches.

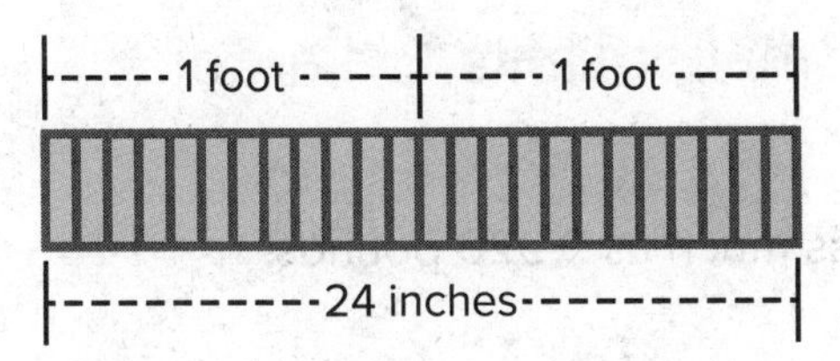

There are 2 whole feet in 24 inches.

Step 2 Find the number of yards.

For every 3 feet, there is 1 yard. There are only 2 feet. Another foot is needed to have 1 whole yard.

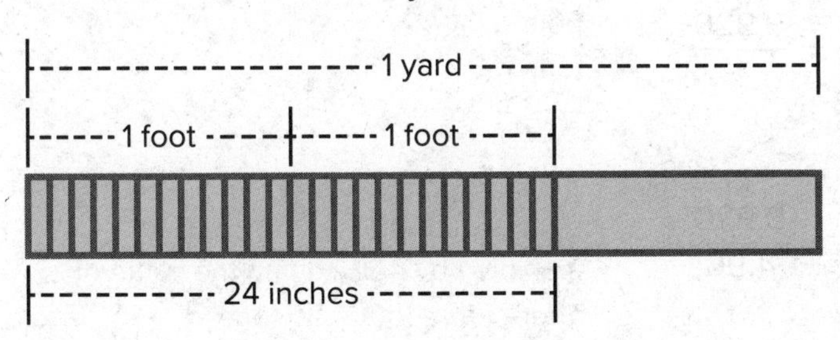

There are only two out of three sections shaded. So, there are 24 inches in $\frac{2}{3}$ yard.

(continued on next page)

Talk About It!

Why might it not always be advantageous to use a bar diagram to convert measurement units? Would you choose to use a bar diagram to convert 126 inches to yards? Why or why not?

Method 2 Use unit ratios and equivalent ratios.

Step 1 Convert 24 inches to feet.

There are 12 inches in every 1 foot. The unit ratio of inches to feet is 12 : 1. Let *f* represent the unknown number of feet.

$$\frac{12}{1} = \frac{24}{f}$$ (inches → top, feet → bottom)

$$\frac{12}{1} \underset{\times 2}{\overset{\times 2}{=}} \frac{24}{2}$$

Because 12 × 2 = 24, multiply 12 by 2 to find the value of *f*. There are 2 feet.

Step 2 Convert 2 feet to yards.

Because there are 3 feet in every 1 yard, and there are only 2 feet, the number of yards is $\frac{2}{3}$.

So, using either method, there are 24 inches in $\frac{2}{3}$ yard.

Think About It!
Will the number of tons be less than, greater than, or equal to 9,920? Explain.

Example 2 Convert Smaller Units to Larger Units

A male hippopotamus can weigh as much as 9,920 pounds.

How much is this weight in tons?

Use unit ratios and equivalent ratios.

There are 2,000 pounds for every 1 ton. The unit ratio of pounds to tons is 2,000 : 1. Let *t* represent the unknown number of tons.

$$\frac{2{,}000}{1} = \frac{9{,}920}{t}$$ (pounds → top, tons → bottom)

$$\frac{2{,}000}{1} \underset{\times 4.96}{\overset{\times 4.96}{=}} \frac{9{,}920}{4.96}$$

Because 2,000 × 4.96 = 9,920, multiply 1 by 4.96 to find the value of *t*. There are about 4.96 tons.

So, the male hippopotamus can weigh as much as 4.96 tons.

Talk About It!
How do you know that the number of tons should be less than 5, but very close to 5?

Check

How many yards are in 54 inches?

Go Online You can complete an Extra Example online.

Apply Soccer Practice

The table shows the amount of drinking water each athlete drinks during one soccer practice. The coach pays $1.75 per quart of bottled water. How much will the coach spend on water for one practice session?

Athlete	Amount (c)
Deon	2
Sierra	1.5
Carmen	3.5
Mia	3
Ella	2

1 What is the task?

Make sure you understand exactly what question to answer or problem to solve. You may want to read the problem three times. Discuss these questions with a partner.

First Time Describe the context of the problem, in your own words.
Second Time What mathematics do you see in the problem?
Third Time What are you wondering about?

2 How can you approach the task? What strategies can you use?

3 What is your solution?

Use your strategy to solve the problem.

4 How can you show your solution is reasonable?

Write About It! Write an argument that can be used to defend your solution.

Talk About It!
Is the number of quarts of water for one practice session less than, greater than, or equal to the number of cups of water for one practice session? Explain.

Check

On Tuesday, Joaquin drank 6 glasses of water each containing 10 fluid ounces. His goal was to drink 2 quarts. How many more fluid ounces does he need to drink in order to reach his goal?

Show your work here

Go Online You can complete an Extra Example online.

Pause and Reflect

What are the advantages of using a bar diagram to convert Customary measurement units? When might it be more advantageous to use unit ratios and equivalent ratios?

Record your observations here

Name ______________________ Period __________ Date __________

Practice

Go Online You can complete your homework online.

Use any strategy to solve each problem.

1. Mrs. Menary made $4\frac{1}{2}$ quarts of lemonade for a school party. How many fluid ounces of lemonade did she make? **(Example 1)**

2. A class walked 2.5 miles for a walk-a-thon. How many yards did the class walk? **(Example 1)**

3. The Martinez family has $\frac{3}{4}$ gallon of orange juice in the refrigerator. How many cups of orange juice are in the refrigerator? **(Example 1)**

4. A grand piano can weigh $\frac{1}{2}$ ton. How many ounces can a grand piano weigh? **(Example 1)**

5. A female hippopotamus can weigh 48,000 ounces. How many tons can a female hippopotamus weigh? **(Example 2)**

6. At soccer practice, Tracey's best kick traveled a distance of 1,200 inches. For how many yards did she kick the ball? **(Example 2)**

7. An elephant can drink up to 6,400 fluid ounces of water a day. How many gallons of water can an elephant drink per day? **(Example 2)**

8. A recipe for ice cream calls for 56 fluid ounces of milk. How many pints of milk are there in the recipe? **(Example 2)**

9. One quart of strawberries weighs about 2 pounds. About how many quarts of strawberries would weigh $\frac{1}{4}$ ton?

Test Practice

10. Open Response A mini fruit juice box contains 4 fluid ounces of juice. You need $2\frac{1}{2}$ quarts of fruit juice. How many mini fruit juice boxes will you need?

Apply

11. At the grocery store, Mr. Arnett allowed each of his children to fill their own bag with trail mix for their hike. The table shows the amount of trail mix for each child. The trail mix costs $4.50 per pound. How much will Mr. Arnett pay for all the trail mix?

Child	Amount of Trail Mix (oz)
Ava	15
Grayson	14
Mason	10
Tyler	17

12. A hockey player needs to shoot a puck 60 yards from his current location to his opponent's goal to score a goal. After the shot, the puck is 48 inches from his opponent's goal. How many feet did the puck travel?

13. There are 60 minutes in one hour and 60 seconds in one minute. Using this information, explain how you could convert 20 miles per hour to feet per second.

14. MP **Identify Structure** When converting from larger units such as quarts to smaller units such as cups, will the number of smaller units be greater than the number of larger units? Explain your reasoning.

15. Give two different measurements that are equivalent to $3\frac{1}{2}$ pints.

16. MP **Find the Error** A student's work for converting 4 gallons to cups is shown. Find the mistake and correct it.

$$\frac{16 \text{ gallons}}{1 \text{ cup}} = \frac{4 \text{ gallons}}{d}$$

So, d is equal to $\frac{1}{4}$ cup.

Understand Rates and Unit Rates

I Can... understand how a rate is related to a ratio, and use ratio and rate reasoning to find a unit rate.

Today's Standards
6.RP.A.2, 6.RP.A.3, 6.RP.A.3.A, 6.RP.A.3.B
MP1, MP2, MP3, MP6

What Vocabulary Will You Learn?
rate
unit price
unit rate

Explore Compare Quantities with Different Units

Online Activity You will use Web Sketchpad to determine how many noodles a machine can make in various amounts of time, if the machine makes the same number of noodles per second.

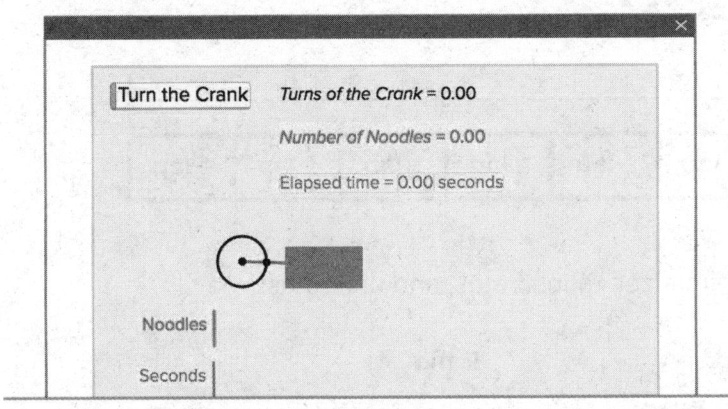

Learn Understand a Rate and a Unit Rate

Luciana ran 4 laps around the track at her middle school in a total of 6 minutes. Suppose she ran at a constant speed. The bar diagram represents the relationship between the number of minutes and the number of laps.

6 minutes			
1 lap	1 lap	1 lap	1 lap

The ratio of the number of minutes to the number of laps is 6 : 4. Because the units, minutes and laps, are different, this kind of ratio is called a rate. A **rate** is a special kind of ratio in which the units are different. The ratio 6 : 4 has the associated rate *6 minutes for 4 laps*.

To find the number of minutes per lap, find the value of each section. Because $6 \div 4 = 1.5$, Luciana ran at a rate of 1.5 minutes per lap.

6 minutes			
1.5 min	1.5 min	1.5 min	1.5 min

This rate is called a unit rate. A **unit rate** is a rate in which the first quantity is compared to 1 unit of the second quantity. The phrase *per* is used to describe unit rates. It means *for each*.

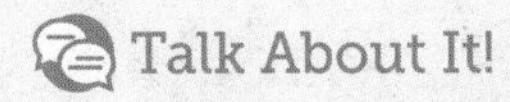

If Luciana's unit rate in minutes per lap is 1.5, how long did it take her to run each lap?

(continued on next page)

Your Notes

![Talk About It!] **Talk About It!**
How does this bar diagram compare to the one on the previous page? Do they represent the same relationship between the two quantities?

Suppose you wanted to find how many laps Luciana can run in 1 minute, at this same rate. The bar diagram represents the relationship between the number of laps and the number of minutes.

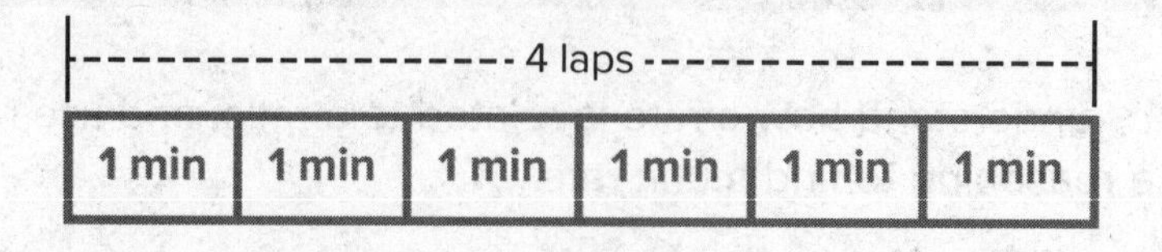

The ratio of the number of laps to the number of minutes is 4 : 6, because Luciana ran 4 laps in 6 minutes. The ratio 4 : 6 has the associated rate *4 laps in 6 minutes*.

To find the number of laps per minute, find the value of each section. Because $4 \div 6 = \frac{4}{6}$, or $\frac{2}{3}$, Melanie ran at a rate of $\frac{2}{3}$ lap per minute.

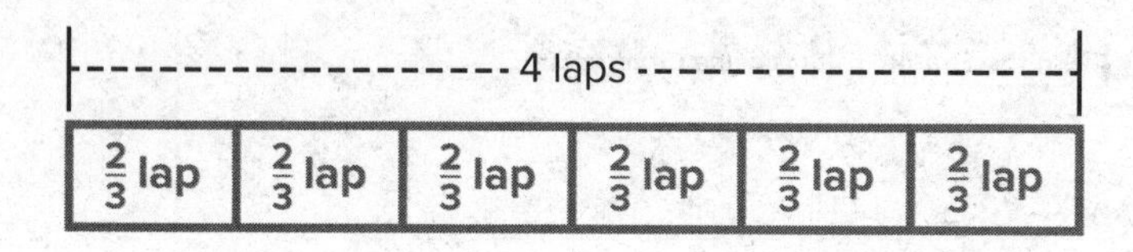

The table summarizes ratios, rates, and unit rates.

Ratio		
Words	**Units**	**Examples**
a comparison between two quantities, in which for every *a* units of one quantity, there are *b* units of another quantity	units can be alike or different	6 laps to 4 laps 6 : 4 4 laps in 6 minutes 4 : 6
Rate		
Words	**Units**	**Examples**
a special kind of ratio in which the units are different	units are different	6 minutes for 4 laps 4 laps in 6 minutes
Unit Rate		
Words	**Units**	**Examples**
a rate in which the first quantity is given for every 1 unit of the second quantity	units are different	1.5 minutes per lap $\frac{2}{3}$ laps per minute

 **Talk About It!**
Which unit rate, minutes per lap or laps per minute, would be helpful if you wanted to predict how many minutes it will take Luciana, at that rate, to run 5 laps? Why?

Example 1 Find a Unit Rate

A scientist studying hummingbirds recorded that a hummingbird flapped its wings 1,590 times in 30 seconds during normal flight.

Assuming a constant rate, how many times did the hummingbird flap its wings per second?

Method 1 Use a ratio table.

Create a ratio table with the given information.

Scale backward to find the number of wing flaps per second.

$\div 30$

Number of Wing Flaps	53	1,590
Number of Seconds	1	30

$\div 30$

Method 2 Use equivalent rates.

Write and solve an equation stating that two rates are equivalent. Let s represent the unknown number of wing flaps per second.

wing flaps → $\frac{s}{1} = \frac{1{,}590}{30}$ ← wing flaps
seconds → ← seconds

$\div 30$

$$\frac{s}{1} = \frac{1{,}590}{30}$$

$\div 30$

Because $30 \div 30 = 1$, divide 1,590 by 30 to find the value of s.

$$\frac{53}{1} = \frac{1{,}590}{30}$$

$1{,}590 \div 30 = 53$; So, $s = 53$.

So, using either method, the hummingbird flapped its wings at a rate of 53 flaps per second.

Think About It!

Why might a bar diagram not be the best method to use to find the unit rate?

Talk About It!

At this rate, how many times would the hummingbird flap its wings in 2 minutes? Justify your response.

Check

Refer to Example 1. The scientist also recorded that the hummingbird took 6,250 breaths over a period of 25 minutes. Assuming a constant rate, how many breaths per minute did the hummingbird take?

Go Online You can complete an Extra Example online.

Learn Unit Price

A grocery store sells a 6-ounce container of yogurt for $0.78. The store also sells a 32-ounce container of the same yogurt for $3.84. To determine which is the better buy – per ounce – find the unit price of each item. The **unit price** is the cost per unit of an item. You can use what you know about unit rates to find a unit price.

6-Ounce Container

Scale backward to find the price per ounce. The unit price is $0.13 per ounce.

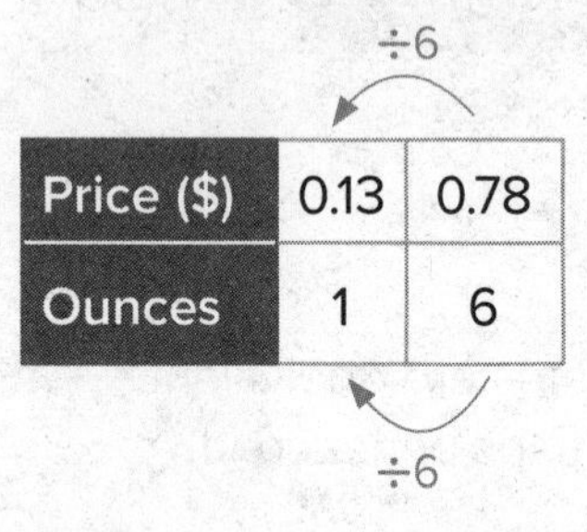

Price ($)	0.13	0.78
Ounces	1	6

32-Ounce Container

Scale backward to find the price per ounce. The unit price is $0.12 per ounce.

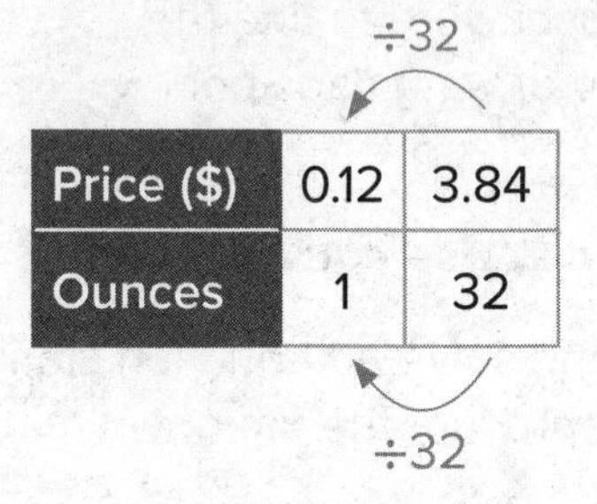

Price ($)	0.12	3.84
Ounces	1	32

Per ounce, the 32-ounce container of yogurt is the better buy, because the unit price is less than that of the 6-ounce container.

Talk About It!

When might it be better to buy the 6-ounce container instead of the 32-ounce container?

Example 2 Find a Unit Price

For Carolina's birthday, her mother took her and four friends to a water park. Carolina's mother can pay either $130 for a 5-pack of student tickets, or $28 for each individual student ticket.

Which ticket payment option has the lesser unit price?

The unit price is given for buying the tickets individually, $28 per ticket. Find the unit price for the 5-pack of student tickets.

Scale backward to find the unit price. The unit price is $26 per ticket.

So, the 5-pack ticket payment option has the lesser unit price because $26 < $28.

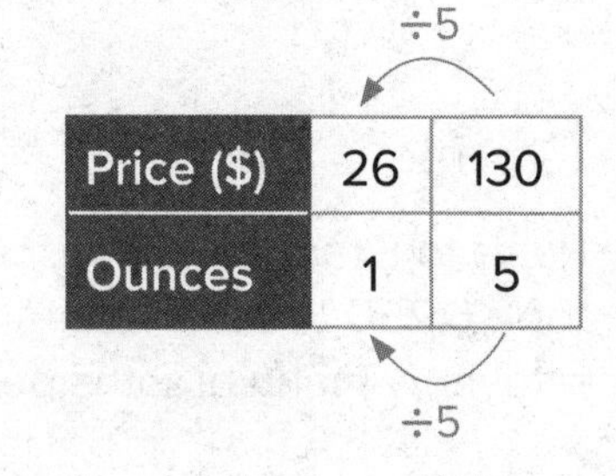

Price ($)	26	130
Ounces	1	5

Check

Refer to Example 2. The water park also offers a 3-pack of student tickets for $82.50. What is the unit price for the 3-pack?

Go Online You can complete an Extra Example online.

Apply Travel

The Martinez family and the Davidson family each drove at a constant rate. The Martinez family drove 260 miles in 4 hours and the Davidson family traveled 305 miles in 5 hours. Which family traveled at a faster rate? How much faster?

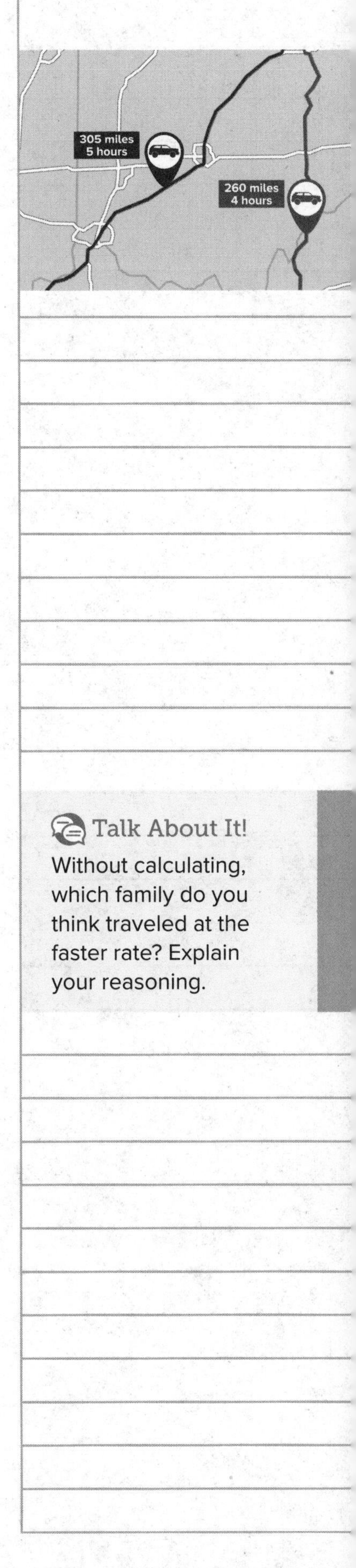

Talk About It!
Without calculating, which family do you think traveled at the faster rate? Explain your reasoning.

1 What is the task?

Make sure you understand exactly what question to answer or problem to solve. You may want to read the problem three times. Discuss these questions with a partner.

First Time Describe the context of the problem, in your own words.
Second Time What mathematics do you see in the problem?
Third Time What are you wondering about?

2 How can you approach the task? What strategies can you use?

3 What is your solution?

Use your strategy to solve the problem.

4 How can you show your solution is reasonable?

Write About It! Write an argument that can be used to defend your solution.

Check

A runner is training for a half marathon. On Wednesday, she ran 6 miles in 40 minutes. On Thursday, she ran 4 miles in 25 minutes. Assume she ran at a constant rate each day. On which day did she run faster? By how much faster did she run?

Go Online You can complete an Extra Example online.

Pause and Reflect

How did what you learned in this lesson relate to a previous lesson or lessons in this module?

Name ________________ Period ________ Date ________

Practice

Go Online You can complete your homework online.

Use any strategy to solve each problem.

1. A hippopotamus can run 6 kilometers in 15 minutes. At this rate, how far can the hippopotamus run in 1 minute? **(Example 1)**

2. Imena earned $261 last week. If she worked 18 hours and earned the same amount each hour, how much was she paid per hour? **(Example 1)**

3. A cat's heart beats approximately 45 times in 15 seconds. At this rate how many times does the cat's heart beat per second? **(Example 1)**

4. Mr. Farley used 4 pounds of hamburger to make 10 hamburger patties of the same size. How many pounds of hamburger did he use per patty? **(Example 1)**

5. At the school festival, Heather can buy 25 game tickets for $10, or she can pay $0.50 per game ticket. Which option has the lesser price per ticket? **(Example 2)**

6. At a toy store, Colton can buy a package of 6 mini footballs for $7.50, or a package of 8 mini footballs for $9.60. Which option has the lesser price per mini football? **(Example 2)**

7. The table shows the options Zoe's mother has for buying tickets to an adventure day camp for Zoe and 5 of her friends. Which option has the lesser cost per student ticket? **(Example 2)**

Adventure Camp Tickets	
Option	**Cost ($)**
6-pack of Student Tickets	126.00
Individual Student Ticket	21.50

Test Practice

8. Multiple Choice Which of the following offers the least price per ounce of shampoo?

Ⓐ $6 for 8 ounces of shampoo

Ⓑ $4 for 5 ounces of shampoo

Ⓒ $8 for 12 ounces of shampoo

Ⓓ $12 for 16 ounces of shampoo

Apply

9. Nolan found two stores that sell filled party favor bags. The table shows his options. Which store has the lesser cost per filled bag? How much less?

Store	Number of Bags	Cost ($)
Party R Us	8	12
Celebrations	12	21

10. The Houck family and Roberts family took trains for their family vacations, traveling at constant rates. The Houck family's train traveled 552 miles in 6 hours and the Roberts family's train traveled 744 miles in 8 hours. Which family's train is traveling at a faster rate? How much faster?

11. Caleb paid $4.50 for 12 bagels. Describe a unit price for bagels that is greater than the unit price Caleb paid.

12. MP **Find the Error** A large box of spaghetti noodles contains 3 pounds and costs $2.40. A student said the unit cost is $1.20 per pound. Is the student correct? Explain.

13. MP **Justify Conclusions** If you travel at a rate of 60 miles per hour, how many minutes will it take you to travel 1 mile? Write and argument that can be used to justify your conclusion.

14. MP **Reason Inductively** Suppose two boxes of cereal contain the same number of ounces but cost different amounts. Without computing, how can you determine which cereal will cost more per ounce of cereal? Explain.

Solve Rate Problems

I Can... solve real-world problems involving rates and unit rates by using bar diagrams, double number lines, and equivalent rates.

Today's Standards
6.RP.A.2, 6.RP.A.3, 6.RP.A.3.B
MP1, MP2, MP3, MP6

Learn Use Bar Diagrams to Solve Rate Problems

Destiny drove 220 miles in 4 hours. Santiago drove 248 miles in 4 hours. At these rates, how many more miles can Santiago drive in 9 hours than Destiny? You can create bar diagrams to solve this rate problem.

Step 1 Construct bar diagrams to represent the rates.

Draw two bars. Each bar represents the number of miles each person drove in 4 hours. Because each person drove 4 hours, divide each bar into 4 equal-size sections. Each section represents 1 hour.

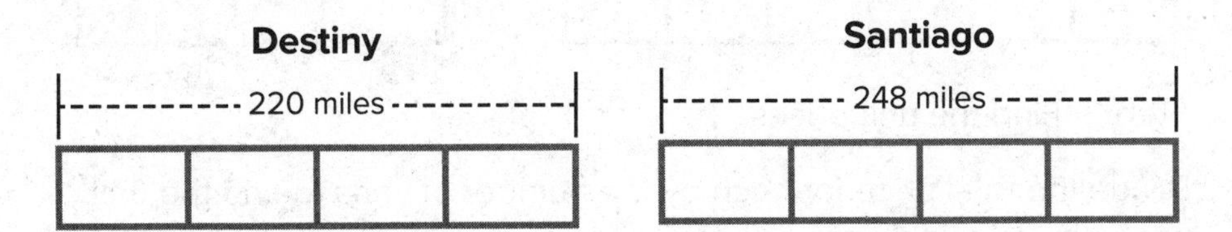

Step 2 Find the unit rates.

Divide the total number of miles each person drove by the number of sections in the diagram to find the unit rate, the number of miles they drove per hour.

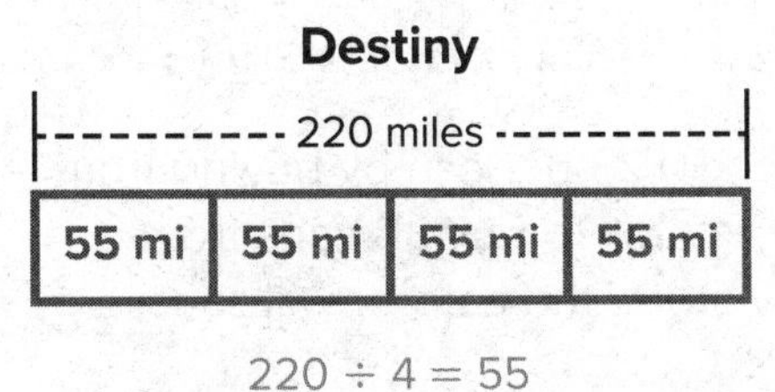

$220 \div 4 = 55$
The unit rate is 55 miles per hour.

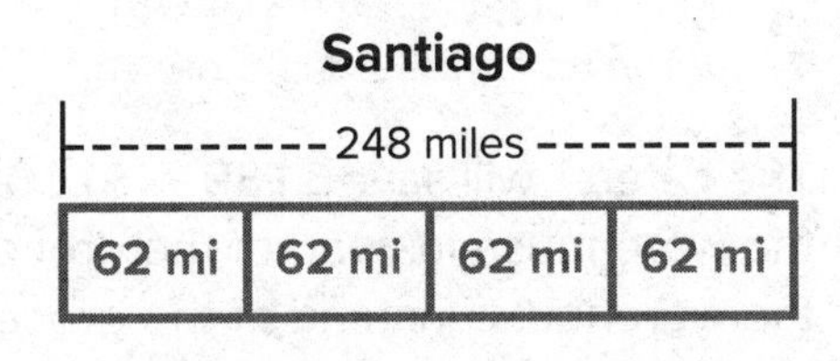

$248 \div 4 = 62$
The unit rate is 62 miles per hour.

Destiny's unit rate is 55 miles per hour. Santiago's unit rate is 62 miles per hour.

Each hour, Santiago can drive $62 - 55$, or 7 miles more than Destiny.

In 9 hours, Santiago can drive 9×7, or 63 miles more than Destiny.

Talk About It!
Can you solve this rate problem another way? Explain.

Your Notes

Think About It!
Why do you need to know the sizes of the cans? Do you need to use that number when solving the problem?

Talk About It!
How can you use estimation to help you solve this problem if you are in a store and do not have access to pencil, paper, or a calculator?

Example 1 Use Bar Diagrams to Solve Rate Problems

A warehouse sells 15-ounce cans of tomato sauce in a case. Each case contains 6 cans and sells for a price of \$9.96. At a local grocery store, three 15-ounce cans of the same brand of tomato sauce are on sale for \$5.67. A caterer needs to buy 36 cans.

How much will the caterer save by buying 36 cans from the warehouse instead of from the grocery store?

Step 1 Construct bar diagrams to represent each situation.

Draw two bars, one to represent the cost of tomato sauce cans at the warehouse, and one to represent the cost of tomato sauce cans at the grocery store. Each section represents one can.

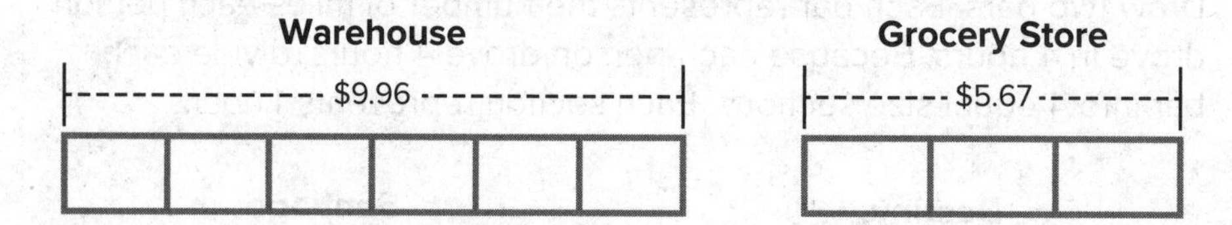

Step 2 Find the unit prices.

Divide the total price for each by the number of cans to find the unit price, the price per can..

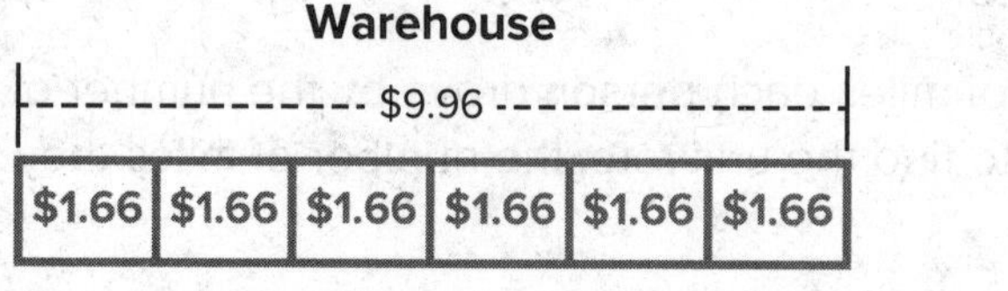

$\$9.96 \div 6 = \$1.66$
The unit price is \$1.66 per can.

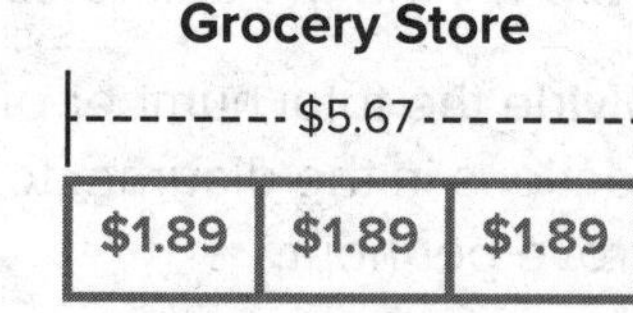

$\$567 \div 3 = \$1.89$
The unit price is \$1.89 per can.

The caterer will save \$1.89 − \$1.66, or \$0.23 per can by buying from the warehouse instead of the grocery store. To buy 36 cans from the warehouse instead of the grocery store, the caterer will save 36 × \$0.23, or ______.

Check

Miranda typed 325 words in 5 minutes, while Joseph typed 295 words in 5 minutes. At these rates, how many more words can Miranda type in 9 minutes than Joseph?

Go Online You can complete an Extra Example online.

Learn Use Double Number Lines and Equivalent Rates to Solve Rate Problems

A veterinarian measured the number of heartbeats of her dog and cat for 4 minutes and recorded the results in the table. At these rates, how many more times does the cat's heart beat in 6 minutes than the dog?

Animal	Heartbeats
Dog	360
Cat	520

Method 1 Use a double number line.

Step 1 Construct a double number line.

In four minutes, the cat's heart beats 520 – 360, or 160 more times than the dog's heart. Draw a double number line to represent this difference.

Difference in Heartbeats

Heartbeats: 0, 160, ?

Minutes: 0, 1, 2, 3, 4, 5, 6

Step 2 Use scaling to find the unit rate.

Scale back to find the difference in heartbeats for 1 minute. Then scale forward to find the difference in heartbeats for 6 minutes.

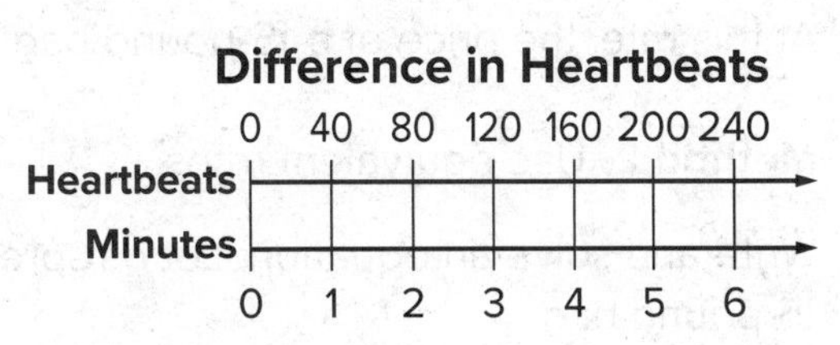

The cat's heart beats 240 more times in 6 minutes than the dog's heart.

Method 2 Use equivalent rates.

Write and solve an equation. Let d represent the difference in heartbeats for 6 minutes. The difference in heartbeats for 4 minutes is 160 beats.

minutes → $\frac{6}{d} = \frac{4}{160}$ ← minutes

difference in heartbeats → ← difference in heartbeats

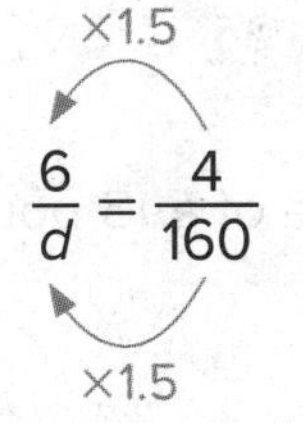

$\frac{6}{d} = \frac{4}{160}$ Because $4 \times 1.5 = 6$, multiply 160 by 1.5.

$\frac{6}{240} = \frac{4}{160}$ $160 \times 1.5 = 240$; So, $d = 240$.

So, using either method, the cat's heart beats 240 more times in 6 minutes than the dog's heart.

Talk About It!

A classmate stated that you can also find each animal's unit rate in heartbeats per minute first. Then multiply each unit rate by 6 minutes to determine the number of heartbeats in 6 minutes for each animal. Finally, subtract to find the difference. Is this method a valid method? Explain.

Example 2 Use Double Number Lines and Equivalent Rates to Solve Rate Problems

A bulk food store sells a 12-pound bag of Red Delicious apples for $18.

At this rate, what is the price of a 15-pound bag of apples?

Method 1 Use a double number line.

Step 1 Construct a double number line.

Draw a double number line to represent the price of a 12-pound bag. Mark equal increments on the bottom number line.

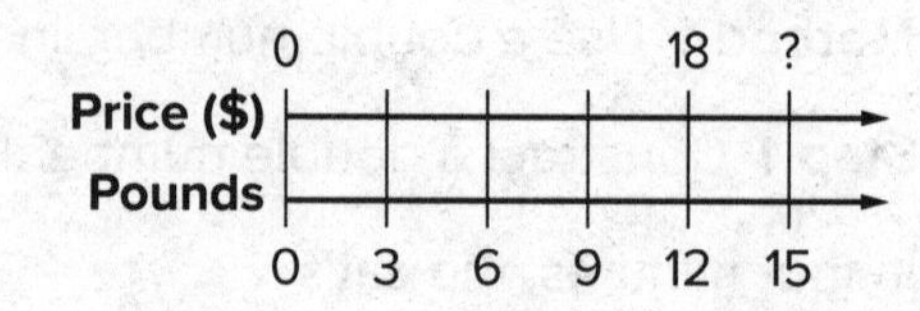

Step 2 Use scaling to find the unit rate.

Scale back to find price for a 3-pound bag. Then scale forward to find the price for a 15-pound bag.

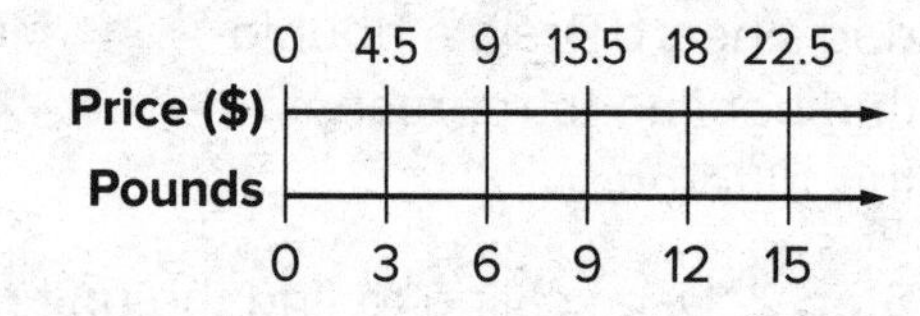

At this rate, the price of a 15-pound bag of apples is $22.50.

Method 2 Use equivalent rates.

Write and solve an equation. Let p represent the price of the 15-pound bag.

pounds → $\frac{15}{p} = \frac{12}{18}$ ← pounds
price ($) → ← price ($)

×1.25

$$\frac{15}{p} = \frac{12}{18}$$

×1.25

Because $12 \times 1.25 = 15$, multiply 18 by 1.25 to find p.

$$\frac{15}{22.5} = \frac{12}{18}$$

$18 \times 1.25 = 22.5$;
So, $p = 22.5$.

So, using either method, the price of a 15-pound bag is $22.50.

Check

The manager of a small bakery determines that an average of 264 loaves of cinnamon raisin bread are sold every 12 weeks. At this rate, about how many loaves of cinnamon raisin bread are sold every 5 weeks?

Go Online You can complete an Extra Example online.

Think About It!

Will the price of a 15-pound bag be less than twice as much as the price for a 12-pound bag? Why or why not?

Talk About It!

The average weight of a Red Delicious apple is about 5 ounces. About how many apples would you expect to be in a 15-pound bag? Explain.

Apply Bike-a-thon

Keshia can ride her bike 15 miles in 90 minutes. She wants to ride in a bike-a-thon that consists of two trail options, a 56-mile trail or a 36-mile trail. At her current rate, how many more hours will it take her to ride 56 miles than 36 miles? If she wants to ride for about 4 hours, which trail should she choose?

1 What is the task?

Make sure you understand exactly what question to answer or problem to solve. You may want to read the problem three times. Discuss these questions with a partner.

First Time Describe the context of the problem, in your own words.
Second Time What mathematics do you see in the problem?
Third Time What are you wondering about?

2 How can you approach the task? What strategies can you use?

3 What is your solution?

Use your strategy to solve the problem.

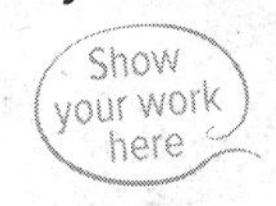

4 How can you show your solution is reasonable?

Write About It! Write an argument that can be used to defend your solution.

Talk About It!

If Keshia raises $1.50 for each mile she rides, how much more money would she raise if she chose the 56-mile trail than the 36-mile trail? Explain.

Check

Martin can run 6 miles in 60 minutes. He wants to run in either one of two upcoming races, a 4-mile race or a 12-mile race. At his current rate, how much longer will it take him to run the 12-mile race than the 4-mile race?

Go Online You can complete an Extra Example online.

Pause and Reflect

What are some problems or situations in which you may have encountered rates, such as a unit price or rate of travel, in your everyday life? How can you use your understanding of ratios and rates to solve everyday problems like these?

Name ______ Period ______ Date ______

Practice

Go Online You can complete your homework online.

Use any strategy to solve each problem.

1. Mr. Anderson is ordering pizzas for a class pizza party. Pizza Place has a special where he can buy 3 large pizzas for $18.75. At Mario's Pizzeria, he can buy 4 large pizzas for $22. If he needs to buy 12 pizzas, how much will he save if he buys the pizzas from Mario's Pizzeria instead of Pizza Place? **(Example 1)**

2. Skylar and Rodrigo each recorded how far they traveled while skateboarding. Skylar traveled 65 feet in 5 seconds and Rodrigo traveled 108 feet in 8 seconds. How much farther did Rodrigo travel per second than Skylar? **(Example 1)**

3. Melissa is buying party favors to make gift bags. Supplies LTD sells a 5-pack of favors for $11.25 and Parties and More sells a 3-pack of favors for $8.25. At these rates, how much more will she spend if she buys 15 favors from Supplies LTD than Parties and More? **(Example 1)**

4. Tara can type 180 words in 4 minutes. At this rate, how many words would you expect her to type in 10 minutes? **(Example 2)**

5. A bakery makes 260 donuts in 4 hours. At this rate, how many donuts can they make in 6 hours? **(Example 2)**

Test Practice

6. **Open Response** While jumping rope, Juan jumped 24 times in 30 seconds. At this rate, how many times will he jump in 50 seconds?

Apply

7. Naomi can run 12 miles in 108 minutes. She is thinking about running in two different races, a 9-mile race and a 13-mile race. At her current rate, how many more minutes will it take her to complete the 13-mile race than the 9-mile race?

8. Leroy wants to buy a new racing bicycle that costs $168. To earn money, he can either do yardwork for his grandmother or babysit his brother and sister. He earns $24 for 3 hours of yardwork and he earns $48 for 4 hours of babysitting. How much longer will it take him to earn the money if he only does yardwork for his grandmother?

9. Billie bikes 9 miles in 45 minutes. At this rate, can she bike 24 miles in 2 hours? Write an argument that can be used to justify your solution.

10. MP **Be Precise** Which method, using a double number line or using equivalent rates, do you prefer to use when solving rate problems? Explain.

11. MP **Persevere with Problems** A fruit stand is selling mandarin oranges for $6 for 4 pounds. A mandarin orange weighs about 2 ounces. There are 16 ounces in a pound. At this rate, how many mandarin oranges can you buy for $9?

12. Create Write and solve a real-world rate problem that you can solve by using a double number line.

Review

Foldables Use your Foldable to help review the module.

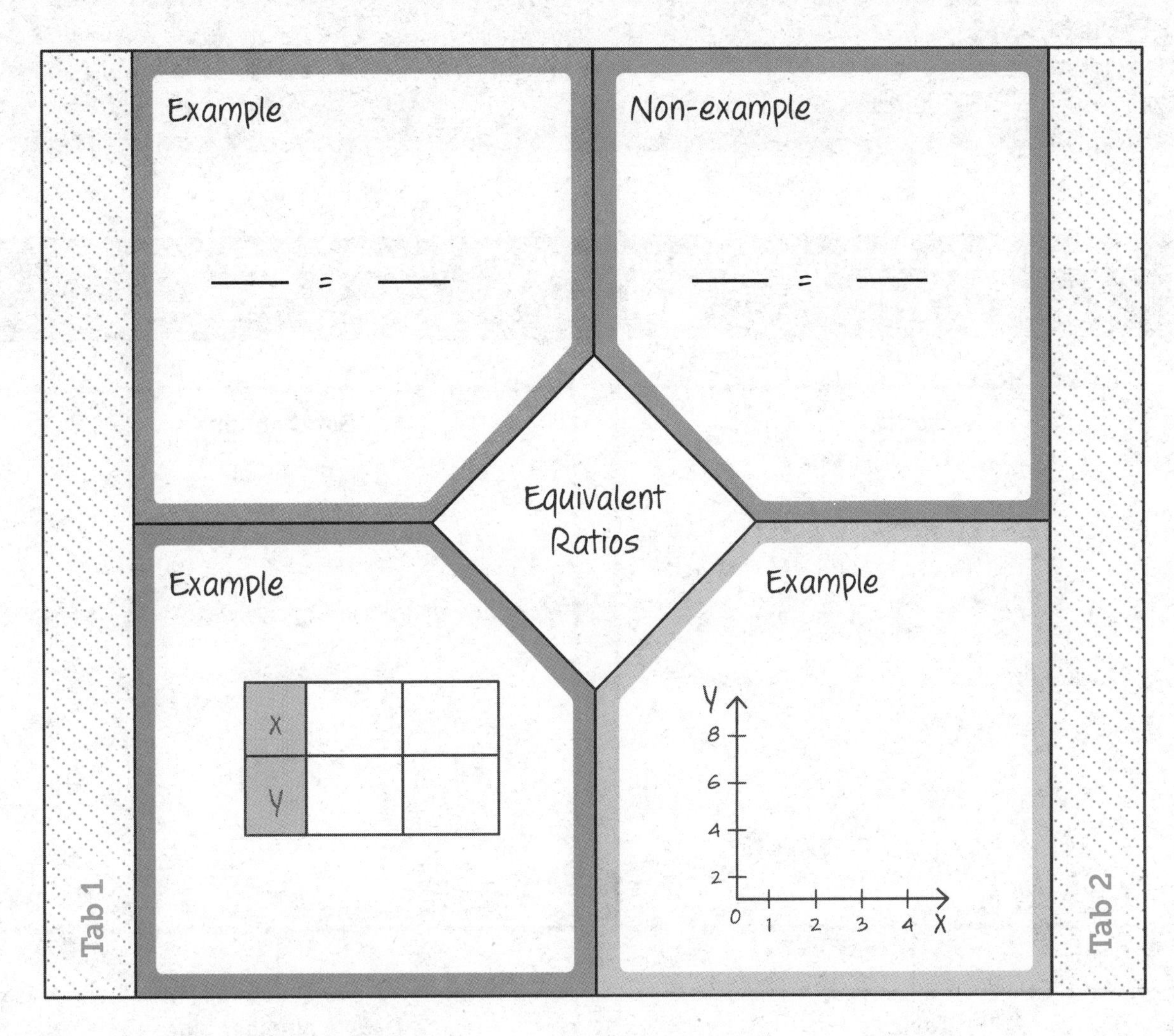

Rate Yourself!   

Complete the chart at the beginning of the module by placing a checkmark in each row that corresponds with how much you know about each topic after completing this module.

Write about one thing you learned.

Write about a question you still have.

Reflect on the Module

Use what you learned about ratios and rates to complete the graphic organizer.

Essential Question

How can you describe how two quantities are related?

Describe how each representation can be used to understand ratios, rates, or unit rates.

Words	Bar Diagrams

Tables	Double Number Lines

Name ________ Period ________ Date ________

Test Practice

1. Equation Editor Jeremy is making a healthy ice cream using only ripe bananas and peanut butter. The recipe makes 4 servings and calls for a ratio 5 bananas to 3 tablespoons of peanut butter. If Jeremy has 30 bananas, how many tablespoons of peanut butter does he need? **(Lesson 1)**

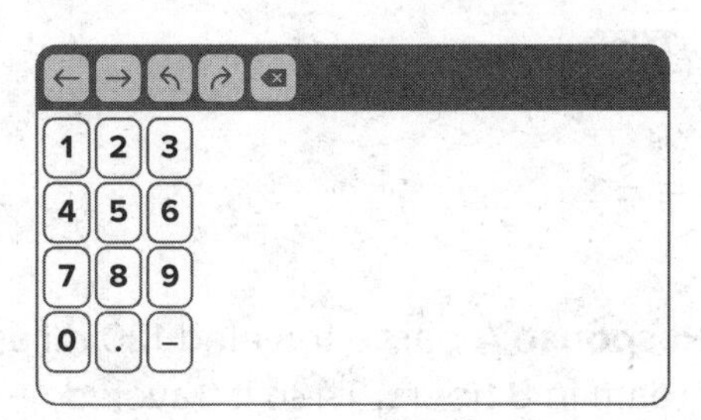

2. Open Response Students at Lincoln Middle School earn $5 for every 4 boxes of cookie dough sold during a fundraiser. Students at Williams Middle School earn $7 for every 6 rolls of wrapping paper sold during their fundraiser. For which fundraiser do students earn the greatest amount of money per item sold? **(Lesson 4)**

3. Multiple Choice A recipe for a punch calls for 12 fluid ounces of orange juice. Reyna needs to make 4 batches of punch for a party. How many quarts of orange juice will Reyna need? **(Lesson 6)**

Ⓐ 0.375 quart

Ⓑ 1.5 quarts

Ⓒ 3 quarts

Ⓓ 6 quarts

4. Table Item Place an X in the column to indicate whether or not Ratio A is equivalent to Ratio B. **(Lesson 2)**

Ratio A	Ratio B	Yes	No
8 questions correct out of 10	4 questions correct out of 5		
15 prizes won in 40 attempts	3 prizes won in 8 attempts		
3 cats for every 6 dogs	1 cat for every 3 dogs		

5. Multiselect Which of the following rates are unit rates? Select all that apply. **(Lesson 7)**

☐ 65 miles per hour

☐ 2 degrees every half hour

☐ 3.2 inches of rain in 2 days

☐ 3 questions for each lesson

6. Open Response The table shows the number of canned goods collected by three different homerooms during a food drive. **(Lesson 2)**

Homeroom	Number of Students	Goods Collected
Mr. Alvarez	25	150
Ms. Jensen	28	154
Mrs. Saunders	27	162

Are the ratios of canned goods per student equivalent between any or all of the classes? Explain your reasoning.

7. Open Response Jessica jogged 4 laps around a track in 9 minutes, Luke jogged 8 laps in 27 minutes. Their rates can be expressed as the ratios $\frac{4 \text{ laps}}{9 \text{ minutes}}$ and $\frac{8 \text{ laps}}{27 \text{ minutes}}$. Are Jessica and Luke's rates equivalent? Explain. (Lesson 7)

8. Grid Kurt uses 3 cups of flour for every 2 cups of sugar in a recipe. Graph the ordered pairs to represent the cups of sugar needed if he uses 3, 6, 9, or 12 cups of flour. (Lesson 3)

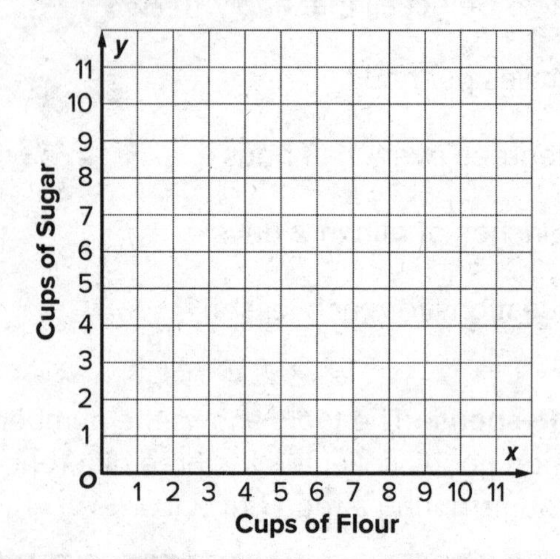

9. Open Response Abigail surveyed 40 students about their favorite kind of movie. The results are shown in the table. If there are 200 students in the school, predict how many more students prefer action movies to scary movies. (Lesson 7)

Type of Movie	Number of Students
Action	14
Animated	3
Comedy	10
Drama	4
Scary	9

10. Multiple Choice Three out of 5 students at Maria's school participate in a school club or sport. There are 175 students at the school. Which of the following shows how equivalent fractions can be used to find the total number of students that participate in a school club or sport? (Lesson 5)

Ⓐ $\frac{3}{5} = \frac{s}{175}$

Ⓑ $\frac{3}{5} = \frac{175}{s}$

Ⓒ $\frac{3}{175} = \frac{s}{5}$

11. Open Response A barge traveled 120 miles downstream in 8 hours. Then it traveled 100 miles upstream in 10 hours. (Lesson 8)

A. How did the rate of speed downstream compare to its rate of speed upstream?

B. What was the difference between the rates of speed?

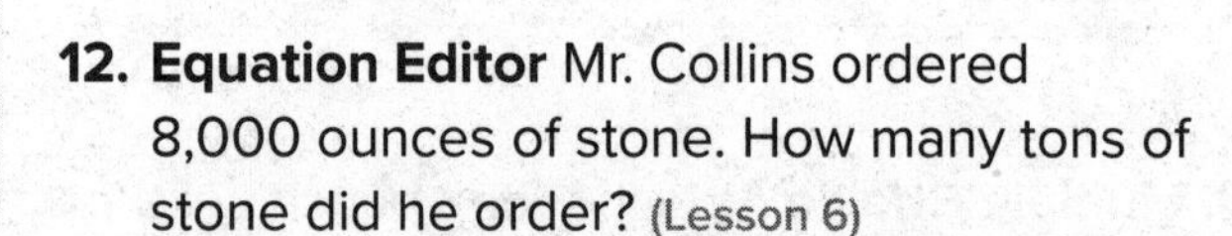

12. Equation Editor Mr. Collins ordered 8,000 ounces of stone. How many tons of stone did he order? (Lesson 6)

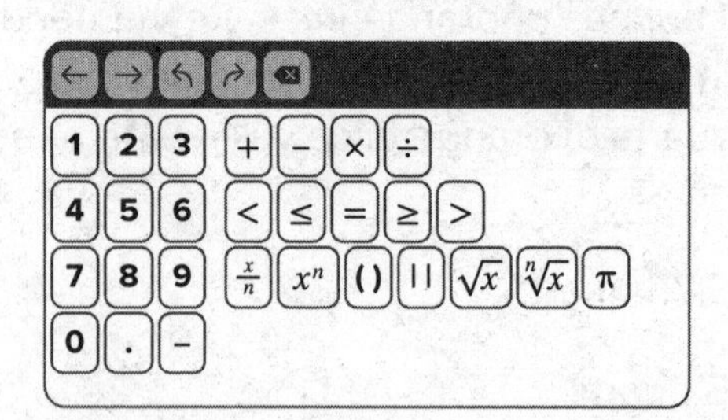

Module 2

Fractions, Decimals, and Percents

Essential Question

How can you use fractions, decimals, and percents to solve everyday problems?

6.RP.A.3, 6.RP.A.3.C
Mathematical Practices: MP1, MP2, MP3, MP5, MP6, MP7

What Will You Learn?

Place a checkmark (✓) in each row that corresponds with how much you already know about each topic **before** starting this module.

KEY
□ — I don't know. ◇ — I've heard of it. ☆ — I know it!

	Before			After		
	□	◇	☆	□	◇	☆
identifying a percent as a rate per 100						
representing percents with 10 × 10 grids and bar diagrams						
writing fractions or mixed numbers as percents						
writing percents as fractions or mixed numbers						
writing decimals as percents						
writing percents as decimals						
finding the percent of a number						
using benchmark percents to estimate the percent of a number						
finding the whole, given a percent and the part of a number						

Foldables Cut out the Foldable and tape it to the Module Review at the end of the module. You can use the Foldable throughout the module as you learn about percents.

What Vocabulary Will You Learn?

Check the box next to each vocabulary term that you may already know.

☐ benchmark percents

☐ percent

Are You Ready?

Study the Quick Review to see if you are ready to start this module.
Then complete the Quick Check.

Quick Review

Example 1

Use part to whole ratios.

The ratio of strawberries to total ingredients in a recipe is 2 to 5. If you have 35 total ingredients, how many are strawberries?

strawberries → $\frac{2}{5} = \frac{s}{35}$ ← strawberries
total ingredients → ← total ingredients

$\frac{2}{5} = \frac{s}{35}$ (×7) Because $5 \times 7 = 35$, multiply 2 by 7 to find the value of s.

$\frac{2}{5} = \frac{14}{35}$ $2 \times 7 = 14$; So, $s = 14$.

So, 14 strawberries are needed to maintain the ratio in the recipe.

Example 2

Use place value to write decimals in word form.

Write each decimal in word form.

0.3 The place value of the last digit, 3, is tenths.

word form: *three tenths*

2.15 The place value of the last digit, 5, is hundredths.

word form: *two and fifteen hundredths*

Quick Check

1. The ratio of cups of borax to total ingredients in a recipe for homemade laundry detergent is 2 : 6. If you need 24 total cups of laundry detergent, how many cups of borax do you need?

2. Write 0.212 in word form.

3. Write 0.145 in word form.

How Did You Do?

Which exercises did you answer correctly in the Quick Check? Shade those exercise numbers at the right.

Understand Percents

I Can... understand the meaning of a percent as a rate per 100, and model percents using 10 × 10 grids and bar diagrams.

Today's Standards
Preparation for 6.RP.A.3, 6.RP.A.3.C
MP2, MP3, MP5, MP7

What Vocabulary Will You Learn?
percent

Learn Use 10 × 10 Grids to Model Percents

A **percent** is a ratio, or rate, that compares a number to 100. *Percent* means *per hundred* and is represented by the symbol %.
For example, 50% means 50 per 100 and is read as *fifty percent*. It represents the ratio 50 : 100, 50 to 100, or $\frac{50}{100}$.

A 10 × 10 grid can be used to model a percent. Because there are 100 squares, each square represents 1%. The 10 × 10 grid shown represents 45% because the ratio of shaded squares to the total number of squares is 45 : 100.

Example	Model
45% means 45 per 100 45 : 100, 45 to 100, or $\frac{45}{100}$ *forty-five percent*	

Other ways to model 45% using a 10 × 10 grid are shown. Note that you do not need to shade the squares in any particular order. As long as the number of shaded squares is 45, you have correctly modeled 45%.

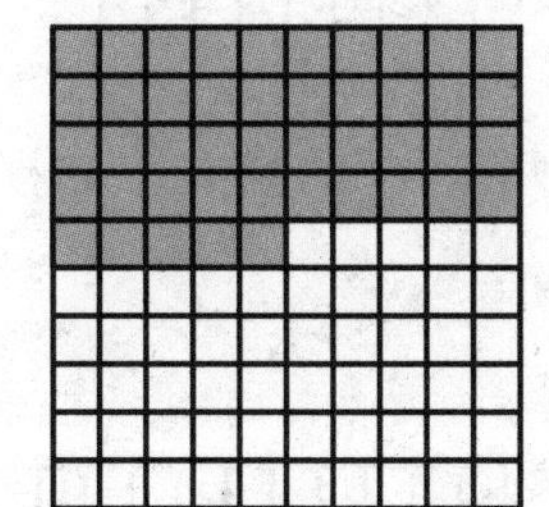

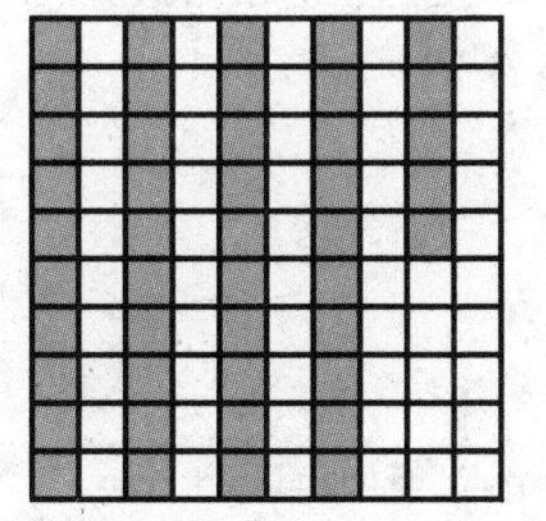

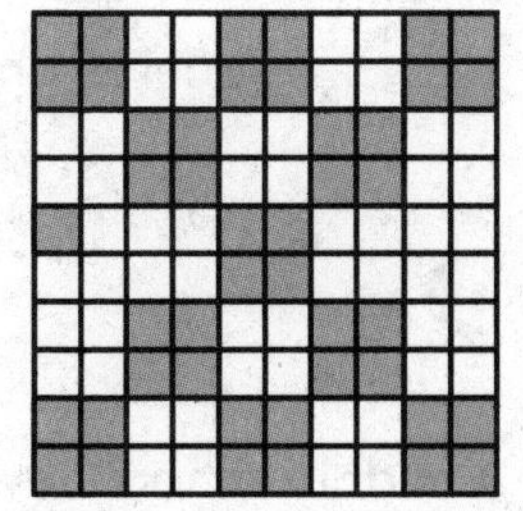

Your Notes

Example 1 Identify the Percent

What percent is represented by the 10 × 10 grid?

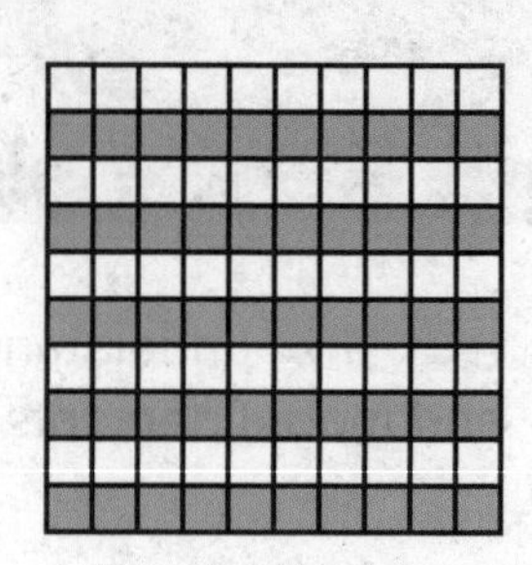

Identify the number of squares shaded. How many squares are shaded? ______

Write the ratio that compares the number of shaded squares to the total number of squares.

The ratio is ______ : 100, ______ to 100, or $\frac{\square}{100}$.

So, the percent represented is ☐%.

Check

What percent is represented by the 10 × 10 grid?

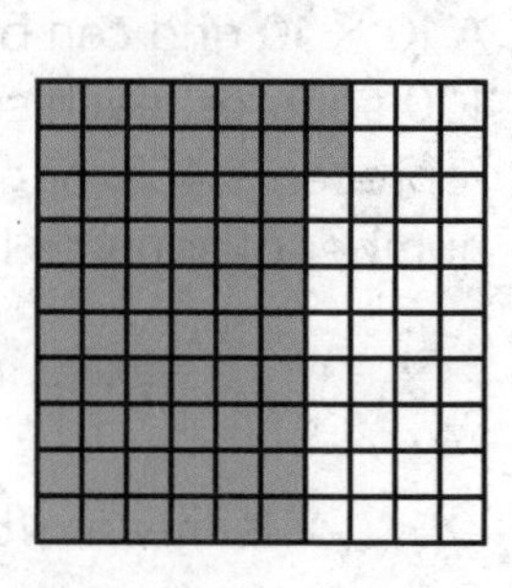

Go Online You can complete an Extra Example online.

Example 2 Model the Percent

In a recent survey, 17% of the people surveyed said that they have a magazine subscription.

Shade the 10 × 10 grid to model 17%.

17% means 17 per 100. There are 100 squares on the 10 × 10 grid. To model 17%, shade ______ squares on the grid.

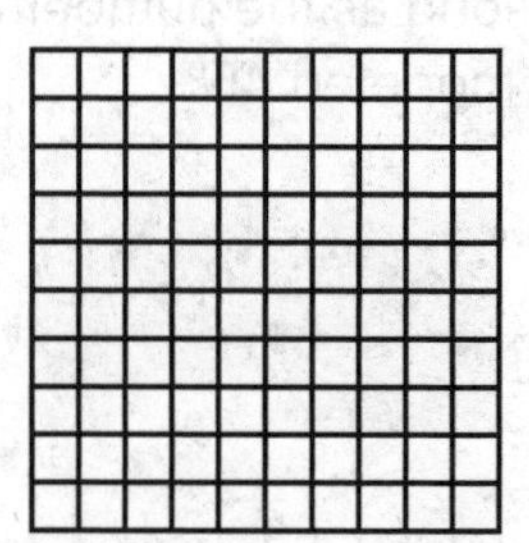

Check

A middle school newspaper surveyed the student body and found that 14% of the students surveyed chose horses as their favorite animal. Shade the 10 × 10 grid to model 14%.

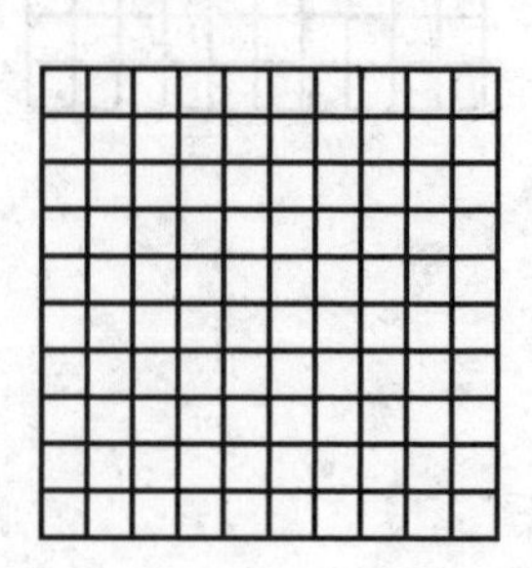

Go Online You can complete an Extra Example online.

Learn Use Bar Diagrams to Model Percents

You can also use bar diagrams to model percents. A bar diagram can be divided into any number of equal-size sections.

To model 10% or a multiple of 10%, you can divide the bar diagram into 10 equal-size sections.

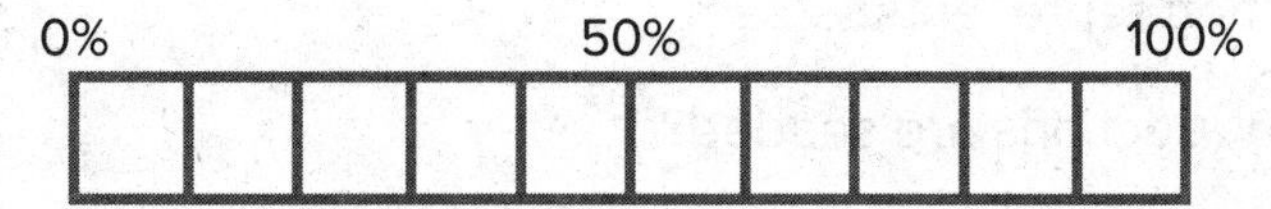

The bar diagrams show representations of several percents that are multiples of 10%.

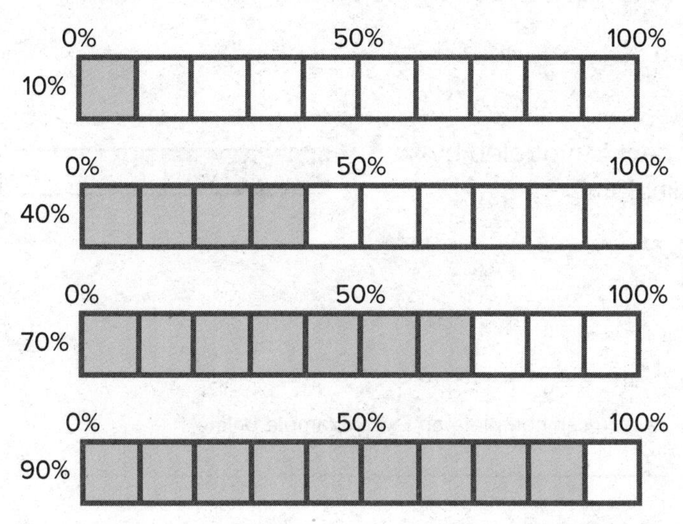

Talk About It!

Describe another way to divide a bar diagram to model 40%.

To model 5% or a multiple of 5%, you can divide the bar diagram into 20 equal-size sections.

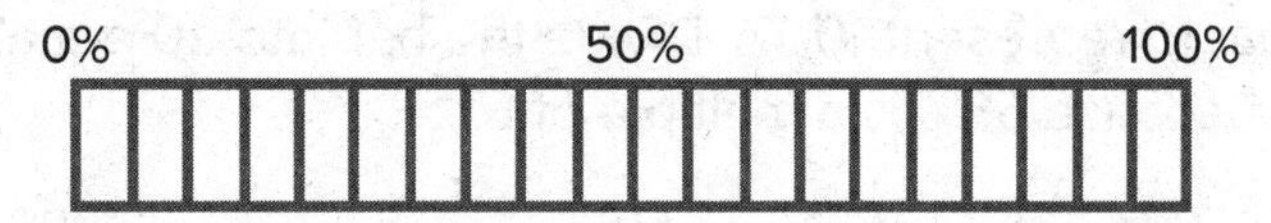

The bar diagrams show representations of several percents that are multiples of 5%.

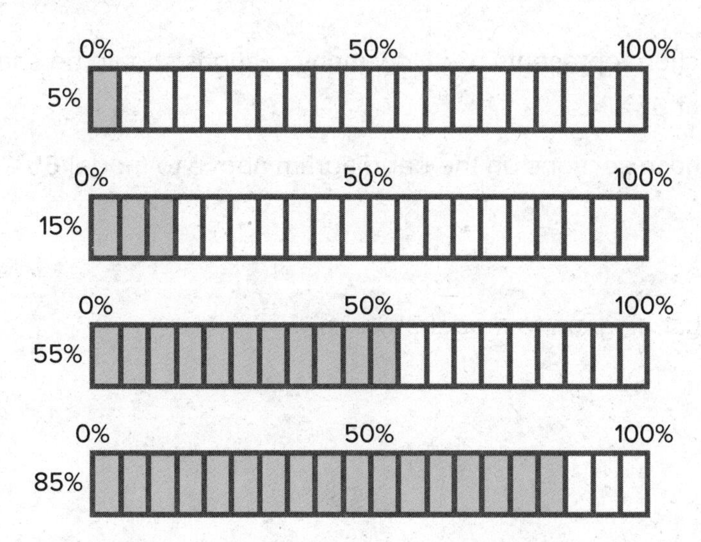

Talk About It!

Why might it not be advantageous to use a bar diagram to model a percent such as 23%?

Example 3 Identify the Percent

What percent is represented by the bar diagram?

0% 50% 100%

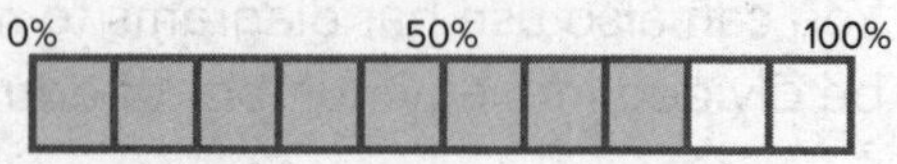

The bar diagram is divided into 10 equal-size sections.

Each section represents _____%.

How many sections are shaded? _____

The total percent represented is _____ × 10%, or _____%.

So, _____% is represented by the bar diagram.

Check

What percent is modeled by the bar diagram?

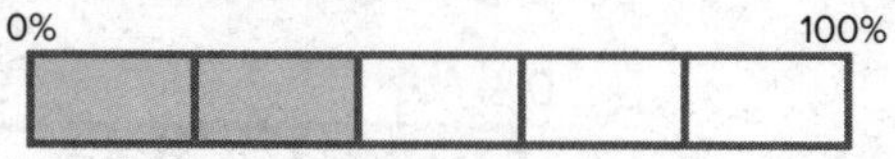

0% 100%

Go Online You can complete an Extra Example online.

Example 4 Model the Percent

Use a bar diagram to model 65%.

Draw a bar to represent 100%. Divide the bar into 20 equal-size sections because 65% is a multiple of 5.

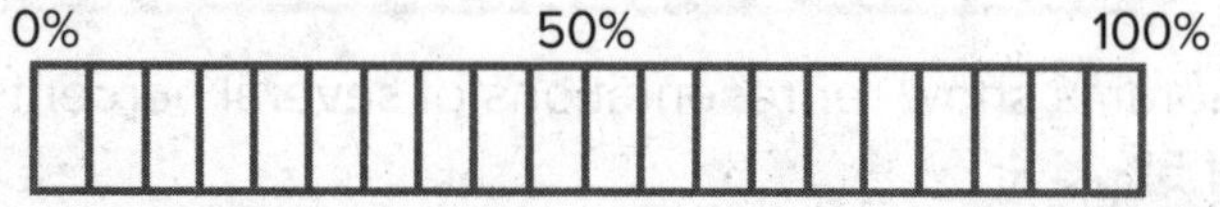

0% 50% 100%

Each section represents 5%. How many sections should be shaded to represent 65%? _____

Shade those sections on the bar diagram above to model 65%.

Check

Draw a bar diagram to model 35%.

Show your work here

Go Online You can complete an Extra Example online.

Name _______________ Period _______ Date _______

Practice

Go Online You can complete your homework online.

For Excecises 1 and 2, identify the percent modeled in each 10 × 10 grid. **(Example 1)**

1. 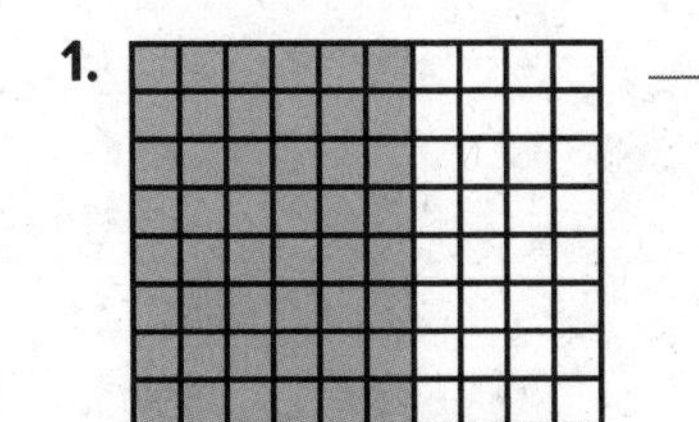 ______

2. 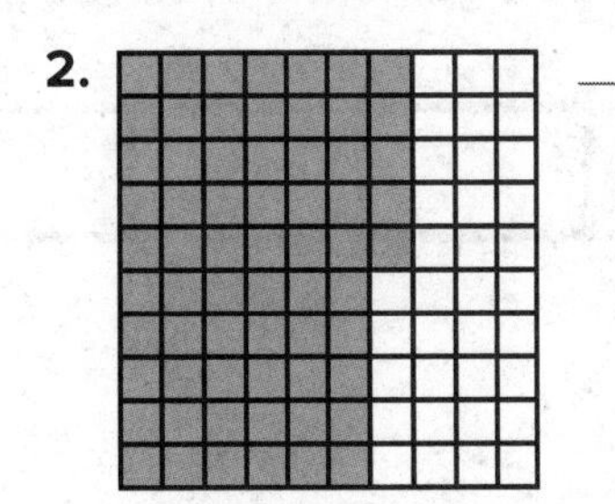 ______

3. In a school survey, 12% of the students surveyed said they like camping. Shade the 10 × 10 grid to model 12%. **(Example 2)**

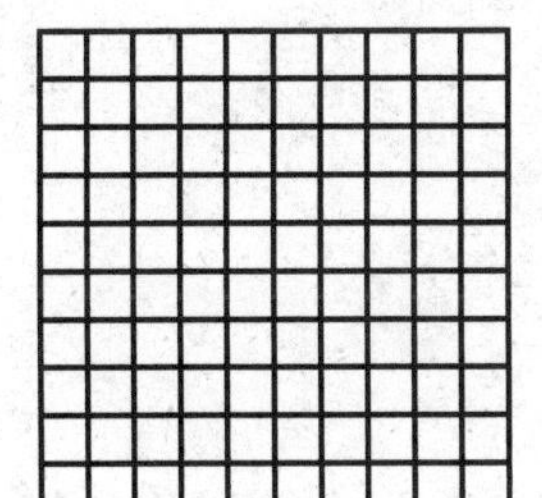

4. Of the students in the lunch line, 9% said they were buying strawberry milk. Shade the 10 × 10 grid to model 9%. **(Example 2)**

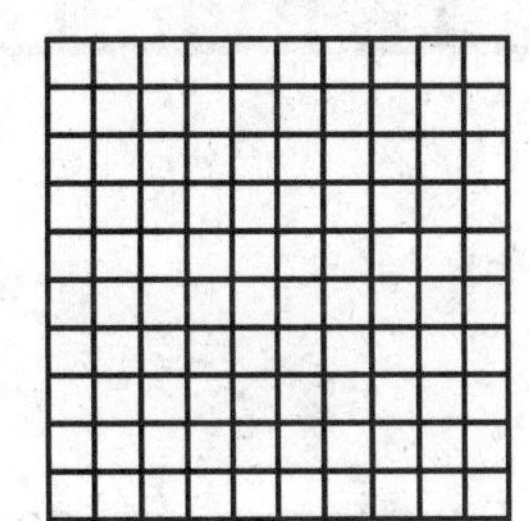

For Exercises 5 and 6, identify the percent modeled by each bar diagram. **(Example 3)**

5. ______

0% 50% 100%

6. ______

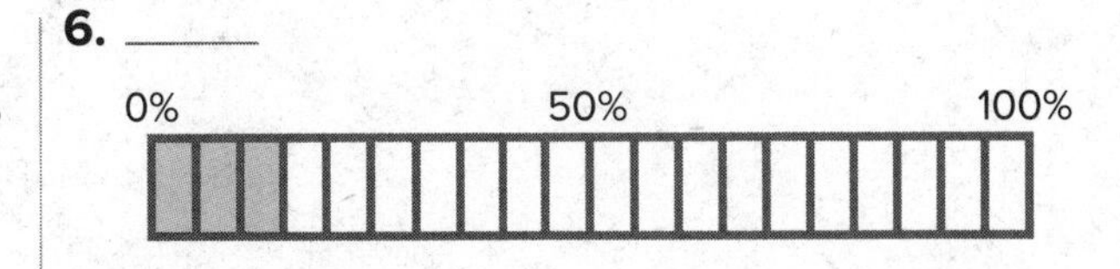

7. Shade the bar diagram to model 25%. **(Example 4)**

0% 50% 100%

Test Practice

8. Open Response How can you use a bar diagram to model 45%?

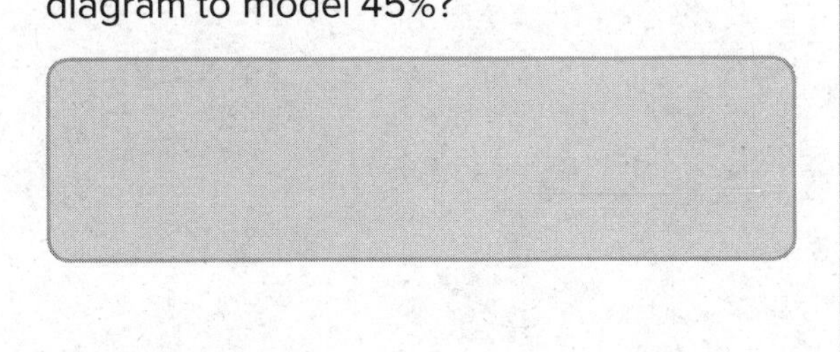

Apply

9. The model shows the percent of students who voted for a tiger as the new school mascot. Did more than 50% of the students *not* vote for a tiger as the mascot? Write an argument that can be used to defend your solution.

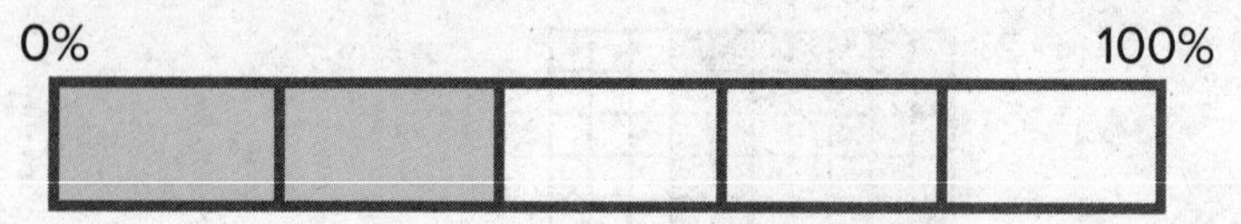

10. The model shows the percent of baseball players on a team who plan to go to a baseball camp on Saturday. Can the coach say that more than 75% of his players are going to the camp? Write an argument that can be used to defend your solution.

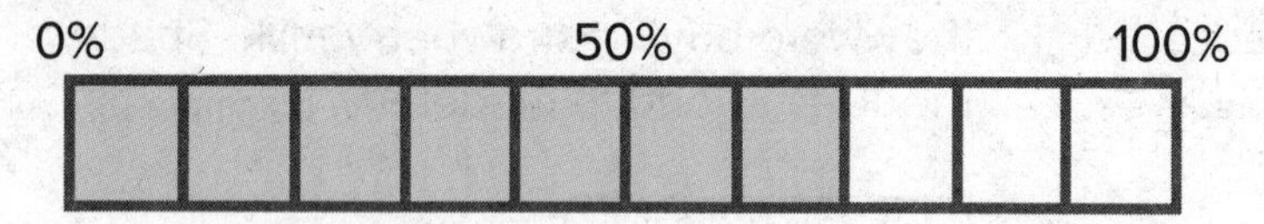

11. MP **Reason Abstractly** Suppose you divide a bar diagram into 25 equal sections and shade 5 sections. What percent is modeled in the diagram? Explain.

12. MP **Find the Error** A student said that to write a percent as fraction, write the number that comes before the percent symbol over a denominator of 100. Is the student correct? Justify your conclusion.

13. MP **Make an Argument** Use an example to explain how you can model percents greater than 100%.

14. **Create** Write a real-world problem that involves a percent less than 50%. Then model the percent.

Lesson 2-2

Percents Greater Than 100% and Less Than 1%

I Can... understand that percents can be greater than 100% or less than 1% and use 10 × 10 grids and bar diagrams to represent them.

Today's Standards
Preparation for 6.RP.A.3, 6.RP.A.3.C
MP2, MP5, MP7

Learn Percents Greater Than 100%

The table shows the total rainfall during April for a certain city for three different years.

Year	April Rainfall (in.)
2017	4.0
2018	3.0
2019	5.0

In 2018, it rained less than it did in 2017. To compare the rainfall in 2018 to that in 2017, use the ratio 3 : 4. Recall that a *percent* is a ratio that compares a number to 100. Use equivalent ratios to show that the rainfall in 2018 was 75% of the rainfall in 2017.

$$\frac{3}{4} = \frac{75}{100}$$

×25 (part → 3, whole → 4; 75/100 → percent)

If the number being compared to 100 is less than 100, then the percent is less than 100%.

In 2019, it rained more than it did in 2017. To compare the rainfall in 2019 to that in 2017, use the ratio 5 : 4. Use equivalent ratios to show that the rainfall in 2019 was 125% of the rainfall in 2017.

$$\frac{5}{4} = \frac{125}{100}$$

×25 (part → 5, whole → 4; 125/100 → percent)

If the number being compared to 100 is greater than 100, then the percent is greater than 100%.

Percents are greater than 100% when the number being compared to 100 is greater than 100. When the percent is greater than 100%, the part is greater than the whole.

Example	Model
125% means 125 per 100 125 : 100, 125 to 100, or $\frac{125}{100}$ *one hundred twenty-five percent*	100% + 25% = 125%

Talk About It!
Suppose the rainfall in 2020 is 5.0 inches. What percent compares the rainfall in 2020 to the rainfall in 2019? Explain why this makes sense.

Your Notes

Example 1 Identify the Percent

What percent is represented by the 10 × 10 grids?

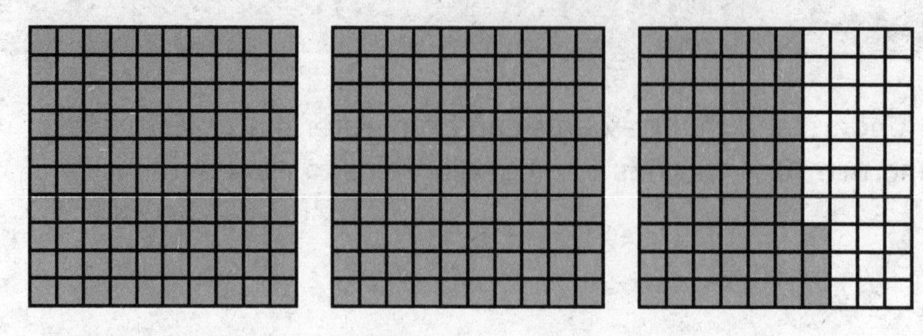

The percent compares the number of shaded squares to 100, because one whole grid contains 100 squares.

How many whole grids are shaded? ______

How many squares are shaded in the third grid? ______

How many squares are shaded altogether? ______

Write the ratio that compares the total number of shaded squares to one whole grid of 100 squares.

The ratio is ______ : 100, ______ to 100, or $\frac{\square}{100}$.

So, the percent represented by the 10 × 10 grids is ____%.

Check

What percent is represented by the 10 × 10 grids?

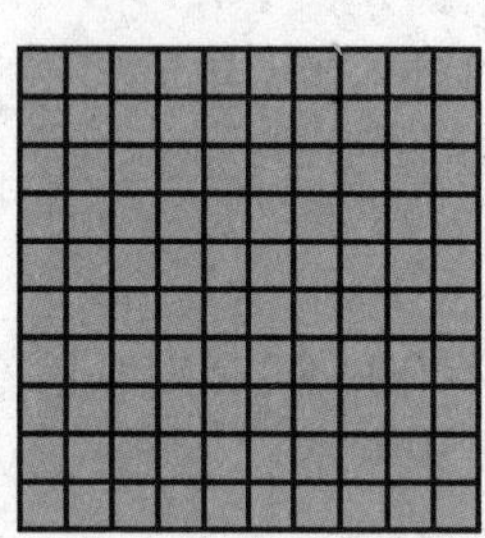

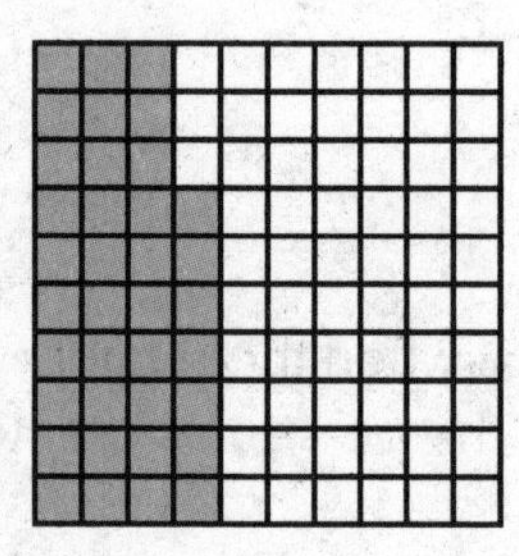

Go Online You can complete an Extra Example online.

Example 2 Model the Percent

At birth, the average kitten weighs 5 ounces. At 3 weeks of age, the average kitten will weigh twice as much as at birth.

Write a percent that compares a kitten's weight at 3 weeks to its weight at birth. Then use 10 × 10 grids to model the percent.

At 3 weeks of age, the kitten will weigh ______ ounces.

10 ounces is twice as much as 5 ounces.

Write a ratio comparing the average kitten's weight at 3 weeks of age to its weight at birth. Use equivalent ratios to show that the average kitten's weight at 3 weeks of age is ______% its weight at birth.

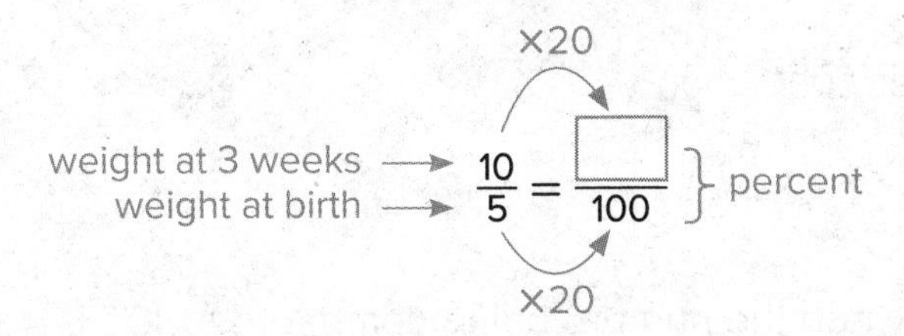

Draw and shade 10 × 10 grids to model 200%.

Check

At birth, a male baby giraffe stands almost 6 feet tall. At 4 years of age, the male giraffe will be about three times as tall as at birth. Write a percent that compares the giraffe's height at 4 years of age to its height at birth. Then draw and shade 10 × 10 grids to model the percent.

Go Online You can complete an Extra Example online.

Think About It!

If a kitten's weight did not change, what percent would compare its unchanged weight to its weight at birth?

Talk About It!

Suppose the veterinarian states that the kitten's weight increased by 100%. Is this claim correct? Why or why not? When talking about the kitten's weight, when is it correct to use 100% and when is it correct to use 200%?

Learn Percents Less Than 1%

Percents can also be less than 1%. Consider the following situation.

The distance from the center of Earth to the surface is also known as the *radius* of Earth. The radius of Earth is about 4,000 miles. The radius of the Sun is about 430,000 miles.

The ratio of Earth's radius to the Sun's radius is 4,000 : 430,000. Use equivalent ratios to show that the radius of Earth is about 0.93% of the Sun's radius. Because 430,000 divided by 4,300 is 100, divide 4,000 by 4,300. Round to the nearest hundredth.

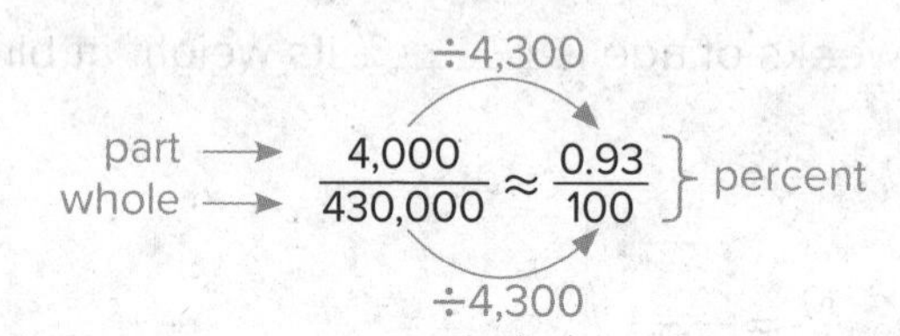

Percents are less than 1% when the number being compared to 100 is less than 1. When the percent is less than 1%, the part is significantly less than the whole. The radius of Earth is significantly less than the radius of the Sun.

Talk About It!

A classmate used a 10 × 10 grid to model 0.93% as shown. What mistake did they make? How does 0.93% compare with 93%?

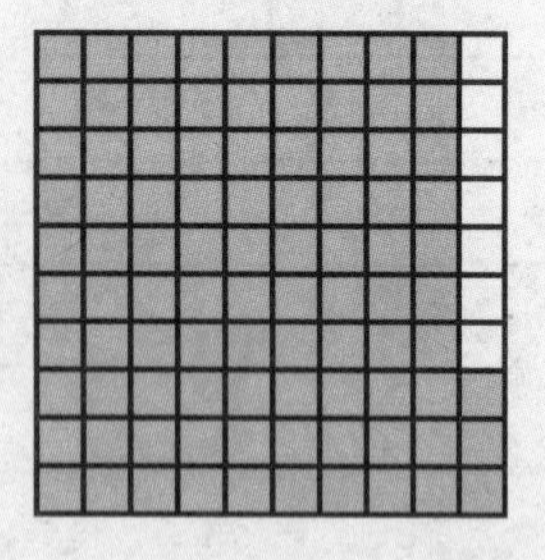

On a 10 × 10 grid, 0.93% is represented by shading 93% of one grid square. One grid square represents 1% and 0.93% is less than 1%. Compared to 100%, 0.93% is significantly less.

Example	Model
0.93% means 0.93 per 100 0.93 : 100, 0.93 to 100, or $\frac{0.93}{100}$ *ninety-three hundredths of a percent*	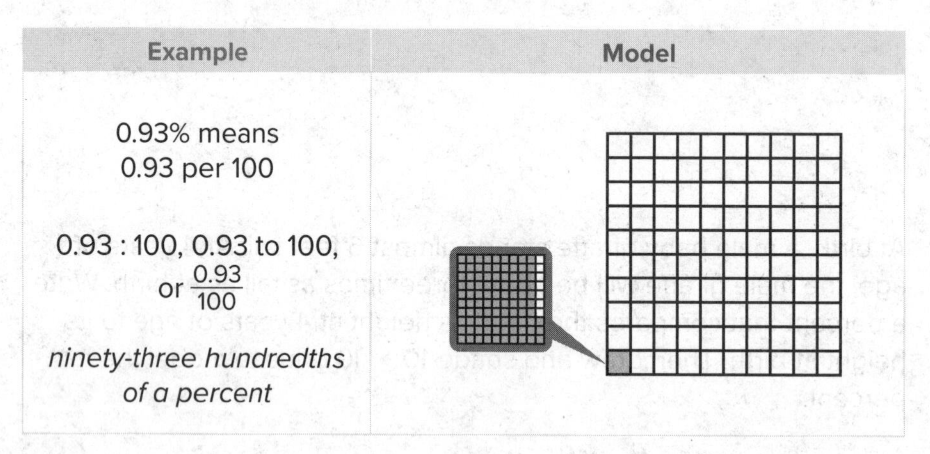

When thinking about how the size of Earth compares to the size of the Sun, it makes sense that Earth's radius is significantly less than the Sun's radius. Earth's radius is a little less than 1% of the Sun's radius.

Example 3 Identify the Percent

What percent is represented by the 10 × 10 grid?

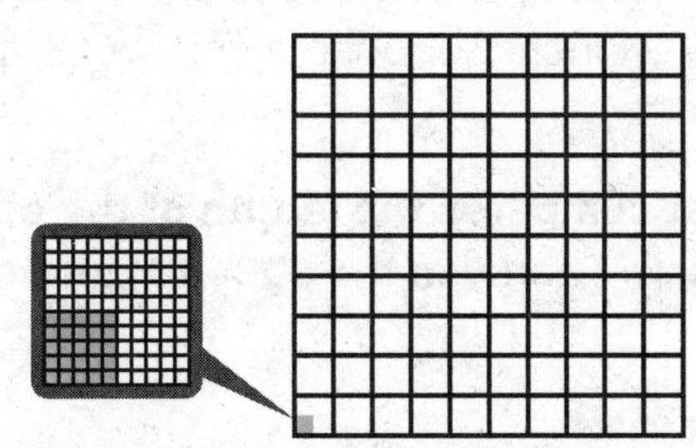

The percent compares the number of shaded squares to 100, because one whole grid contains 100 squares.

Less than 1 grid square is shaded on the 10 × 10 grid. The close-up reveals that one-fourth, $\frac{1}{4}$, or 0.25, of one grid square is shaded.

Write the ratio that compares the total number of shaded squares to one whole grid of 100 squares.

The ratio is 0.25 : 100, 0.25 to 100, or $\frac{0.25}{100}$.

So, the percent represented by the 10 × 10 grid is _______%.
Another way to write this percent is $\frac{1}{4}$%.

Check

What percent is represented by the 10 × 10 grid?

Go Online You can complete an Extra Example online.

Think About It!
How do you know that the percent represented is less than 1%?

Talk About It!
A friend states that the percent represented by the 10 × 10 grid is 25%. How can you use reasoning to explain to your friend that this is incorrect?

Example 4 Model the Percent

Think About It!

Without calculating the percent, how does the length of the plankton compare to the length of the jellyfish?

The diet of a jellyfish consists primarily of plankton, which are tiny organisms living in the ocean. One species of plankton has an average length of 0.04 inch. Suppose a certain jellyfish has a length of 8 inches.

Write a percent that compares the length of the plankton to the length of the jellyfish. Then use the 10 × 10 grid to model the percent.

Step 1 Write a ratio comparing 0.04 inch to 8 inches.

Use equivalent ratios to show that the plankton's length is ______% the length of the jellyfish.

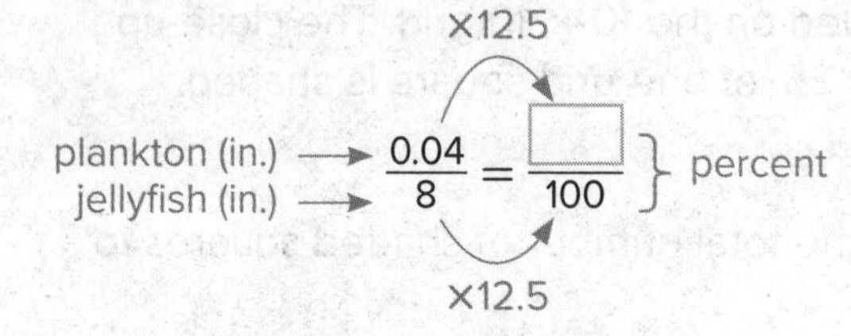

Step 2 Shade the 10 × 10 grid.

To model 0.5%, shade half of one percent by shading half of one grid square.

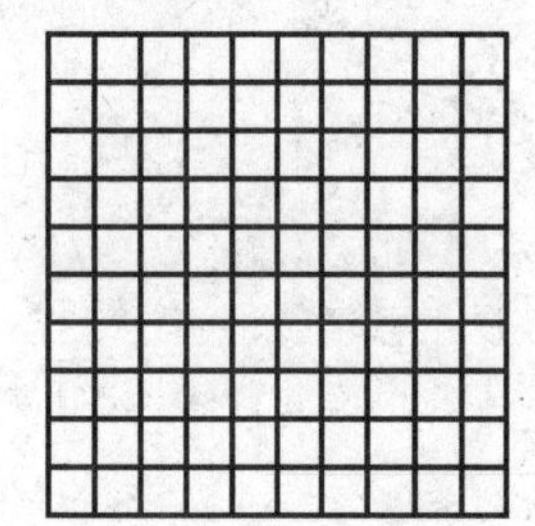

Talk About It!

What might be a common error that someone might make when shading 0.5% on the 10 × 10 grid?

Check

The average weight of a brown bear is about 1,000 pounds. Suppose a large stuffed bear weighs 2.5 pounds. Write a percent to compare the weight of the stuffed animal to the weight of the brown bear. Then use the 10 × 10 grid to model the percent.

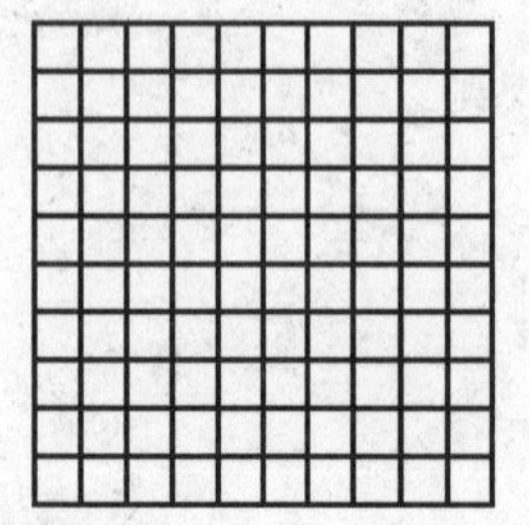

Go Online You can complete an Extra Example online.

Name ______________________ Period ____________ Date ______________

Practice

Go Online You can complete your homework online.

Identify the percent modeled in the 10 × 10 grids. (Examples 1 and 3)

1. 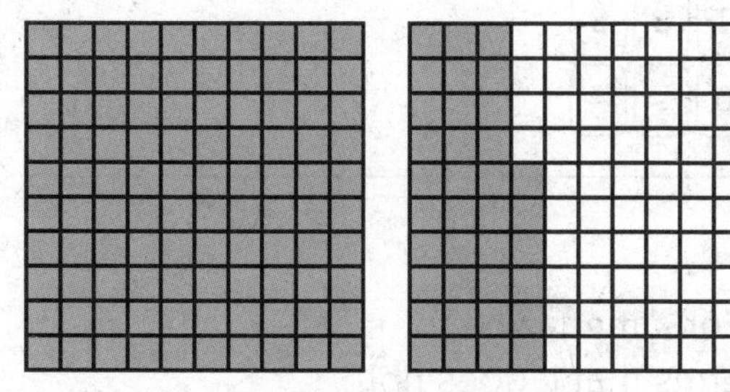

2. 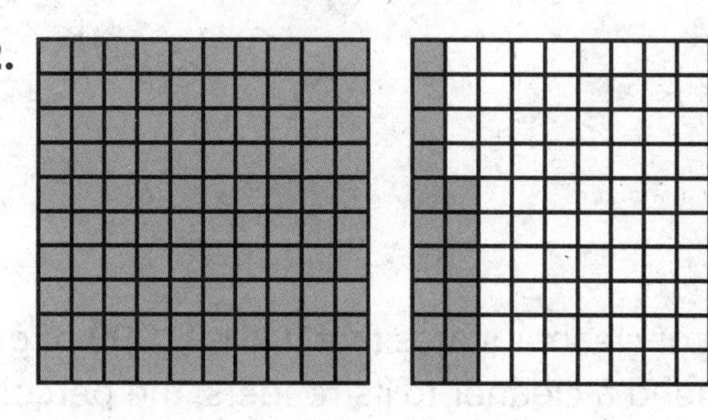

3. 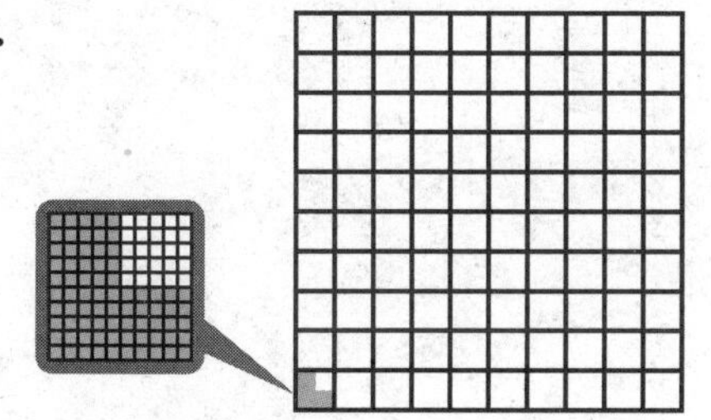

4. 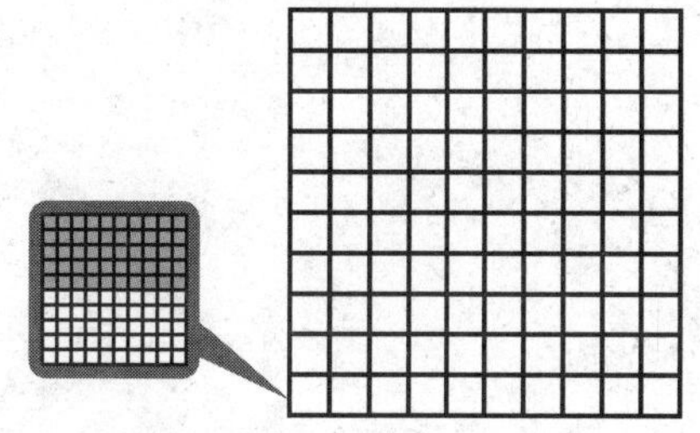

5. The size of a large milkshake is 1.4 times the size of a medium milkshake. Write a percent that compares the size of the large milkshake to the size of the small milkshake. Then draw and shade 10 × 10 grids to model the percent. (Example 2)

6. The Freedom Tower is 1,776 feet tall. Mr. Feeman's students are building a replica of the tower for a class project that will stand 4.44 feet tall. Write a percent that compares the height of the replica to the height of the actual tower. Then shade the 10 × 10 grid to model the percent.

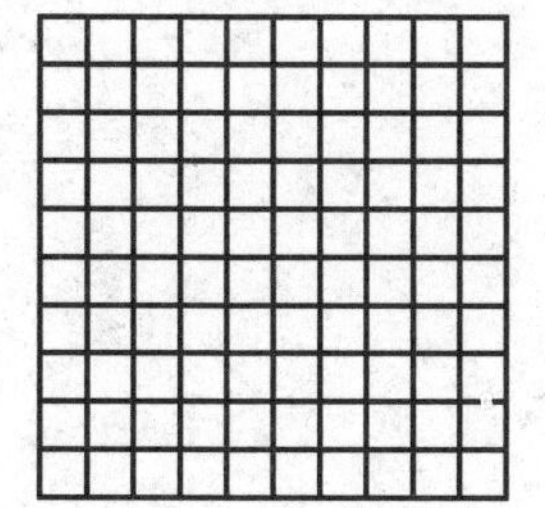

7. Equation Editor The tire pressure on Kate's bicycle decreased by 0.85% of its original pressure. What is 0.85% written as a decimal?

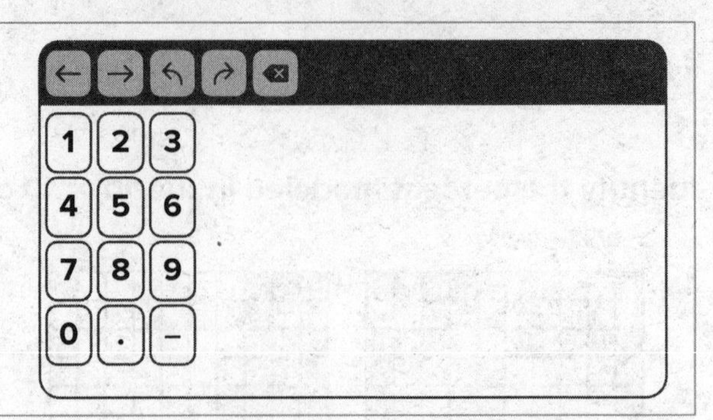

Apply

8. A bottle of cleaner states that it kills 0.999 of germs. For a magazine to recommend a cleaner to its readers, the percent of germs that it does not kill cannot exceed 1%. Would this cleaner be recommended by the magazine? Write an argument that can be used to defend your solution.

9. MP **Persevere with Problems** The top running speed of a giraffe is 250% of the top speed of a squirrel. If a squirrel's top running speed is 12 miles per hour, find the speed of a giraffe.

10. MP **Reason Inductively** Any number that can be written as a fraction is referred to as a *rational number*. Is a percent a rational number? Explain your reasoning.

11. MP **Find the Error** A student said that to represent 0.2% with a 10 × 10 grid, you shade 2 squares in the grid. Find the student's error and correct it.

12. Create Write a real-world problem involving a percent that is greater than 100% or a percent that is less than 1%. Then solve the problem.

Relate Fractions, Decimals, and Percents

I Can... relate fractions, decimals, and percents by using place-value reasoning and understanding a percent as a ratio that compares a number to 100.

Today's Standards
Preparation for 6.RP.A.3, 6.RP.A.3.C
MP1, MP2, MP3

Explore Percents and Ratios

Online Activity You will use 10 × 10 grids to understand the relationship between percents and ratios.

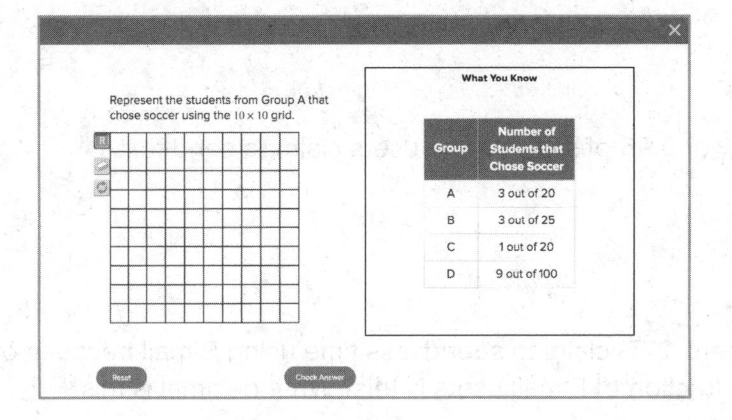

Learn Relate Percents to Fractions and Decimals

By definition, a percent is a ratio that compares a number to 100. The percent 35% compares 35 to 100 as the ratio 35 : 100. In fraction form, this ratio is $\frac{35}{100}$ which means *thirty-five hundredths*. You can use the definition of percent, equivalent ratios, and place-value reasoning to write percents as both fractions and decimals.

Write 35% as a fraction.

$35\% = \frac{35}{100}$ Definition of percent

$= \frac{7}{20}$ Find an equivalent ratio. Divide both 35 and 100 by 5.

As a fraction, $35\% = \frac{35}{100}$, or $\frac{7}{20}$.

Write 35% as a decimal.

$35\% = \frac{35}{100}$ Definition of percent

$= 0.35$ $\frac{35}{100}$ means *thirty-five hundredths*

As a decimal, 35% = 0.35.

Your Notes

Think About It!

What is the first step to writing a percent as a fraction?

Talk About It!

When writing a fraction as a percent, why do you find an equivalent ratio with a denominator of 100?

Example 1 Write Percents as Fractions and Decimals

In a recent survey, about 95% of smartphone users claim to send text messages.

What fraction of smartphone users is this? What decimal is this?

Part A Write 95% as a fraction.

$95\% = \frac{95}{100}$ Definition of percent

$= \frac{19}{20}$ Find an equivalent ratio. Divide both 95 and 100 by 5.

Part B Write 95% as a decimal.

$95\% = \frac{95}{100}$ Definition of percent

$= 0.95$ $\frac{95}{100}$ means *ninety-five hundredths*

So, about $\frac{19}{20}$ or 0.95 of smartphone users claim to send text messages.

Check

Of E-mail users, 22% claim to spend less time using E-mail because of spam. What fraction of E-mail users is this? What decimal is this?

Go Online You can complete an Extra Example online.

Learn Relate Fractions to Percents and Decimals

You can also write fractions as percents and decimals. Suppose you are given the fraction $\frac{3}{20}$. Use your understanding of equivalent ratios, the definition of percent, and place-value reasoning to write $\frac{3}{20}$ as a percent and as a fraction.

Write $\frac{3}{20}$ as a percent.

$\frac{3}{20} \overset{\times 5}{=} \frac{15}{100}$ Find an equivalent ratio with 100 as the denominator. Because $20 \times 5 = 100$, multiply 3 by 5.

$= 15\%$ Definition of percent

As a percent, $\frac{3}{20} = 15\%$.

(continued on next page)

Write $\frac{3}{20}$ as a decimal.

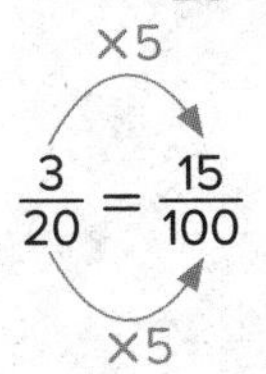

Find an equivalent ratio with 100 as the denominator. Because $20 \times 5 = 100$, multiply 3 by 5.

$= 0.15$ $\quad \frac{15}{100}$ means *fifteen hundredths*

As a decimal, $\frac{3}{20} = 0.15$.

Consider the fraction $\frac{9}{15}$. How can you write this fraction as a percent, knowing that there is no whole number by which you can multiply 15 to obtain 100?

Go Online Watch the animation to learn how to write $\frac{9}{15}$ as a percent.

The animation shows that you can simplify the fraction first, and then find an equivalent ratio with a denominator of 100. To *simplify* a fraction, divide both the numerator and denominator by the same number. By simplifying a fraction, you are finding an equivalent ratio. In this case, find an equivalent ratio with a denominator that is a factor of 100.

Write $\frac{9}{15}$ as a percent.

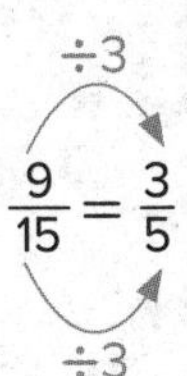

Find an equivalent ratio with 5 as the denominator because 5 is a factor of both 100 and 15. Because $15 \div 3 = 5$, divide 9 by 3.

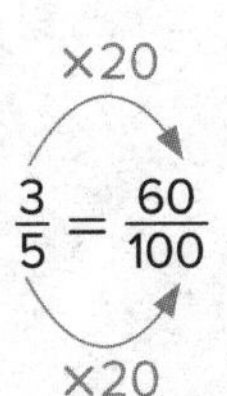

Find an equivalent ratio with 100 as the denominator. Because $5 \times 20 = 100$, multiply 3 by 20.

$= 60\%$ $\quad$ Definition of percent

As a percent, $\frac{9}{15} = 60\%$.

Go Online You can complete an Extra Example online.

Talk About It!

A classmate claims that you can always write a fraction as a decimal by dividing the numerator by the denominator. Is this a valid method? Why or why not?

Talk About It!

A classmate wrote the decimal form of $\frac{9}{15}$ as 0.6. Another classmate wrote the decimal form as 0.60. Who is correct? Why?

Think About It!

A classmate claims that $\frac{6}{8}$ is less than 60%, because $\frac{6}{8} = \frac{60}{80}$, and the denominator 80 is less than 100. Is this reasoning correct? Why or why not?

Example 2 Write Fractions as Percents and Decimals

Write the fraction $\frac{6}{8}$ as a percent and as a decimal.

Part A Write $\frac{6}{8}$ as a percent.

Find an equivalent ratio with a denominator of 100. There is no whole number by which you can multiply 8 by to obtain 100. So, first simplify the fraction.

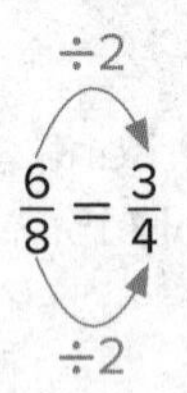

$\frac{6}{8} = \frac{3}{4}$ (÷2, ÷2) Find an equivalent ratio with 4 as the denominator because 4 is a factor of both 100 and 8. Because $8 \div 2 = 4$, divide 6 by 2.

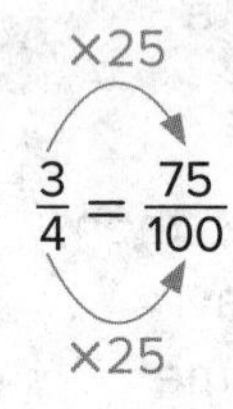

$\frac{3}{4} = \frac{75}{100}$ (×25, ×25) Find an equivalent ratio with 100 as the denominator. Because $4 \times 25 = 100$, multiply 3 by 25.

$= 75\%$ Definition of percent

Part B Write $\frac{6}{8}$ as a decimal.

As a percent, $\frac{6}{8} = 75\%$. Write 75% as a decimal.

$75\% = 0.75$ $75\% = \frac{75}{100}$, which means *seventy-five hundredths*

As a percent, $\frac{6}{8} = 75\%$. As a decimal, $\frac{6}{8} = 0.75$.

Talk About It!

Now that you know that $\frac{6}{8} = 75\%$, what are some other fraction-percent equivalencies with denominators of 8? Explain how you can use reasoning to find them.

Check

Write $\frac{4}{16}$ as a percent and as a decimal.

 **Go Online** You can complete an Extra Example online.

Example 3 Write Mixed Numbers as Percents

The cheetah is the fastest land mammal in the world. The peregrine falcon is the fastest bird in the world. The peregrine falcon's top speed is $2\frac{1}{10}$ times as fast as the top speed of a cheetah.

What percent represents this value?

Step 1 Write the mixed number as an improper fraction.

The fraction $2\frac{1}{10}$ is a mixed number that consists of a whole number part, 2, and a fractional part, $\frac{1}{10}$.

$2\frac{1}{10} = 2 + \frac{1}{10}$ Write the mixed number as a sum.

$= \frac{10}{10} + \frac{10}{10} + \frac{1}{10}$ $2 = 1 + 1$ and $1 = \frac{10}{10}$

$= \frac{21}{10}$ Add.

Step 2 Find an equivalent ratio with 100 as a denominator.

$\frac{21}{10} = \frac{210}{100}$ (×10, ×10) Find an equivalent ratio with 100 as the denominator.

Because $10 \times 10 = 100$, multiply 21 by 10.

$= 210\%$ Definition of percent

So, the peregrine falcon's top speed is ______% that of a cheetah's top speed.

Think About It!

Is the top speed of the falcon greater than 200% that of the cheetah? How do you know?

Talk About It!

How can you use mental math to express $2\frac{1}{10}$ as a percent?

Check

When blue whales feed, they can take in $1\frac{1}{25}$ times their body weight in food and water in one single gulp. What percent of their body weight is this?

Go Online You can complete an Extra Example online.

Learn Relate Decimals to Percents and Fractions

You can use place-value reasoning and equivalent ratios to write decimals as percents and fractions. A decimal with its last nonzero digit in the tenths place can be written as a fraction with a denominator of 10.

$0.7 = \frac{7}{10}$ 0.7 means *seven tenths*

$= \frac{70}{100}$, or 70% Find an equivalent ratio with a denominator of 100. Multiply both 7 and 10 by 10.

As a fraction, $0.7 = \frac{7}{10}$. As a percent, $0.7 = 70\%$

A decimal with its last nonzero digit in the hundredths place can be written as a fraction with a denominator of 100.

$0.34 = \frac{34}{100}$, or 34% 0.34 means *thirty-four hundredths*

As a fraction, $0.34 = \frac{34}{100}$, or $\frac{17}{50}$. As a percent, $0.34 = 34\%$.

A decimal with its last nonzero digit in the thousandths place can be written as a fraction with a denominator of 1,000.

$0.125 = \frac{125}{1,000}$ 0.125 means *one hundred twenty-five thousandths*

$= \frac{12.5}{100}$, or 12.5% Find an equivalent ratio with a denominator of 100. Divide both 125 and 1,000 by 10.

As a fraction, $0.125 = \frac{125}{1,000}$, or $\frac{1}{8}$. As a percent $0.125 = 12.5\%$.

Talk About It!

When might it be advantageous to simplify the fraction $\frac{125}{1,000}$ to $\frac{1}{8}$? When might it be more advantageous to leave the fraction as $\frac{125}{1,000}$?

Example 4 Write Decimals as Percents and Fractions

Write 0.025 as a percent and as a fraction.

$0.025 = \frac{25}{1,000}$ 0.025 means *twenty-five thousandths*

$= \frac{2.5}{100}$ To write 0.025 as a percent, find an equivalent ratio with a denominator of 100. $0.025 = 2.5\%$

$= \frac{1}{40}$ To write 0.025 as a fraction, find an equivalent ratio by simplifying the original fraction $\frac{25}{1,000}$. $0.025 = \frac{1}{40}$

As a percent, $0.025 = 2.5\%$. As a fraction, $0.025 = \frac{25}{1,000}$ or $\frac{1}{40}$.

Check

Write 1.4 as a percent and as a mixed number.

Go Online You can complete an Extra Example online.

Apply School

The table shows the percent of time Allison spent studying each of her school subjects last week. The total time spent studying is 100%. What fraction of the time was spent studying math and history?

Subject	Percent
Math	?
Science	13
Language Arts	11
History	?
Reading	20
Music	16

1 What is the task?

Make sure you understand exactly what question to answer or problem to solve. You may want to read the problem three times. Discuss these questions with a partner.

First Time Describe the context of the problem, in your own words.
Second Time What mathematics do you see in the problem?
Third Time What are you wondering about?

2 How can you approach the task? What strategies can you use?

3 What is your solution?

Use your strategy to solve the problem.

4 How can you show your solution is reasonable?

Write About It! Write an argument that can be used to defend your solution.

Go Online watch the animation.

Talk About It!

Based on the information in the table alone, is it possible to determine the fraction of time Allison spent studying math? Explain.

Check

Ramiro's garden is shown. What percent of the total area of the garden do the cucumbers cover?

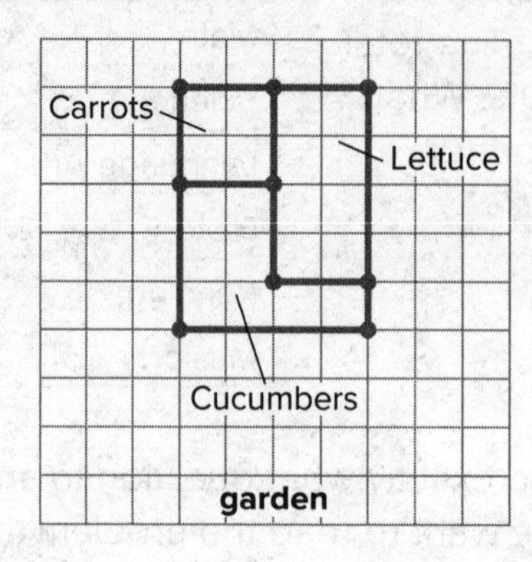

Go Online You can complete an Extra Example online.

Foldables It's time to update your Foldable, located in the Module Review, based on what you learned in this lesson. If you haven't already assembled your Foldable, you can find the instructions on page FL1.

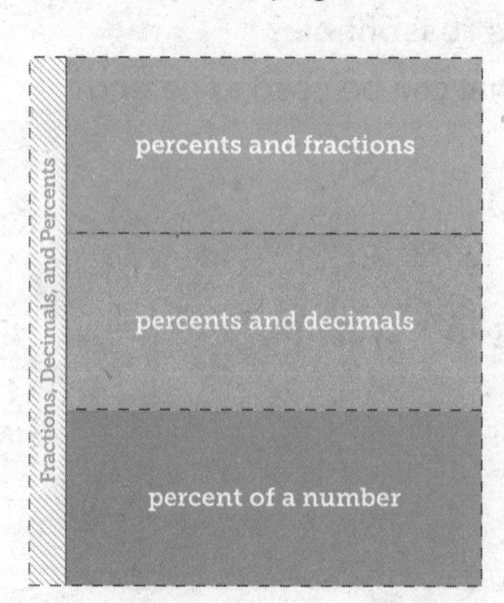

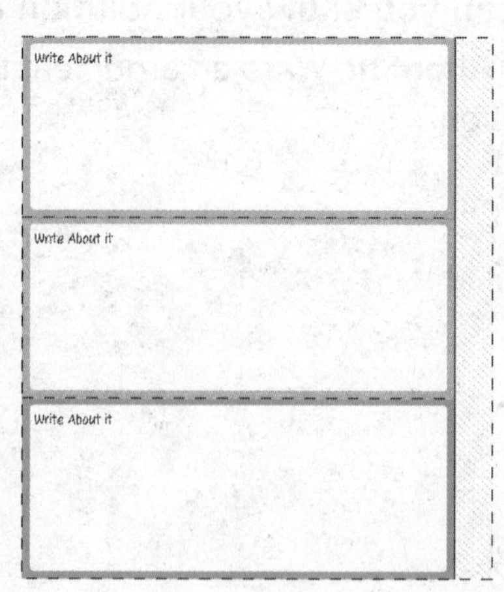

Name ______________________ Period __________ Date __________

Practice

Go Online You can complete your homework online.

Write each percent as a fraction and as a decimal. (Example 1)

1. 45%

2. 72%

3. 80%

Write each fraction as a percent and as a decimal. (Examples 2 and 3)

4. $\frac{3}{20}$

5. $1\frac{3}{4}$

6. $\frac{5}{8}$

Write each decimal as a percent and as a fraction. (Example 4)

7. 0.89

8. 0.82

9. 0.65

10. About 0.41 of Hawaii's total area is water. Write 0.41 as a fraction and as a percent.

11. Over the course of the basketball season, Zane's free throw average went up by 30%. Write 30% as a fraction and as a decimal.

12. There are 25 students in Muriel's class. Write a percent to represent the number of students that have brown eyes. Then write the percent as a fraction and as a decimal.

Eye Color	Number of Students
Blue	6
Brown	10
Green	7
Hazel	2

Test Practice

13. Multiselect Which of the following are equivalent to 85%? Select all that apply.

- ☐ 0.85
- ☐ $\frac{85}{100}$
- ☐ 0.8
- ☐ $\frac{17}{20}$
- ☐ 85

Apply

14. The table shows the results of a recent survey of sixth grade students at Potter Middle School about their favorite sports. What fraction of the students chose football or soccer?

Sport	Percent
Baseball	14
Football	20
Lacrosse	12
Soccer	35
Softball	8
Volleyball	11

15. The table shows the percent of each type of pet owned by pet owners in a neighborhood. The total percent is equal to 100%. What fraction of the pets owned were cats and dogs?

Pet	Percent
Bird	4
Cat	?
Dog	?
Fish	14
Hamster	10
Snake	2

16. MP **Justify Conclusions** Determine if the following statement is *true* or *false*. Justify your conclusion.
Any decimal that ends with a digit in the hundredths place can be written as a fraction with a denominator that is divisible by both 2 and 5.

17. MP **Reason Inductively** A sixth-grade class was surveyed about their favorite kind of drink. The results are shown in the table. Did chocolate milk and lemonade receive more than 50% of the votes? Explain.

Type of Drink	Percent (decimal)
Chocolate Milk	0.22
Iced Tea	0.05
Lemonade	0.24
Orange Juice	0.18
Sports Drink	0.31

18. MP **Persevere with Problems** Explain how you can write $25\frac{2}{5}\%$ as a decimal.

19. MP **Identify Structure** When writing a fraction as a percent, how can you tell if the percent will be less than 100%, equal to 100%, or greater than 100%?

Find the Percent of a Number

I Can... find the percent of a number by reasoning about percent as a rate per 100 and by using bar diagrams, equivalent ratios, double number lines, and ratio tables.

Today's Standards
6.RP.A.3, 6.RP.A.3.C
MP1, MP2, MP3, MP5, MP6, MP7

Explore Percent of a Number

Online Activity You will use 10 × 10 grids and bar diagrams to represent the percent of a number.

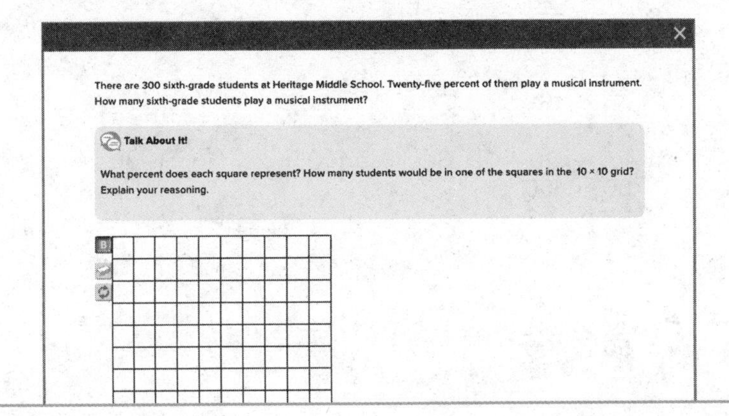

Learn Find the Percent of a Number

Fifty people were surveyed and asked to vote on their favorite flavor of sherbet. The results are shown in the table.

Flavor	Percent
Lemon	20
Orange	26
Peach	14
Watermelon	40

To find the number of people who prefer lemon, you can use a bar diagram. The bar is separated into 10 equal-size sections. The whole is 50 total people, so each section represents 50 ÷ 10, or 5 people. The percent is 20% and the part is 10 people (two sections of 5 people each). The bar diagram shows that 20% of 50 is 10. In context, 10 people, out of the 50 surveyed, which is 20%, prefer lemon.

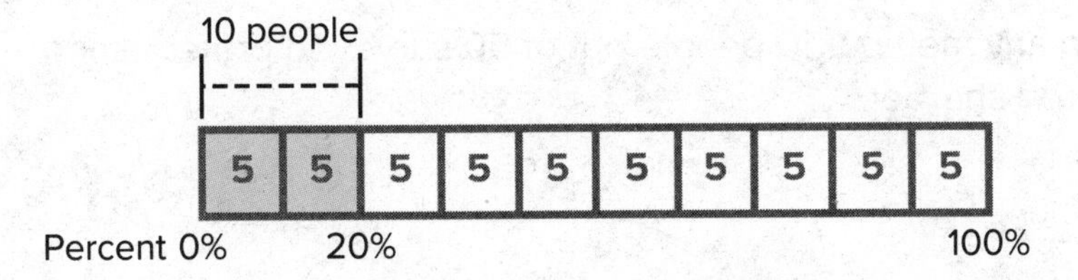

Talk About It!
Why is the bar divided into 10 sections? Is there a different way you can divide the bar to solve the same problem? Explain.

(continued on next page)

Your Notes

You can also use a ratio table to find 20% of 50. You know 100% of 50 is 50. You need to find 20% of 50. Scale back to find 20% of 50 by dividing both 100 and 50 by 5.

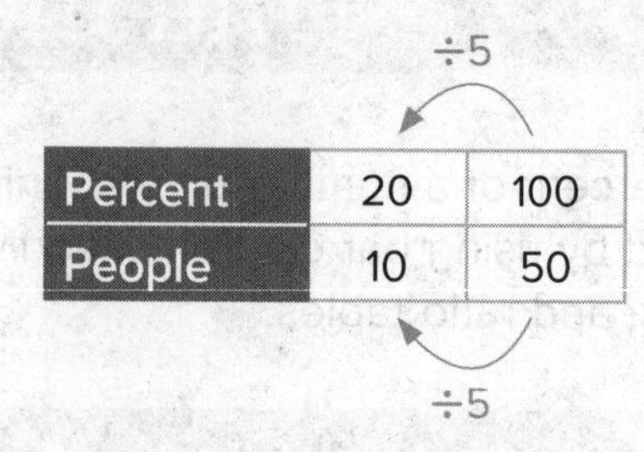

Another method you can use to find 20% of 50 is to write and solve an equation stating the ratios are equivalent. Let n represent the number of people who prefer lemon.

lemon → total surveyed → $\frac{n}{50} = \frac{20}{100}$ } percent

÷2

$\frac{n}{50} = \frac{20}{100}$ Because $100 \div 2 = 50$, divide 20 by 2.

÷2

$\frac{10}{50} = \frac{20}{100}$ $100 \div 2 = 50$; So, $n = 10$.

So, 10 people prefer lemon.

In addition to bar diagrams, ratio tables, and equivalent ratios, you can also use a double number line to find the percent of a number.

Draw a double number line. The bottom number line represents the percent, so use increments of 10 to draw tick marks and label the percents. The top number line represents the whole, so label the tick mark that corresponds with 100% on the bottom number line with 50. Since there are 10 increments, the value of each tick mark on the top number line increases by $50 \div 10$, or 5 units.

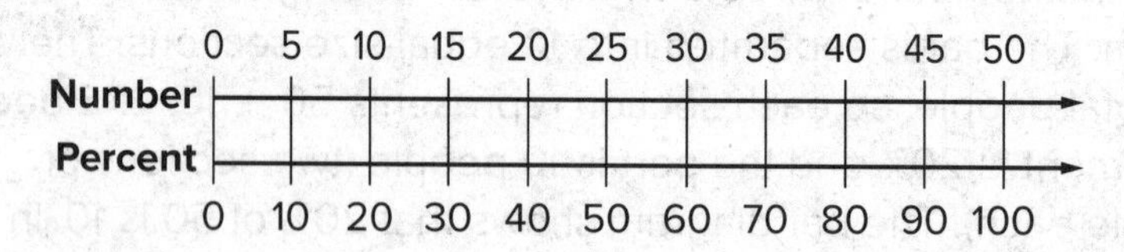

The double number line shows that 20% corresponds to 10 people.

Using any method, 10 people out of 50 surveyed prefer lemon flavored sherbet.

Talk About It!

Which representation helps you to visualize the problem? Can you think of a situation in which it might not be advantageous to use that representation?

Example 1 Find the Percent of a Number

The graph shows the types of snacks that students at York Middle School bring with them to school. Suppose there are 300 students at the school.

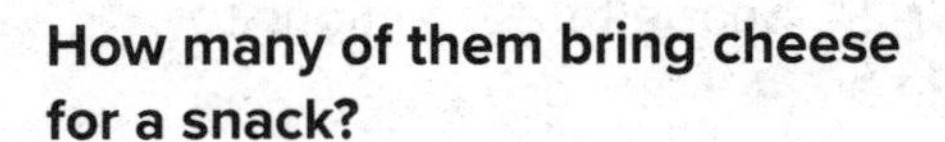

How many of them bring cheese for a snack?

First, identify the part, the whole, and the percent. The part is unknown. The whole is 300. The percent is 15%.

Method 1 Use the rate per 100 and mental math.

The percent is 15%. This means, that for every 100 students, 15 of them bring cheese for a snack. This is the rate per 100.

$15 + 15 + 15$	There are three 100s in 300. For each 100, 15 students bring cheese as a snack.
$= 3 \times 15$	Write repeated addition as multiplication.
$= 45$	Multiply. 45 students bring cheese as a snack.

Method 2 Use equivalent ratios.

Write and solve an equation stating the ratios are equivalent. Let n represent the number of students who bring cheese as a snack.

cheese → total students → $\frac{n}{300} = \frac{15}{100}$ } percent

$\frac{n}{300} \overset{\times 3}{=} \frac{15}{100}$ Because $100 \times 3 = 300$, multiply 15 by 3.

$\frac{45}{300} = \frac{15}{100}$ $15 \times 3 = 45$; So, $n = 45$.

So, using either method, ______ students bring cheese as a snack.

Check

Approximately 11% of the U.S. population is left-handed. If there are 700 students at a middle school, about how many students are expected to be left-handed?

Go Online You can complete an Extra Example online.

Think About It!

A classmate claims that because 15% is a little over 10% and 10% of 300 is 30, that 15% of 300 will be a little over 30. Do you think this reasoning is correct? Why or why not?

Talk About It!

How can you use a bar diagram to find 15% of 300?

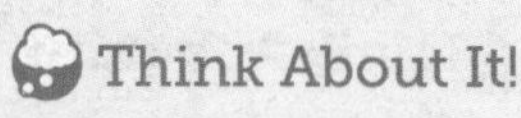

Think About It!

Is 30% of 240 less than, greater than or equal to 120? How do you know?

Talk About It!

Now that you know 30% of 240, use mental math to find 60% of 240, 90% of 240, and 15% of 240.

Example 2 Find the Percent of a Number

What is 30% of 240?

The part is unknown. The whole is 240. The percent is 30%.

Method 1 Use a bar diagram.

Draw a bar diagram with 10 equal-size sections. The whole is 240, so each section represents $240 \div 10$ or 24. Shade three sections to represent 30%. So, 30% of 240 is $24 + 24 + 24$, or 72.

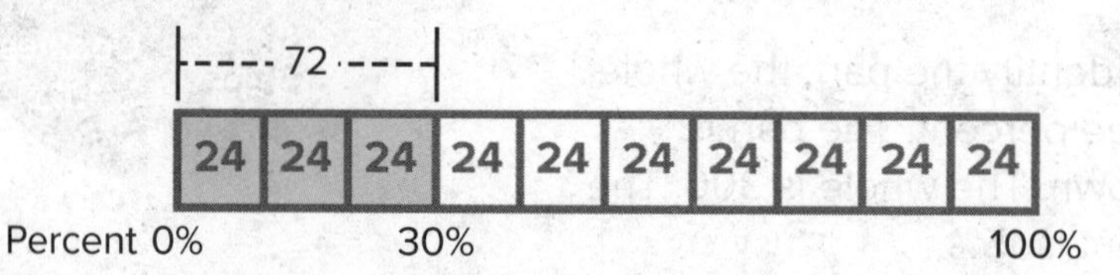

Method 2 Use a double number line.

Draw a double number line. The bottom number line represents the percent, so use increments of 10 to draw tick marks and label the percents. The top number line represents the whole, so label the tick mark that corresponds with 100% on the bottom number line with 240. Since there are 10 increments, the value of each tick mark on the top number line increases by $240 \div 10$, or 24 units. So, 30% on the bottom number line corresponds with 72 on the top number line.

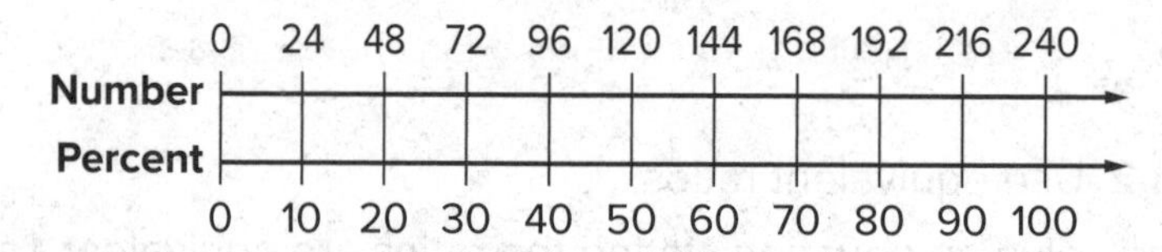

Method 3 Use equivalent ratios.

Write and solve an equation stating the ratios are equivalent. Let n represent the unknown part.

part →, whole → $\frac{n}{240} = \frac{30}{100}$ } percent

$\times 2.4$

$\frac{n}{240} = \frac{30}{100}$ Because $100 \times 2.4 = 240$, multiply 30 by 2.4.

$\times 2.4$

$\frac{72}{240} = \frac{30}{100}$ $30 \times 2.4 = 72$; So, $n = 72$.

So, using any method, 30% of 240 is ______.

Check

What is 70% of 580? Use any strategy.

Go Online You can complete an Extra Example online.

Example 3 Find the Percent of a Number

What is 145% of 320?

The part is unknown. The whole is 320. The percent is 145%.

Method 1 Use a ratio table.

You know that 100% of 320 is 320. You need to find 145% of 320. Use a ratio table to scale back from 100% to 1%. Then scale forward from 1% to 145%.

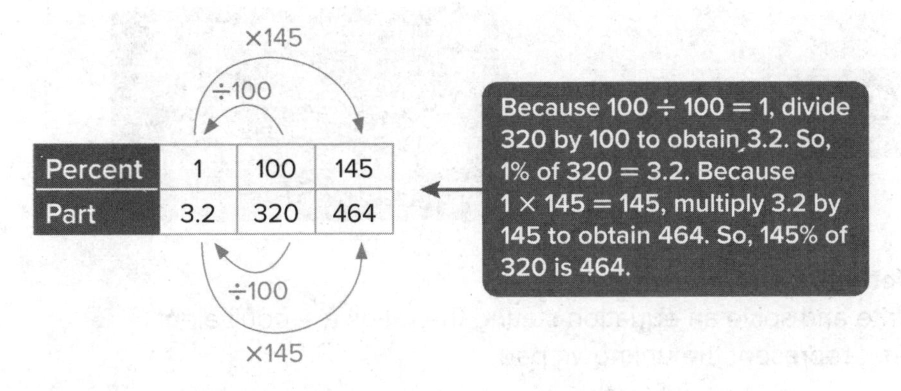

Percent	1	100	145
Part	3.2	320	464

Method 2 Use equivalent ratios.

Write and solve an equation stating the ratios are equivalent. Let n represent the unknown part.

part → whole → $\frac{n}{320} = \frac{145}{100}$ } percent

×3.2

$\frac{n}{320} = \frac{145}{100}$ Because $100 \times 3.2 = 320$, multiply 145 by 3.2.

×3.2

$\frac{464}{320} = \frac{145}{100}$ $145 \times 3.2 = 464$; So, $n = 464$.

So, using either method, 145% of 320 is ______.

Check

What is 275% of 4? Use any strategy.

Go Online You can complete an Extra Example online.

Think About It!

Is 145% of 320 less than, greater than, or equal to 320? How do you know?

Talk About It!

Compare the part, 464, to the whole, 320. Does it make sense that 464 is greater than 320? Why or why not?

Think About It!

Why might it not be advantageous to use a bar diagram to find 0.25% of 58?

Talk About It!

Compare the part, 0.145, to the whole, 58. Does it make sense that 0.145 is significantly less than 58? Why or why not?

Example 4 Find the Percent of a Number

What is 0.25% of 58?

The part is unknown. The whole is 58. The percent is 0.25%.

Method 1 Use a ratio table.

You know that 100% of 58 is 58. You need to find 0.25% of 58. Use a ratio table to scale back from 100% to 1%. Then scale back again from 1% to 0.25%.

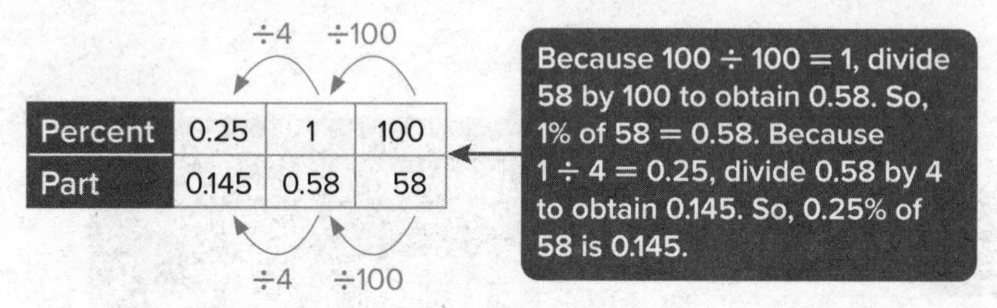

Percent	0.25	1	100
Part	0.145	0.58	58

Method 2 Use equivalent ratios.
Write and solve an equation stating the ratios are equivalent. Let n represent the unknown part.

part → whole → $\frac{n}{58} = \frac{0.25}{100}$ } percent

$\frac{n}{58} = \frac{0.25}{100}$ (×0.58) Because $100 \times 0.58 = 58$, multiply 0.25 by 0.58.

$\frac{0.145}{58} = \frac{0.25}{100}$ $0.25 \times 0.58 = 0.145$; So, $n = 0.145$.

So, using either method, 0.25% of 58 is ______.

Check

What is 0.55% of 220? Use any strategy.

Go Online You can complete an Extra Example online.

Apply Book Fair!

Students were asked which night they planned on attending the book fair. The results of the survey are shown in the table. Twenty percent of the students who planned to attend on Wednesday attended on Thursday instead. Twenty-five percent of the students who planned to attend on Thursday attended on Wednesday instead. Which day, Wednesday or Thursday, had a greater actual attendance? By how many students?

Day	Number of Students
Monday	55
Tuesday	80
Wednesday	70
Thursday	112
Friday	65

1 What is the task?

Make sure you understand exactly what question to answer or problem to solve. You may want to read the problem three times. Discuss these questions with a partner.

First Time Describe the context of the problem, in your own words.
Second Time What mathematics do you see in the problem?
Third Time What are you wondering about?

2 How can you approach the task? What strategies can you use?

3 What is your solution?

Use your strategy to solve the problem.

4 How can you show your solution is reasonable?

Write About It! Write an argument that can be used to defend your solution.

Talk About It!
Would the solution be the same if 25% of the students who planned to attend Wednesday attended on Thursday, instead of 20%? Explain.

Check

Five hundred students were asked what color they prefer for the new school colors. The results are shown in the table. How many more students prefer blue than black?

Color	Percent
Yellow	7
Blue	36
Orange	15
Red	12
Black	30

Go Online You can complete an Extra Example online.

Pause and Reflect

Create a graphic organizer that shows your understanding of how you can use the following methods to find the percent of a number.

- bar diagram
- ratio table
- double number line
- equivalent ratios

Name ______________________ Period ____________ Date ______________

Practice

Go Online You can complete your homework online.

Use any strategy to solve each problem.

1. The graph shows the career interests of the students at Linda's school. Suppose there are 400 students at the school. How many of them want to be an athlete? **(Example 1)**

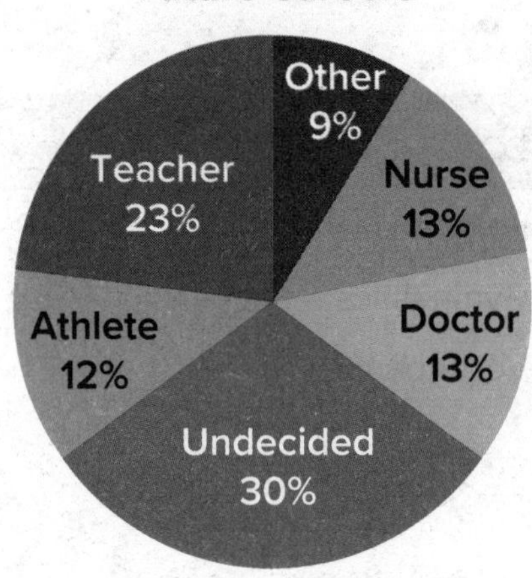

2. The graph shows the favorite activities of campers at a summer camp. Suppose there are 300 campers at the camp. How many campers favor fishing? **(Example 1)**

Favorite Camp Activities

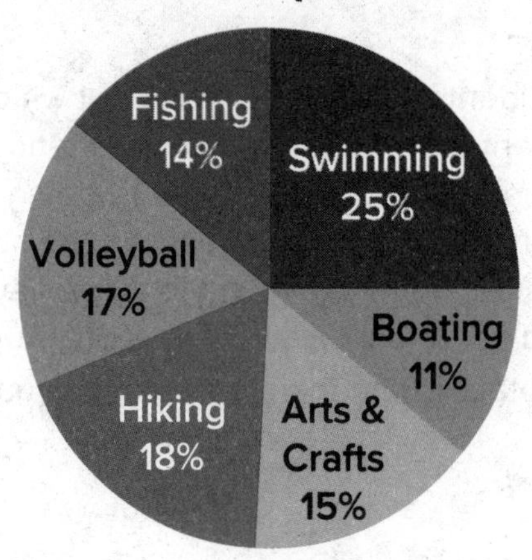

Use any method to find the percent of each number. **(Examples 2–4)**

3. 15% of 240 = ________

4. 65% of 180 = ________

5. 40% of 20 = ________

6. 250% of 82 = ________

7. 150% of 44 = ________

8. 320% of 65 = ________

9. 0.15% of 350 = ________

10. 0.4% of 168 = ________

11. 0.25% of 360 = ________

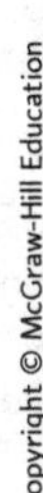

Test Practice

12. Open Response Kenzie is putting the family vacation videos onto a flash drive. The flash drive can hold 200 minutes of video. Kenzie has used 45% of the memory space already. How many minutes of the flash drive has she already used?

Apply

13. Students were asked which night they planned on going to the school festival. The results of the survey are shown in the table. If 18% of the students did not go on Friday, and 15% of the students did not go on Saturday, how many more students went on Friday than on Saturday?

Night	Number of Students
Friday	550
Saturday	480

14. Students were surveyed about which school athletic event they were planning to attend this week. Of the students who said they were going to the football game, 25% did not attend. Of the students who stated they were going to the volleyball game, 20% did not attend. How many more students went to the football game than the volleyball game?

Event	Number of Students
Football Game	120
Gymnastics Meet	95
Volleyball Game	80

15. MP **Persevere with Problems** Olive is going to buy a scooter that costs $95. The sales tax rate is 8.5%. What is the total cost of the scooter including tax to the nearest cent?

16. MP **Justify Conclusions** Is 18% of 30 the same as 30% of 18? Justify your conclusion.

17. MP **Identify Structure** How can you find 40% of 150 using mental math? Explain.

18. Be Precise Explain how the part of a whole can be greater than the whole itself. Use an example.

Estimate the Percent of a Number

I Can... estimate the percent of a number by using benchmark percents and rounding.

Today's Standards
6.RP.A.3, 6.RP.A.3.C
MP1, MP2, MP3, MP5

What Vocabulary Will You Learn?
benchmark percents

Learn Estimate the Percent of a Number

You learned how to find the percent of a number, such as 27% of 40, by reasoning about percent as a rate per 100 and by using bar diagrams, equivalent ratios, double number lines, and ratio tables. The equivalent ratios show that 27% of 40 is 10.8.

÷2.5

part → whole → $\frac{10.8}{40} = \frac{27}{100}$ } percent

÷2.5

Sometimes, it is not necessary to calculate the exact percent of a number. It may be sufficient to approximate, or estimate, the percent of a number. These situations can occur when estimating how much of a tip to leave on a restaurant bill, or estimating how much an item will cost after a percent discount.

When estimating the percent of a number, you can use benchmark percents. **Benchmark percents** are common percents, such as 10%, 20%, 25%, and their multiples. You can often perform mental calculations using benchmark percents.

The bar diagram shows the benchmark percent 25%, its multiples, and its corresponding fractional values.

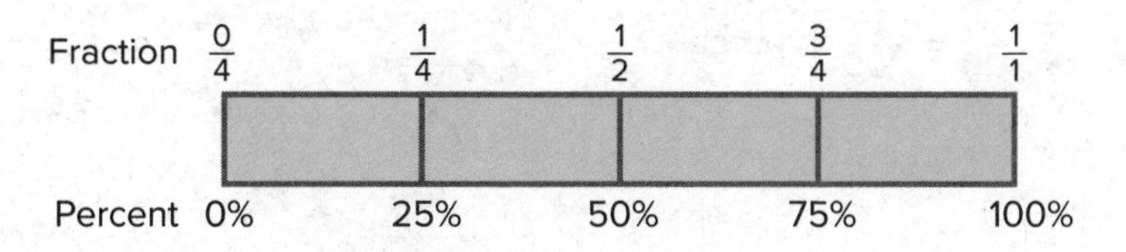

Suppose you wanted to estimate 27% of 40. You can use the benchmark percent 25% because 27% is close to 25%.

27% of 40 ≈ 25% of 40	27% is close to the benchmark percent 25%.
$\approx \frac{1}{4}$ of 40	25% of 40 is $\frac{1}{4}$ of 40.
≈ 10	$\frac{1}{4}$ of 40 is 10. So, 27% of 40 ≈ 10.

Because 10 is close to 10.8, the estimated part of the whole is close to the part of the whole.

Talk About It!
Why is the estimated part, 10, less than the actual part, 10.8?

(continued on next page)

Your Notes

Talk About It!

Compare and contrast 30% of 40 and the estimate you found on the previous page, 25% of 40. Which one is closer to the actual value, 27% of 40? Why?

Talk About It!

How can you use the benchmark percent 10% to find 30% of 40?

Some other benchmark percents you can use are 20%, 10%, and their multiples. The bar diagrams show the benchmark percents 20%, 10%, their multiples, and corresponding fractional values.

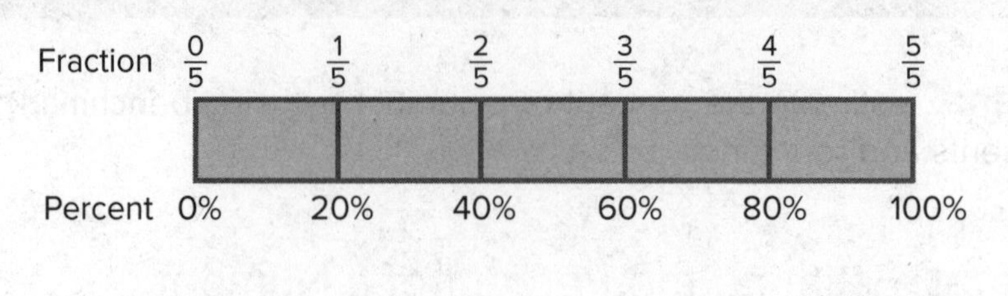

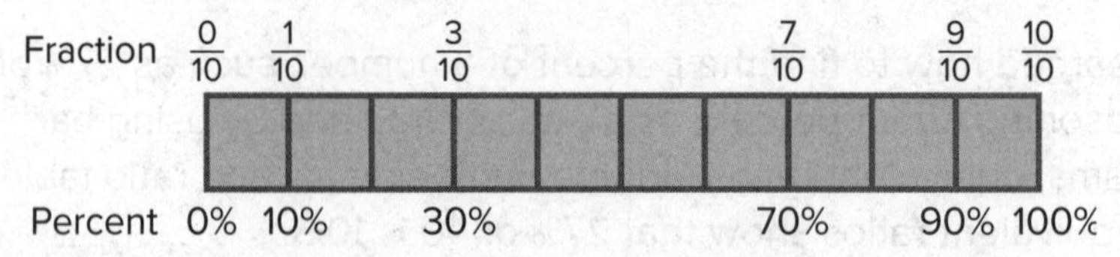

You can also use rounding to estimate the percent of a number. When estimating 27% of 40, you might round 27% to 30% and find 30% of 40 by using equivalent ratios. The equivalent ratios show that 30% of 40 is 12. So, 27% of 40 is about 12.

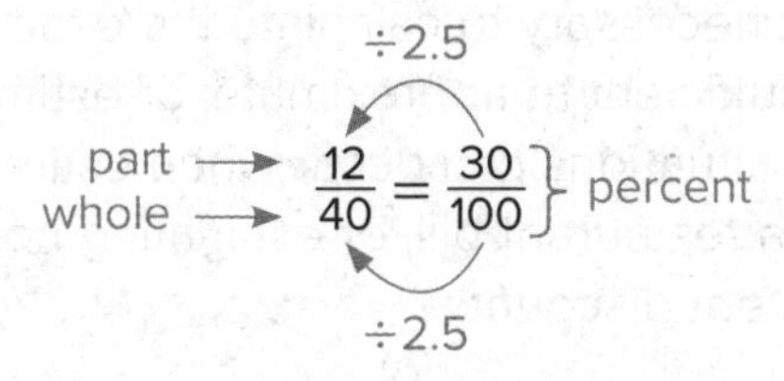

Sometimes, you might find it beneficial to also round the whole when estimating the percent of a number. Suppose you want to estimate 27% of 22. You can round 22 to 20 and round 27% to 25%, and then estimate 25% of 20 by using the methods shown in this Learn.

÷5

part → $\frac{x}{20} = \frac{25}{100}$ } percent

whole →

÷5

Because $100 \div 5 = 20$, divide 25 by 5.

$\frac{5}{20} = \frac{25}{100}$ $\quad 25 \div 5 = 5$; So, $x = 5$.

So, 27% of 22 is approximately 5.

Example 1 Estimate the Percent of a Number

Marita and five of her friends went out to dinner. Their total bill was \$47.45, and they would like to tip 18% of the bill.

About how much money should they leave as a tip?

Use the benchmark percent 20% because 18% is close to 20%. Round \$47.45 to \$50.

18% of \$47.45 ≈ 20% of \$50   18% is close to the benchmark percent 20%.

Method 1 Use a bar diagram.

The bar diagram shows that 20% of \$50 is \$10.

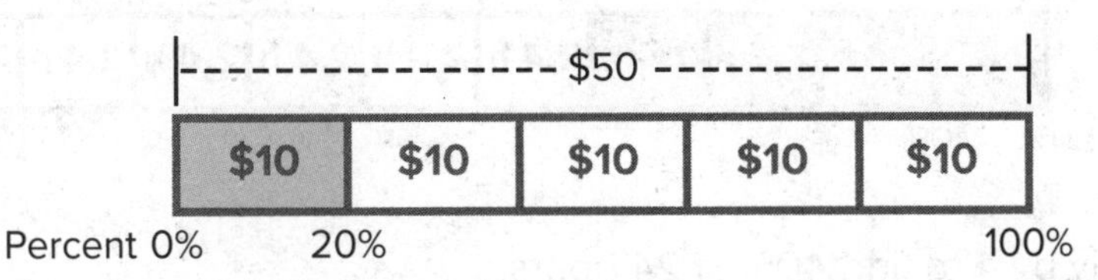

Method 2 Use equivalent ratios.

Let n represent the unknown part.

part → whole → $\frac{n}{50} = \frac{20}{100}$ } percent

$\frac{n}{50} = \frac{20}{100}$ (÷2 on both)   Because 100 ÷ 2 = 50, divide 20 by 2.

$\frac{10}{50} = \frac{20}{100}$ 20 ÷ 2 = 10; So, $n = 10$.

So, using either method, 18% of \$47.45 is about ______. Marita and her friends should leave a \$10 tip.

Check

Of the 78 teenagers at a youth camp, 63% have birthdays in the spring. About how many teenagers have birthdays in the spring?

Go Online You can complete an Extra Example online.

Think About It!

Is 18% of \$47.45 less than, greater than, or equal to \$5? How do you know?

Talk About It!

A classmate rounded \$47.45 to \$48 and found 20% of \$48 to be \$9.60. Is this a valid strategy? Explain. Which rounding strategy is closer to the actual value? Why might someone choose to round to \$50 instead of \$48?

Example 2 Estimate the Percent of a Number

Most pet birds spend about 41% of the day sleeping.

About how many hours a day do they spend sleeping?

You need to estimate 41% of 24, because there are 24 hours in a day. Because 41% is close to 40%, 41% of 24 ≈ 40% of 24.

Method 1 Use the benchmark percent 10%.

Draw a bar diagram with 10 equal-size sections. Each section represents 10%. The value of each section is 24 ÷ 10 or 2.4 hours. So, 10% of 24 hours is 2.4 hours.

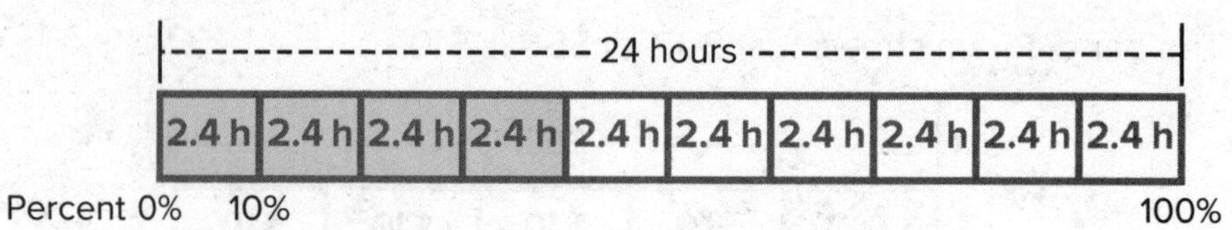

Multiply by 4 to find 40% of 24 hours.

40% of 24	= 4(10% of 24)	40% = 4(10%)
	= 4(2.4)	10% of 24 = 2.4
	= 9.6	Multiply.

Method 2 Use the benchmark percent 20%.

Draw a bar diagram with 5 equal sections. Each section represents 20%. The value of each section is 24 ÷ 5 or 4.8 hours. So, 20% of 24 hours is 4.8 hours.

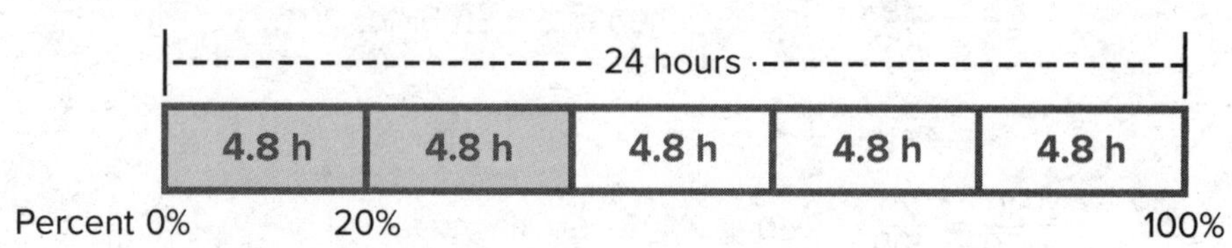

Multiply by 2 to find 40% of 24 hours.

40% of 24	= 2(20% of 24)	40% = 2(20%)
	= 2(4.8)	20% of 24 = 4.8
	= 9.6	Multiply.

So, using either method, 41% of 24 hours is about ______ hours. Pet birds spend about 9.6 hours a day sleeping.

Check

Estimate 76% of 122. Use any strategy.

Go Online You can complete an Extra Example online.

Think About It!

Do pet birds spend less than, greater than, or equal to 12 hours a day sleeping? Explain.

Talk About It!

Why might it be more advantageous to use the benchmark percent 10% than 20%?

Apply Financial Literacy

Sabrina takes her car to the car wash and chooses the Gold Star service that includes a wash, wax, and interior cleaning. This service normally costs $53.99, but is on special for $5.00 off. She must also pay a 6% sales tax, which is applied to the discounted price, and then added to find the total price. Estimate the total amount Sabrina paid at the car wash.

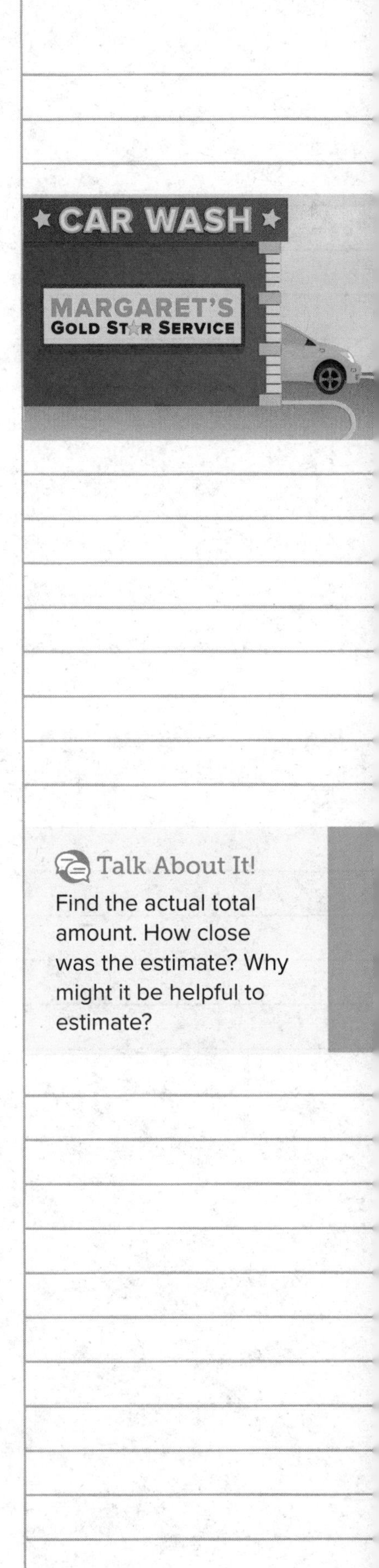

1 What is the task?

Make sure you understand exactly what question to answer or problem to solve. You may want to read the problem three times. Discuss these questions with a partner.

First Time Describe the context of the problem, in your own words.
Second Time What mathematics do you see in the problem?
Third Time What are you wondering about?

2 How can you approach the task? What strategies can you use?

3 What is your solution?

Use your strategy to solve the problem.

4 How can you show that your solution is reasonable?

Write About It! Write an argument that can be used to defend your solution.

Talk About It!
Find the actual total amount. How close was the estimate? Why might it be helpful to estimate?

Check

There were 48,500 people at an amusement park on Monday. Forty-two percent of the people wanted to ride the new roller coaster. Twenty-three percent of those people decided not to ride the coaster because the line was too long. About how many people waited in line for the new roller coaster that day?

Show your work here

Go Online You can complete an Extra Example online.

Pause and Reflect

Describe a situation in which you have estimated the percent of a number in your everyday life, or describe a situation in which you might do so in the future.

Record your observations here

Name ______________________ Period __________ Date __________

Practice

Go Online You can complete your homework online.

For Exercises 1–11, estimate each percent. Show your estimation. (Examples 1 and 2)

1. 51% of 62 ≈ _____

2. 26% of 78 ≈ _____

3. 39% of 198 ≈ _____

4. 78% of 148 ≈ _____

5. 19% of 103≈ _____

6. 98% of 59 ≈ _____

7. Emilia and her three sisters went to the movies. The total cost of the movie tickets and snacks was $38.75. Emilia paid 23% of the total cost. About how much money did she pay?

8. Karl earned $188 last month doing chores after school. If 68% of the money he earned was from doing yard work, about how much did Karl earn doing yard work?

9. The concession stand at a football game served 288 customers. Of those customers, about 77% bought a hot dog. About how many customers bought a hot dog?

10. In a recent season, the Chicago Cubs won 64% of the 161 regular season games they played. About how many games did they win?

11. The table shows how the 515 students at West Middle School get to school. About how many of the students walk to school?

Way to School	Number of Students
Bus	53%
Car	21%
Walk	26%

Test Practice

12. Open Response Carolyn's homeroom sold 207 magazines. Of the magazines sold, 28% were fashion magazines. About how many fashion magazines were sold?

Apply

13. Paul takes his dog to the groomer and selects the deluxe grooming package. He has a coupon for $10 off any grooming service. He must pay an 8% sales tax, which is applied to the discounted price, and then added to find the total price. Estimate the total amount Paul paid the dog groomer.

Grooming Package	Cost ($)
Regular	48.99
Deluxe	58.75

14. A toy store purchases a video game system for $192 and marks up the price of the video game system by 176%. The store is having a sale where everything is 20% off the sticker price. Estimate the final price of the video game system.

15. There were 59,500 people who attended a football game. Twenty-four percent of the people received a voucher for a free water bottle. Six percent of those people never claimed their water bottle. About how many people claimed their water bottle?

16. MP **Reason Inductively** Zeb wants to buy a fishing pole regularly priced at $64. It is on sale for 60% off. Zeb estimates that he will save 60% of $60, or $36. Will the actual amount saved be more or less than $36? Explain.

17. Explain how you can estimate 39% of $197.

18. MP **Justify Conclusions** A store is having a 40% off sale. If you have $38, will you have enough money to buy an item that regularly sells for $65.99? Write an argument to justify your conclusion.

Find the Whole

I Can... find the whole, given the part and the percent by using bar diagrams, ratio tables, double number lines, and equivalent ratios.

Today's Standards
6.RP.A.3, 6.RP.A.3.C
MP1, MP2, MP3, MP5, MP6, MP7

Learn Find the Whole

Sixty percent of the sixth graders at Jackson Middle School play a sport. If 114 sixth graders play a sport, how many sixth graders are there in the school?

You are given the part, 114 students, and the percent, 60%. You need to find the whole. In other words, 60% of what number is 114?

You can use bar diagrams, ratio tables, double number lines, and equivalent ratios to find the whole.

Method 1 Use a bar diagram.

Sixty is a multiple of 10 and 10 is a factor of 100. Draw a bar diagram with 10 equal-size sections of 10% each, because $10 \times 10 = 100$. Shade 6 sections to represent 60%. Label the shaded sections as 114 students, because 60% of the whole is 114.

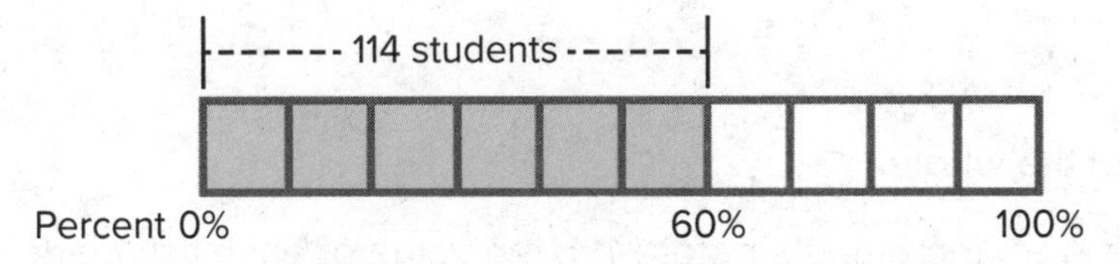

Each section represents the same number of students. There are 6 shaded sections. Divide 114 by 6 to find the number of students represented by each section.

$114 \div 6 = 19$ Divide. Each section represents 19 students.

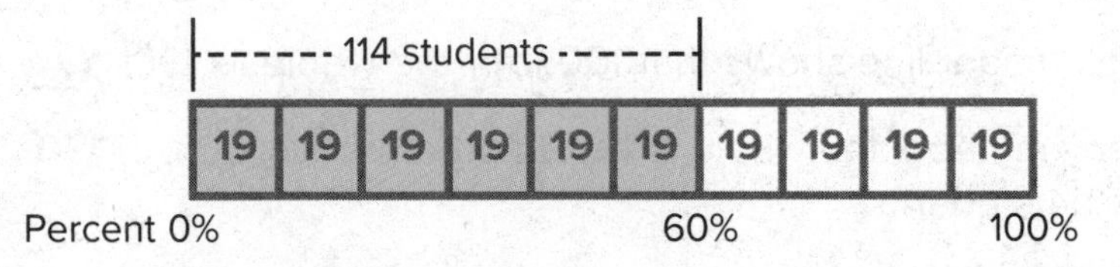

Because each section represents 19 students and there are 10 total sections, multiply 19 by 10 to find the total number of students.

$19 \times 10 = 190$ Multiply. The whole is 190 students.

So, 60% of 190 is 114. There are 190 sixth graders at the school.

Talk About It!
How can you use the bar diagram to find the number of sixth graders who do *not* play a sport?

(continued on next page)

Your Notes

Method 2 Use a ratio table.

You know that 60% of some number is 114. Use a ratio table to scale back from 60% to 10%. Then scale forward from 10% to 100%.

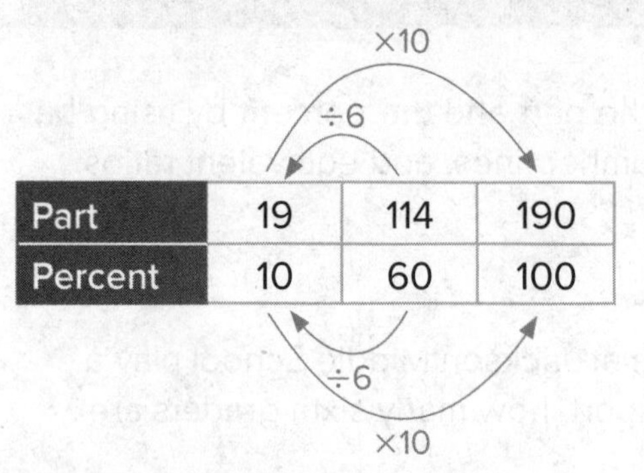

Part	19	114	190
Percent	10	60	100

Because $60 \div 6 = 10$, divide 114 by 6 to obtain 19. Because $10 \times 10 = 100$, multiply 19 by 10 to obtain 190. So, 60% of 190 is 114.

Method 3 Use a double number line.

Step 1 Draw a double number line.

Draw a double number line. The bottom number line represents the percent, so use increments of 10 to draw tick marks and label the percents. The top number line represents the whole, so label the tick mark that corresponds with 60% on the bottom number line with 114.

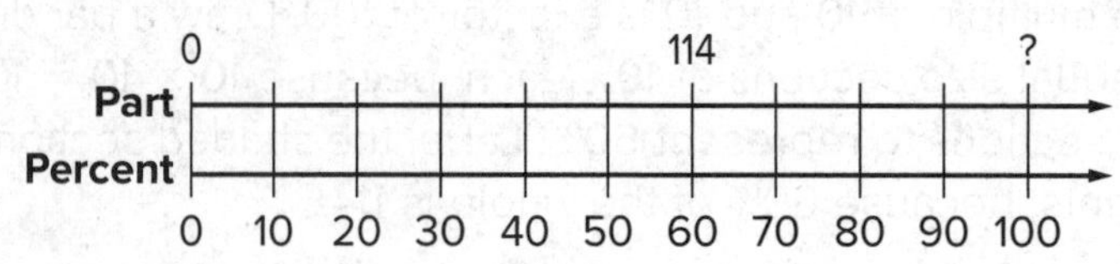

Step 2 Find the whole.

Since there are 6 increments before 114, the value of each tick mark on the top number line increases by $114 \div 6$, or 19 units.

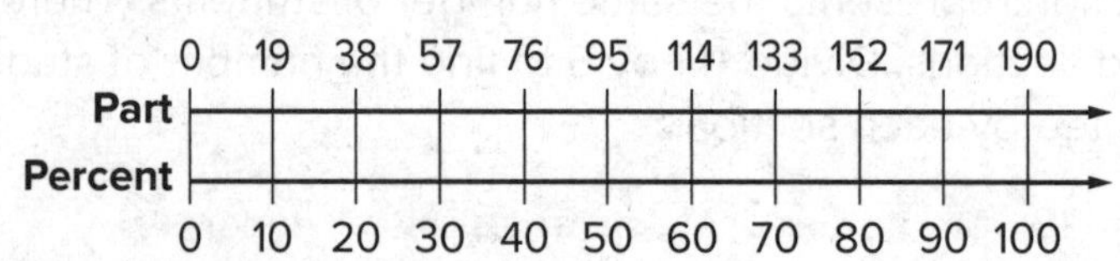

The double number line shows that 100%, or the whole, is 190.

So, using any method, the whole is 190. In other words, 60% of 190 students is 114 students.

Talk About It!

A classmate let w represent the unknown whole and set up two equivalent ratios as shown. Is this method valid? Why might this method not be the most advantageous one to use in this case?

$$\frac{114}{w} = \frac{60}{100}$$

Example 1 Find the Whole

Country music makes up 75% of Landon's music library.

If he has downloaded 90 country music songs, how many songs does Landon have in his music library?

The part is 90 country music songs. The percent is 75%. The whole, the number of songs he has in his library, is the unknown.

Method 1 Use a bar diagram.
Draw a bar diagram with 4 equal-size sections of 25% each. Shade 3 sections to represent 75%. Label the shaded sections as 90 songs.

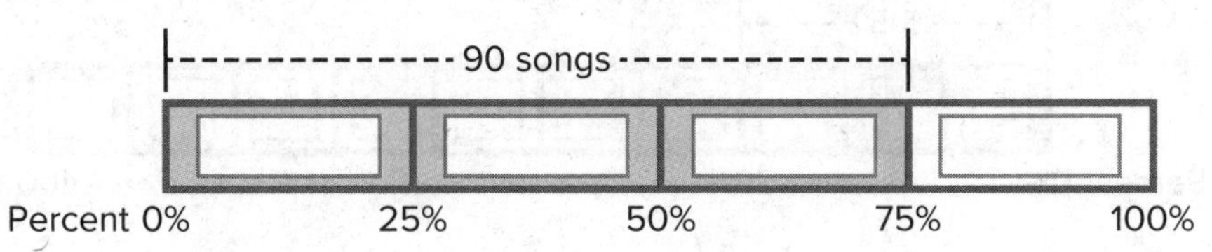

How many songs are represented by each section? ____________

Label each section on the bar diagram.

How many songs are represented by the whole? ____________

Method 2 Use equivalent ratios.
Let w represent the whole.

part → percent → $\frac{90}{w} = \frac{75}{100}$ } percent

$\frac{90}{w} = \frac{3}{4}$ Simplify $\frac{75}{100}$ as $\frac{3}{4}$.

×30

$\frac{90}{120} = \frac{3}{4}$ Because $3 \times 30 = 90$, multiply 4 by 30 to obtain 120. So, $w = 120$.

×30

So, using either method, Landon has 120 songs in his music library.

Check

In the first year of ownership, a new car lost 20% of its value. If the car lost $4,200 of its value, how much did the car originally cost? Use any strategy.

Go Online You can complete an Extra Example online.

Think About It!

A classmate claims that because 75% is less than 100, Landon should have more than 90 music songs in his library. Do you think this reasoning is correct? Why or why not?

Talk About It!

Explain why setting up the equation relating the equivalent ratios was advantageous to use in this example.

Example 2 Find the Whole

Marissa saved $15 because she bought a sweater that was on sale for 30% off.

What was the original price of the sweater?

The part is $15. The percent is 30%. The whole is the unknown.

Method 1 Use a bar diagram.
Draw a bar diagram with 10 equal-size sections of 10% each. Shade 3 sections to represent 30%. Label the shaded sections as $15.

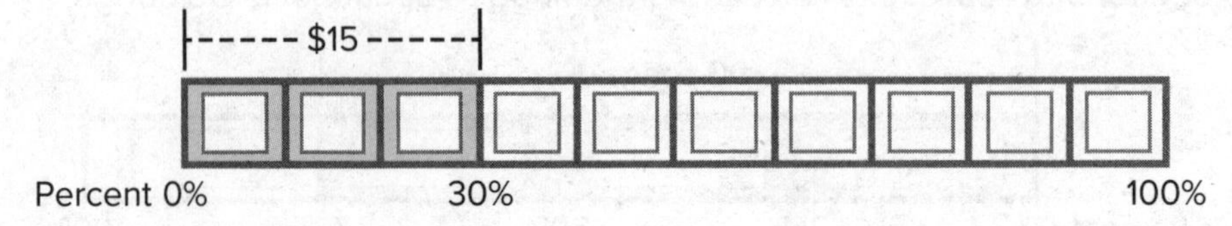

How much money is represented by each section? ________

Label each section on the bar diagram.

How much money is represented by the whole? ________

Method 2 Use a double number line.

Step 1 Draw a double number line.

Label the part, 15, with its corresponding percent, 30%.

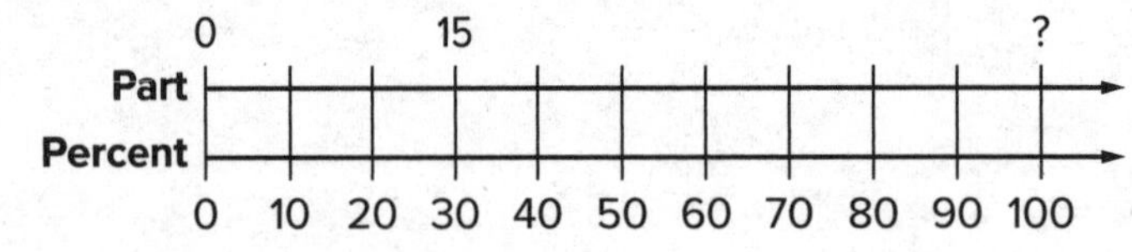

Step 2 Find the whole.

The value of each tick mark on the top number line increases by 15 ÷ 3, or 5 units. The number line shows that the whole, or 100%, is $50.

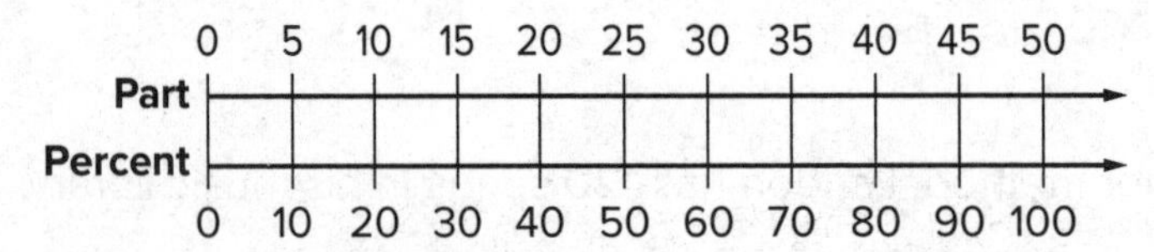

So, using either method, the original cost of the sweater was $50.

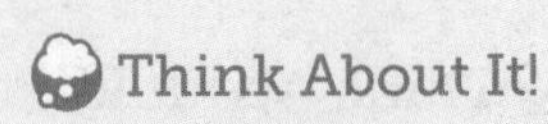

Think About It!

Is the whole less than, greater than, or equal to $15? How do you know?

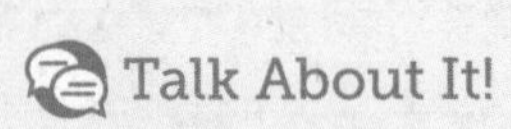

Talk About It!

Choose another strategy, such as a ratio table or an equation relating two equivalent ratios, to solve this problem. Compare and contrast the methods.

Check

Kai calculates that he spends 15% of a school day in science class. If he spends 75 minutes in science class, how many minutes does Kai spend in school each day?

Go Online You can complete an Extra Example online.

Apply Sales

The table shows the percentage of each type of popcorn flavor at a specialty food store. A store clerk put all of the bags of cinnamon popcorn and cheese popcorn in a display in the front of the store. If the clerk put 60 bags in the front, how many bags of popcorn does the store have in all? If the store sells all of the bags of popcorn for $4.75 per bag, how much will the store earn in sales?

Flavor	Percent
Kettle Corn	60
Cinnamon	15
Caramel	10
Cheese	15

Go Online watch the animation.

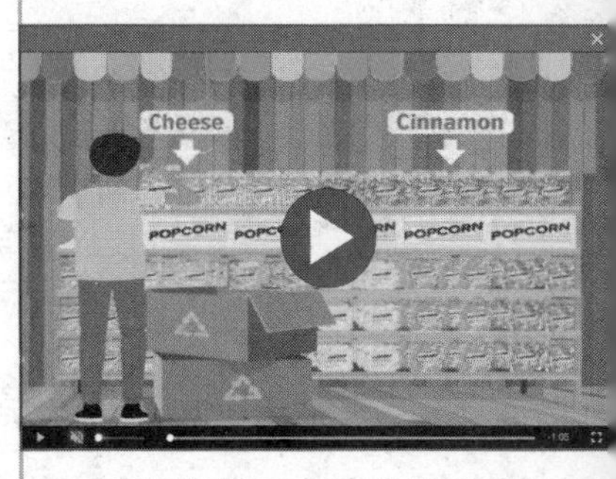

1 What is the task?

Make sure you understand exactly what question to answer or problem to solve. You may want to read the problem three times. Discuss these questions with a partner.

First Time Describe the context of the problem, in your own words.
Second Time What mathematics do you see in the problem?
Third Time What are you wondering about?

2 How can you approach the task? What strategies can you use?

3 What is your solution?

Use your strategy to solve the problem.

Talk About It!
How much more will the store earn in sales for selling all of the bags of kettle corn popcorn than caramel popcorn? Describe two different ways to solve this problem.

4 How can you show that your solution is reasonable?

Write About It! Write an argument that can be used to defend your solution.

Check

The table shows the percent of each type of puzzle in a toy store. During a sale, the store sold all of the 300-piece and 500-piece puzzles. If they sold 120 puzzles, how many puzzles did the store have before the sale? If they sell all of the puzzles for $8.19 per puzzle, how much will the store make in sales?

Number of Pieces	Percent of Stock
300	50
500	30
750	15
1,000	5

Go Online You can complete an Extra Example online.

Pause and Reflect

Create a graphic organizer that shows your understanding of how you can use the following methods to find the whole, given the part and the percent.

- bar diagram
- ratio table
- double number line
- equivalent ratios

Name ______________________ Period ____________ Date ____________

Practice

Go Online You can complete your homework online.

Use any strategy to solve each problem. (Examples 1 and 2)

1. Yolanda's club requires that 80% of the members be present for any vote. If at least 20 members must be present to have a vote, how many members does the club currently have?

2. Action movies make up 25% of Sara's DVD collection. If she has 16 action DVDs, how many DVDs does Sara have in her collection?

3. Marcus saved $10 because he bought a baseball glove that was on sale for 40% off. What was the original price of the baseball glove?

4. Of the students in the marching band, 55% plan to go to the school dance. If there are 110 students going to the dance, how many students are in the marching band?

5. Melcher used 24% of the memory card on his digital camera while taking pictures at a family reunion. If Melcher took 96 pictures at the family reunion, how many pictures can the memory card hold?

6. Mallorie has $12 in her wallet. If this is 20% of her monthly allowance, what is her monthly allowance?

7. The table shows the number of minutes Tim has for lunch and study hall. He calculates that these two periods account for 18% of the minutes he spends at school. How many minutes does he spend at school?

Period	Number of Minutes
Lunch	45
Study Hall	45

Test Practice

8. Open Response The number of sixth grade students accounts for 35% of the total number of students enrolled in middle school. There are 245 sixth grade students. How many students are enrolled in the middle school?

Apply

9. The table shows the percent of each type of lunch choice a cafeteria had available on Friday. There are a total of 270 cheese and pepperoni pizza lunches. How many lunches did they have in all on Friday? If the cafeteria sells all the lunches at $3.50 per lunch, how much money will the cafeteria earn?

Lunch	Percent
Cheese Pizza	50
Pepperoni Pizza	40
Fried Chicken	10

10. The volleyball team was selling snack bags to raise money for new uniforms. The table shows the percentage of each type of bag. There are 210 bags of pretzels and cheese puffs. How many snack bags did they have in all? If the team sells all the snack bags at $1.75 per bag, how much money will they raise?

Snack	Percent
Cheese Puffs	10
Corn Chips	15
Popcorn	25
Potato Chips	30
Pretzels	20

11. MP **Be Precise** Of the number of sixth grade students at a middle school, 120 prefer online magazines over print magazines. Of the number of seventh grade students, 140 prefer online magazines. A student said that this means a greater percent of seventh grade students prefer online magazines than sixth grade students. Is the student correct? Use precise mathematical language to explain your reasoning.

12. MP **Use Math Tools** In a photography club, 48% of members are girls. If there are 26 members who are girls, explain how you can use mental math to estimate the total number of members in the photography club?

13. Create Write and solve a real-world problem where you use equivalent ratios to find the whole.

14. If 10% of x is 100, how can you find the value of x?

Review

Foldables Use your Foldable to help review the module.

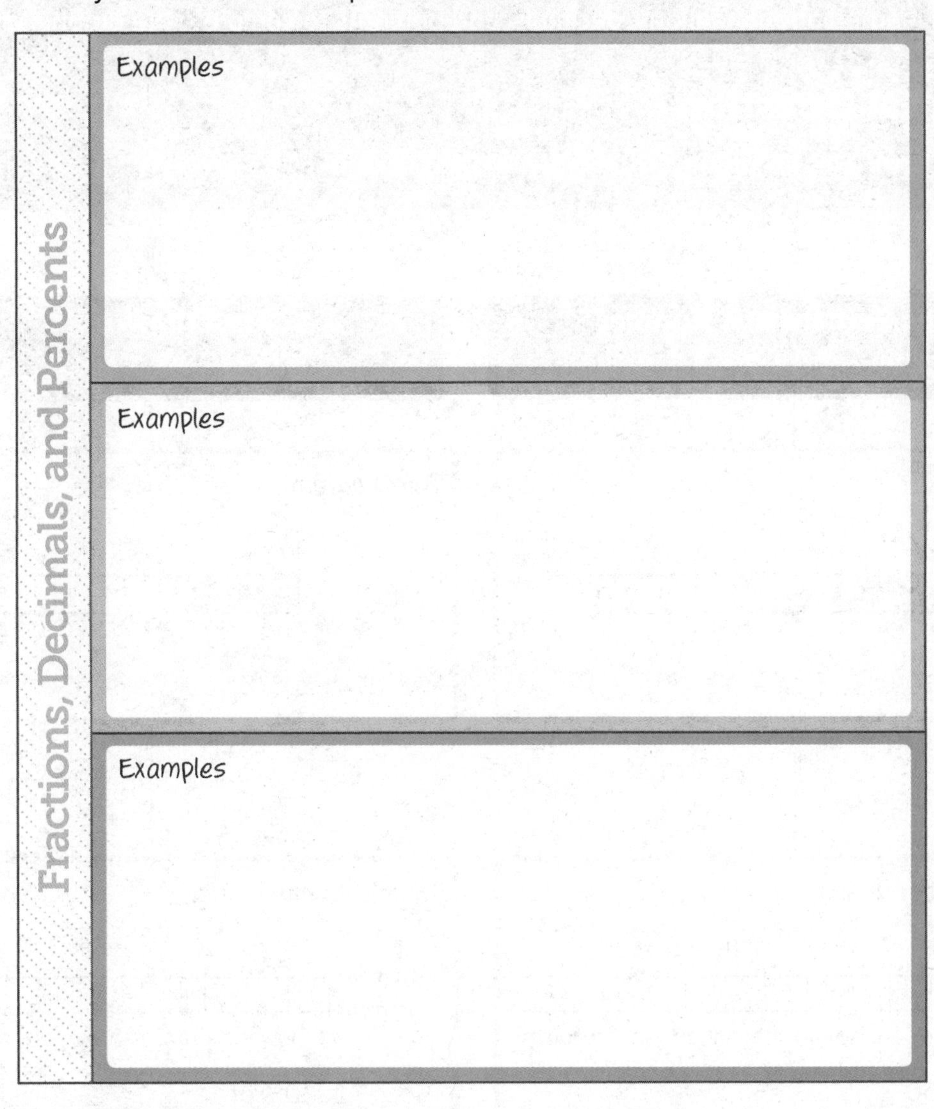

Rate Yourself!

Complete the chart at the beginning of the module by placing a checkmark in each row that corresponds with how much you know about each topic after completing this module.

Write about one thing you learned.

Write about a question you still have.

Reflect on the Module

Use what you learned about fractions, decimals, and percents to complete the graphic organizer.

Essential Question

How can you use fractions, decimals, and percents to solve everyday problems?

Find the Percent of a Number

What is 60% of 60?

Bar Diagram:

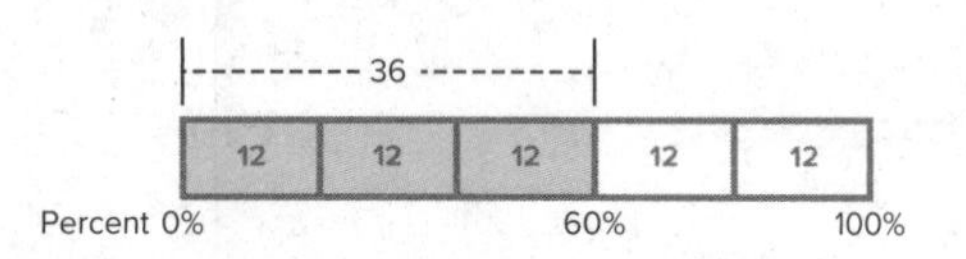

So, 60% of 60 is ______.

Double Number Line:

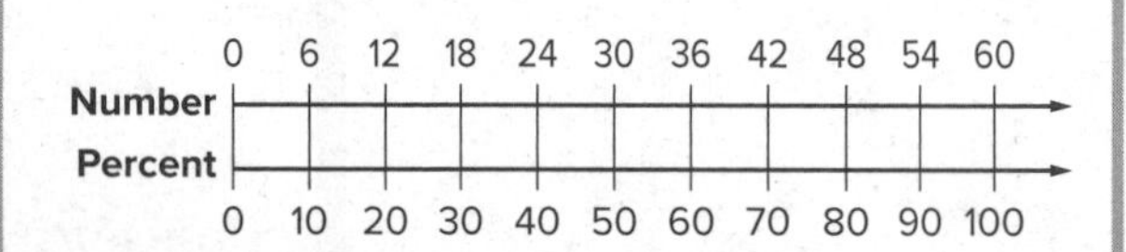

So, 60% of 60 is ______.

Equivalent Ratios:

Part → Whole → $\frac{x}{60} = \frac{60}{100}$ } Percent

×0.6

$\frac{36}{60} = \frac{60}{100}$ Because $100 \times 0.6 = 60$, multiply 60 by 0.6.

×0.6

So, 60% of 60 is ______.

Find the Whole

27 is 30% of what?

Bar Diagram:

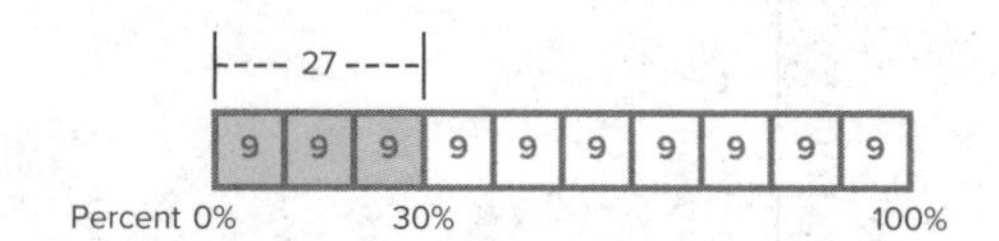

So, 27 is 30% of ______.

Double Number Line:

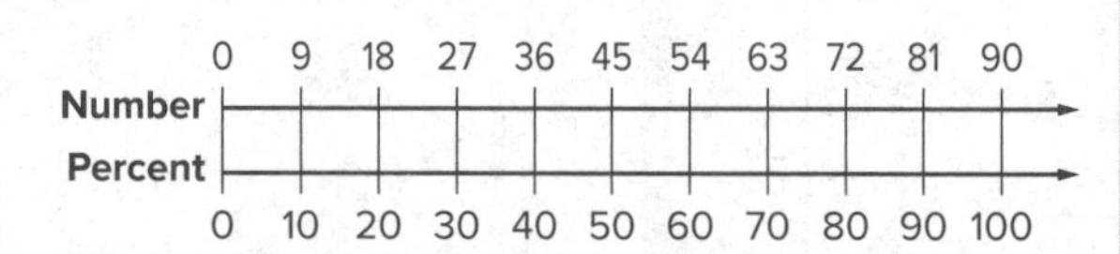

So, 27 is 30% of ______.

Equivalent Ratios:

Part → Whole → $\frac{27}{x} = \frac{30}{100}$ } Percent

×0.9

$\frac{27}{90} = \frac{30}{100}$ Because $30 \times 0.9 = 27$, multiply 100 by 0.9.

×0.9

So, 27 is 30% of ______.

Name ______________________ Period __________ Date __________

Test Practice

1. Multiple Choice What is 2.6% written as a decimal? **(Lesson 2)**

(A) 0.26

(B) 0.026

(C) 26

(D) 260

2. Equation Editor At a baking competition, 0.5 dishes were cooked by girls, $\frac{3}{10}$ were cooked by boys, and $\frac{1}{5}$ were cooked by adults. What fraction of the dishes were cooked by boys and girls? **(Lesson 1)**

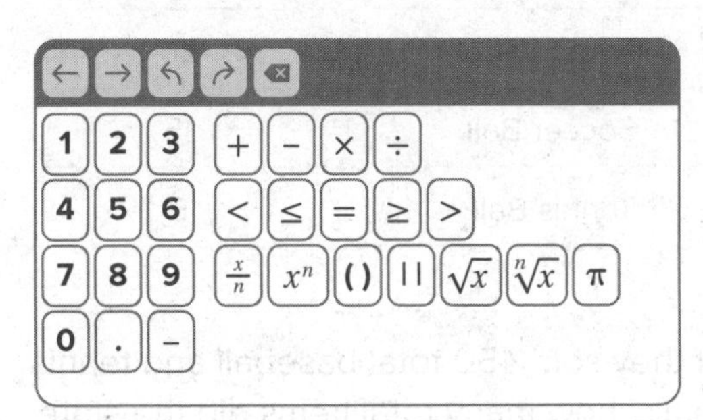

3. Open Response Vineisha earned 22 out of 20 points on her science quiz over the phases of the moon due to an extra credit question. What percent did she earn on the quiz? **(Lesson 2)**

4. Open Response Refer to the grid shown below. **(Lesson 2)**

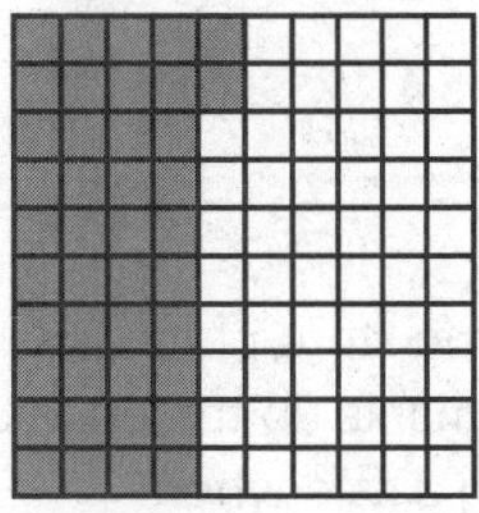

A. What percent of the grid is shaded?

B. Write your answer from part A as a fraction and a decimal.

5. Multiselect Which number forms below are equivalent to 0.28? Select all that apply. **(Lessons 1 and 4)**

☐ 28%

☐ $\frac{28}{100}$

☐ $\frac{14}{50}$

☐ $\frac{7}{25}$

6. Open Response At a food festival, $\frac{3}{8}$ of the dishes were from China. Another 12.5% of the dishes were from Japan. What percent of the dishes were from other countries? **(Lesson 3)**

7. Open Response A basketball player made 40% of the shots she attempted. If she made 32 baskets, how many shots did she attempt? (Lesson 6)

8. Multiple Choice A clothing store purchases sweatshirts for $26 and marks up the price by 146%. The store is having a sale where everything is on sale for 20% off. Choose the most reasonable estimate for the final price of a sweatshirt. (Lesson 4)

Ⓐ $7.60

Ⓑ $30.40

Ⓒ $38.00

Ⓓ $45.60

9. Open Response Three hundred students were surveyed about their favorite subject. The results are shown in the table below. How many more students prefer science than math? (Lesson 4)

Subject	Percent
Language Arts	15
Math	24
Science	33
Social Studies	21
Elective	7

10. Open Response The original price of a DVD is $11. The sale price is 30% off the original price. What is the sale price of the DVD? (Lesson 4)

11. Open Response The table shows the percent of total sales for each type of ball sold at a sports equipment store in one week. (Lesson 6)

Type of Ball	Percent
Baseball	25
Basketball	35
Football	20
Soccer Ball	15
Tennis Ball	5

A. If they sold 450 total baseball and tennis balls, how many total items did the store sell in one week?

B. If each item is sold for $10.95, how much did the store have in sales?

12. Open Response Twenty-one students in Michael's classroom are wearing jeans. There are 25 students in his class. Michael says that 80% of his class is wearing jeans. Is Michael correct? Explain your reasoning. (Lesson 4)

Module 3

Compute with Multi-Digit Numbers and Fractions

Essential Question

How are operations with fractions and decimals related to operations with whole numbers?

6.NS.A.1, 6.NS.B.2, 6.NS.B.3
Mathematical Practices: MP1, MP2, MP3, MP4, MP5, MP6, MP7, MP8

What Will You Learn?

Place a checkmark (✓) in each row that corresponds with how much you already know about each topic **before** starting this module.

KEY
▢ — I don't know. ◊ — I've heard of it. ☆ — I know it!

	Before			After		
	▢	◊	☆	▢	◊	☆
dividing multi-digit numbers						
adding and subtracting multi-digit decimals						
multiplying multi-digit decimals						
dividing multi-digit decimals						
finding reciprocals						
dividing whole numbers by fractions						
dividing fractions by fractions						
dividing fractions by whole numbers						

Foldables Cut out the Foldable and tape it to the Module Review at the end of the module. You can use the Foldable throughout the module as you learn about computing with multi-digit numbers and fractions.

What Vocabulary Will You Learn?

Check the box next to each vocabulary term that you may already know.

- ☐ dividend
- ☐ divisor
- ☐ Inverse Property of Multiplication
- ☐ multiplicative inverse
- ☐ quotient
- ☐ reciprocal

Are You Ready?

Study the Quick Review to see if you are ready to start this module. Then complete the Quick Check.

Quick Review

Example 1

Multiply whole numbers.

Find 13 × 15.

$$\begin{array}{r} 13 \\ \times\ 15 \\ \hline 65 \\ +130 \\ \hline 195 \end{array}$$

65 — Multiply the ones.
+130 — Multiply the tens.
195 — Add.

Example 2

Divide whole numbers.

Find 323 ÷ 17.

$$\begin{array}{r} 19 \\ 17\overline{)323} \\ -17 \\ \hline 153 \\ -153 \\ \hline 0 \end{array}$$

Divide the tens.
Divide the ones.

Quick Check

1. Find 19 × 51.

2. Find 49 × 23.

3. Find 539 ÷ 11.

4. Find 432 ÷ 16.

How Did You Do?

Which exercises did you answer correctly in the Quick Check? Shade those exercise numbers at the right.

① ② ③ ④

Divide Multi-Digit Whole Numbers

I Can... use the standard algorithm to divide multi-digit numbers when solving problems.

Today's Standards
6.NS.B.2
MP1, MP2, MP4, MP6, MP8

What Vocabulary Will You Learn?
dividend
divisor
quotient

Learn Divide Multi-Digit Numbers

When one number is divided by another, the result is called a **quotient**. The **dividend** is the number that is divided and the **divisor** is the number used to divide the dividend.

Label each part of the division expression with the terms *quotient*, *dividend*, and *divisor*.

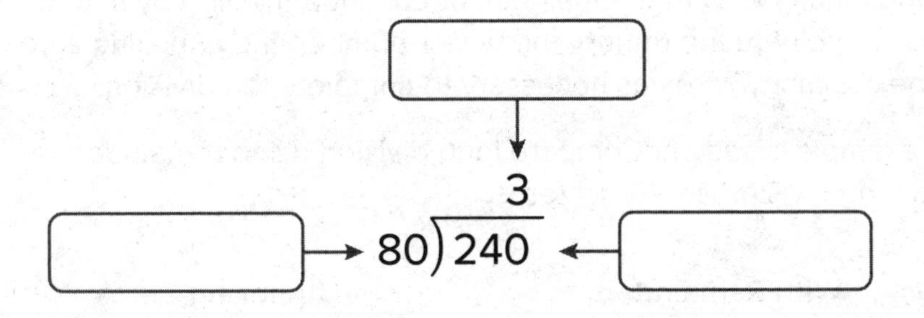

Example 1 Divide Multi-Digit Numbers

Find 25,740 ÷ 12.

```
     2,145
12)25,740
  −24
    17
   −12
     54
    −48
      60
     −60
       0
```

Divide each place value position from left to right.

So, 25,740 ÷ 12 is ______.

Talk About It!
How can you check to see if the quotient is correct?

Your Notes

Check

Find 868 ÷ 14.

Show your work here

Go Online You can complete an Extra Example online.

Learn Divide Multi-Digit Numbers

If two numbers do not divide evenly, you can write the quotient as a whole number with a remainder, or continue dividing by adding a decimal point to the right of the whole number and annexing zeros. Annex as many zeros as necessary to complete the division.

An example is shown. Compare long division using remainders and long division annexing zeros.

With Remainders

$$\begin{array}{r} 32 \text{ R}20 \\ 25\overline{)820} \\ -75 \\ \hline 70 \\ -50 \\ \hline 20 \end{array}$$

Annexing Zeros

$$\begin{array}{r} 32.8 \\ 25\overline{)820.0} \\ -75 \\ \hline 70 \\ -50 \\ \hline 200 \\ -200 \\ \hline 0 \end{array}$$

Recall that a remainder can be written as a fraction with the remainder in the numerator and the dividend in the denominator. To check that 32 R 20 is equal to 32.8, first write the remainder as a fraction and then convert the fraction to a decimal.

$$32 \text{ R } 20 = 32\frac{20}{25}$$
$$= 32\frac{20}{25} \text{ or } 32.8$$

The solution to 820 ÷ 25 is 32 with a remainder of 20, or 32.8.

Talk About It!
How do you know that 820 and 820.0 are equivalent?

Example 2 Divide Multi-Digit Numbers

Find 5,272 ÷ 64.

```
        82.375
64)5,272.000
   -5 12          Divide each place value from left to right.
                  Multiply 8 × 64, then subtract.
     152
    -128          Multiply 2 × 64, then subtract.
      240         There is a remainder. Annex a zero.
     -192         Multiply 3 × 64, then subtract.
       480        Annex a zero and continue dividing.
      -448        Multiply 7 × 64, then subtract.
        320       Annex a zero and continue dividing.
       -320       Multiply 5 × 64, then subtract.
          0       The remainder is 0.
```

So, 5,272 ÷ 64 is ________.

Think About It!
How will you set up the long division?

Talk About It!
How do you know when you are done dividing?

Check

Find 16,047 ÷ 60.

Go Online You can complete an Extra Example online.

Example 3 Divide Multi-Digit Numbers

Find 5,287 ÷ 340.

```
        15.55
340)5,287.00      Divide each place value from left to right.
   −3 40          Multiply 1 × 340, then subtract.
    1 887
   −1 700         Multiply 5 × 340, then subtract.
      1870
     −1700        Multiply 5 × 340, then subtract.
       1700
      −1700       Multiply 5 × 340, then subtract.
          0       The remainder is 0.
```

So, 5,287 ÷ 340 is ________.

Check

Find 4,562 ÷ 25.

Go Online You can complete an Extra Example online.

Apply Fundraising

The table shows the number of cookies donated for the school bake sale. The cookies were placed into bags with a dozen cookies in each bag. How many bags of a dozen cookies were available to sell?

Bake Sale Cookies	
Type	**Number**
Chocolate Chip	125
Oatmeal	60
Peanut Butter	245
Sugar	116

1 What is the task?

Make sure you understand exactly what question to answer or problem to solve. You may want to read the problem three times. Discuss these questions with a partner.

First Time Describe the context of the problem, in your own words.
Second Time What mathematics do you see in the problem?
Third Time What are you wondering about?

2 How can you approach the task? What strategies can you use?

3 What is your solution?

Use your strategy to solve the problem.

4 How can you show your solution is reasonable?

Write About It! Write an argument that can be used to defend your solution.

Talk About It!
Why is the final answer given as a whole number when the quotient is a decimal?

Check

There are 24 seats in each row of the middle school auditorium. The table shows the number of students from each grade who attended a concert. If the students fill each row in the auditorium, how many rows would be needed for all of the students?

Grade	Number of Students
Sixth	310
Seventh	256
Eighth	262

Go Online You can complete an Extra Example online.

Pause and Reflect

Have you ever wondered when you might use the concepts you learn in math class? What are some everyday scenarios in which you might use what you learned today?

Name ______ Period ______ Date ______

Practice

Go Online You can complete your homework online.

Find each quotient. (Examples 1–3)

1. 52,080 ÷ 15 = ______

2. 38,480 ÷ 26 = ______

3. 648 ÷ 18 = ______

4. 3,409 ÷ 14 = ______

5. 8,890 ÷ 40 = ______

6. 3,120 ÷ 64 = ______

7. 6,750 ÷ 240 = ______

8. 4,415 ÷ 800 = ______

9. 5,777 ÷ 160 = ______

10. The table shows the distances between major cities. Mr. Santiago has a flight from Los Angeles to Toronto. If the plane travels at 520 miles per hour, how long is the flight?

New York to Paris	3,636 miles
Los Angeles to Toronto	2,171 miles

Test Practice

11. Equation Editor What is the value of the expression 3,082 ÷ 23?

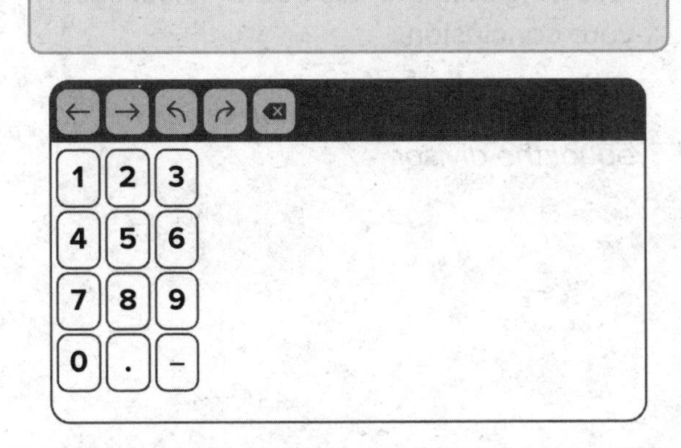

Apply

12. The table shows the number of each type of greeting card a gift shop had remaining at the end of the year. The store created blind bags with 15 random cards in each bag to sell. How many complete blind bags of cards were they able to make?

Card Type	Number of Cards
Anniversary	163
Birthday	258
Get Well	98
Thank You	47

13. The table shows the number of each type of seed packet a garden center had remaining at the end of summer. Bags were created with 20 random seed packets in each to sell at their summer sale. How many complete bags of seeds can be created?

Seed Type	Number of Packets
Aster	40
Daisy	95
Pansy	160
Sunflower	125
Wildflower	70

14. Use the digits 9, 6, and 3 one time each in the following multi-digit division problem. Then rewrite the problem.

☐,☐00 ÷ ☐0 = 40

15. MP **Persevere with Problems** If the divisor is 60, what is the least four-digit dividend that would not have a remainder?

16. MP **Justify Conclusions** Determine if the following statement is *true* or *false*. Justify your conclusion.

The remainder in a division problem can equal the divisor.

17. How can you check that your quotient is correct when dividing multi-digit whole numbers?

Compute With Multi-Digit Decimals

I Can... solve problems by using the standard algorithms for addition, subtraction, multiplication, and division to compute with multi-digit decimals.

Today's Standards
6.NS.B.3
MP1, MP2, MP3, MP4, MP6, MP7

Learn Add and Subtract Multi-Digit Decimals

You have already added and subtracted decimals to the hundredths place. You can apply the same rules when adding and subtracting decimals to the thousandths place, even when the place values are different. First, align the decimal points, then annex zeros until the place values are the same.

Find 45.16 + 21.384.

$$\begin{array}{r} 45.160 \\ +\ 21.384 \\ \hline \end{array}$$

Align the decimal points and annex a zero.

$$\begin{array}{r} 45.160 \\ +\ 21.384 \\ \hline 66.544 \end{array}$$

Add as with whole numbers.

45.16 + 21.384 = ☐

The steps show 32.94 − 15.386.

$$\begin{array}{r} 32.940 \\ -\ 15.386 \\ \hline \end{array}$$

Align the decimal points and annex a zero.

$$\begin{array}{r} 32.940 \\ -\ 15.386 \\ \hline 17.554 \end{array}$$

Subtract as with whole numbers.

32.94 − 15.386 = ☐

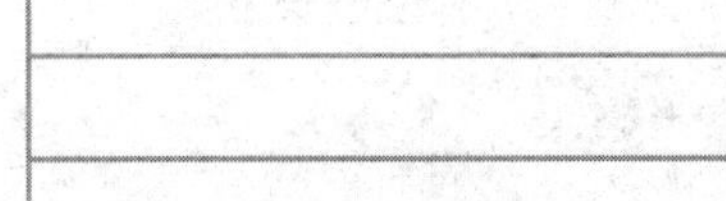

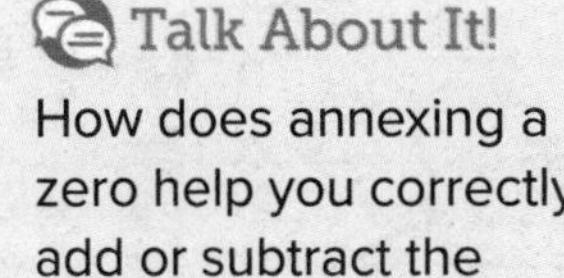

Talk About It!

How does annexing a zero help you correctly add or subtract the numbers?

Your Notes

Example 1 Add Multi-Digit Decimals

Find 23.498 + 14.93. Check the solution.

Make an estimate. Round to the nearest whole number.

23.498 + 14.93 ≈ ☐ + ☐ or ☐

Find the sum.

$$\begin{array}{r} 23.498 \\ +\ 14.930 \\ \hline 38.428 \end{array}$$

Align the decimal points and annex a zero.

Add. Place the decimal point in the answer.

So, 23.498 + 14.93 is ________.

Check the solution.

Compare the solution to the estimate:

☐ ≈ ☐

Talk About It!

Why is estimation useful when solving problems involving multi-digit decimals?

Check

Find 356.72 + 142.4.

Go Online You can complete an Extra Example online.

Example 2 Subtract Multi-Digit Decimals

Find 163.45 − 85.374. Check the solution.

Make an estimate. Round to the nearest ten.

163.45 − 85.374 ≈ ☐ − ☐ or ☐

Find the difference.

$$\begin{array}{r} 163.450 \\ -\ 85.374 \\ \hline 78.076 \end{array}$$

Align the decimal points and annex a zero.

Subtract. Place the decimal point in the answer.

So, 163.45 − 85.374 is ______.

Check the solution.

Compare the solution to the estimate:

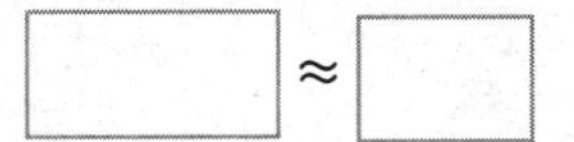

Check

Find 356.2 − 142.25.

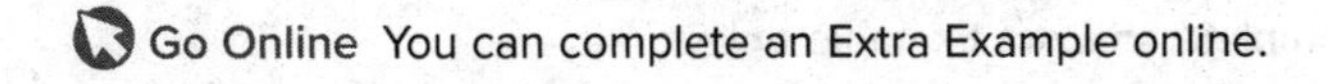

Example 3 Subtract Multi-Digit Decimals

Find 25 – 17.46. Check the solution.

Make an estimate. Round to the nearest whole number.

$25 - 17.46 \approx$ ☐ – ☐ or ☐

Find the difference.

$$\begin{array}{r} 25.00 \\ -\ 17.46 \\ \hline 7.54 \end{array}$$

Align the decimal points and annex zeros.

Subtract. Place the decimal point in the answer.

So, 25 – 17.46 is ________.

Check the solution.

Compare the solution to the estimate:

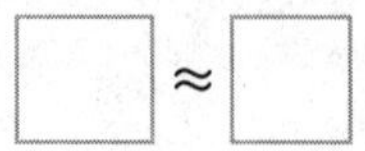

Check

Find 34 – 9.14.

Go Online You can complete an Extra Example online.

Learn Multiply Decimals

When multiplying a decimal by a decimal, multiply as with whole numbers. To place the decimal point in the product, find the sum of the number of decimal places in each factor. The product has the same number of decimal places. If there are not enough decimal places in the product, annex zeros to the left of the first non-zero digit.

Find 0.014 × 3.7.

```
   0.014   ←— three decimal places
 ×   3.7   ←— one decimal place
 -------
      98
 +   420
 -------
  0.0518   ←— Annex a zero to make four decimal places.
```

So, 0.014 × 3.7 is ________.

Pause and Reflect

Are you ready to move on to the next Example? If yes, what have you learned that you think will help you? If no, what questions do you still have? How can you get those questions answered?

Record your observations here

Talk About It!

When you add or subtract decimals, you need to align the decimal point. In multiplication, the decimal points are not aligned. Why don't you need to align the decimal points when multiplying?

Example 4 Multiply Multi-Digit Decimals

Find 0.067 × 1.42. Check your solution.

Make an estimate.

0.067 × 1.42 ≈ ☐ × ☐ or ☐

Find the product.

0.067	Write the problem.
× 1.42	Multiply as with whole numbers.
134	Multiply. 2 × 67 = 134
268	Multiply. 4 × 67 = 268
+ 67	Multiply. 1 × 67 = 67
0.09514	Add.

How many decimal places are in the original problem? ________

How many zeros need to be annexed? ________

So, 0.067 × 1.42 is ________.

Check the solution.

Compare the solution to the estimate:

☐ ≈ ☐

Talk About It!

Should the product of a number and 1.42 be larger or smaller than the original number? Explain your reasoning.

Check

Find 14.7 × 11.361.

Show your work here

Go Online You can complete an Extra Example online.

Learn Divide Decimals

When dividing by decimals, it is easier to complete the division when the divisor is a whole number. To change the divisor into a whole number, multiply both the divisor and the dividend by the same power of 10.

Line up the numbers by place value as you divide. Annex zeros in the dividend to continue dividing after the decimal point.

Find the quotient of 0.006 ÷ 0.12.

$0.12\overline{)0.006}$ Multiply the dividend and divisor by 100.

$$\begin{array}{r} 0.05 \\ 12\overline{)0.60} \\ -0 \\ \hline 0\,6 \\ -0\,0 \\ \hline 0\,60 \\ -60 \\ \hline 0 \end{array}$$

Rewrite the division problem as 0.6 ÷ 12, then divide as with whole numbers.

Annex a zero and continue to divide.

So, 0.006 ÷ 0.12 is ________.

Talk About It!

Use number patterns to explain why you can rewrite 0.006 ÷ 0.12 as 0.6 ÷ 12.

Talk About It!

Why is the quotient larger than the dividend?

Pause and Reflect

Are you ready to move on to the next Example? If yes, what have you learned that you think will help you? If no, what questions do you still have? How can you get those questions answered?

Example 5 Divide Multi-Digit Decimals

Think About It!

How will you set up the long division? By what will you need to multiply both values to eliminate the decimal point in the divisor?

Find 60.927 ÷ 0.012.

Find the quotient.

$0.012\overline{)60.927}$ Write the problem.

$0.012\overline{)60.927}$ Multiply the dividend and divisor by 1,000 to eliminate the decimal point in the divisor.

$$
\begin{array}{r}
5077.25 \\
12\overline{)60927.00} \\
-60 \\
\hline
092 \\
-84 \\
\hline
87 \\
-84 \\
\hline
30 \\
-24 \\
\hline
60 \\
-60 \\
\hline
0
\end{array}
$$

Divide until there is a remainder of 0.

So, 60.927 ÷ 0.012 is ________.

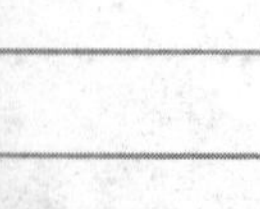

Talk About It!

Why is the quotient so much larger than the dividend?

Check

Find 2.943 ÷ 0.27.

Go Online You can complete an Extra Example online.

Apply Shopping

The table shows the cost of produce per pound at a farmer's market. Mr. Gonzalez bought 0.75 pound of pears and 3.5 pounds of plums. If Mr. Gonzalez paid for his fruit with a $10 bill, how much change will he receive?

Produce	Cost per Pound ($)
Pears	0.98
Oranges	1.29
Carrots	1.18
Plums	1.49

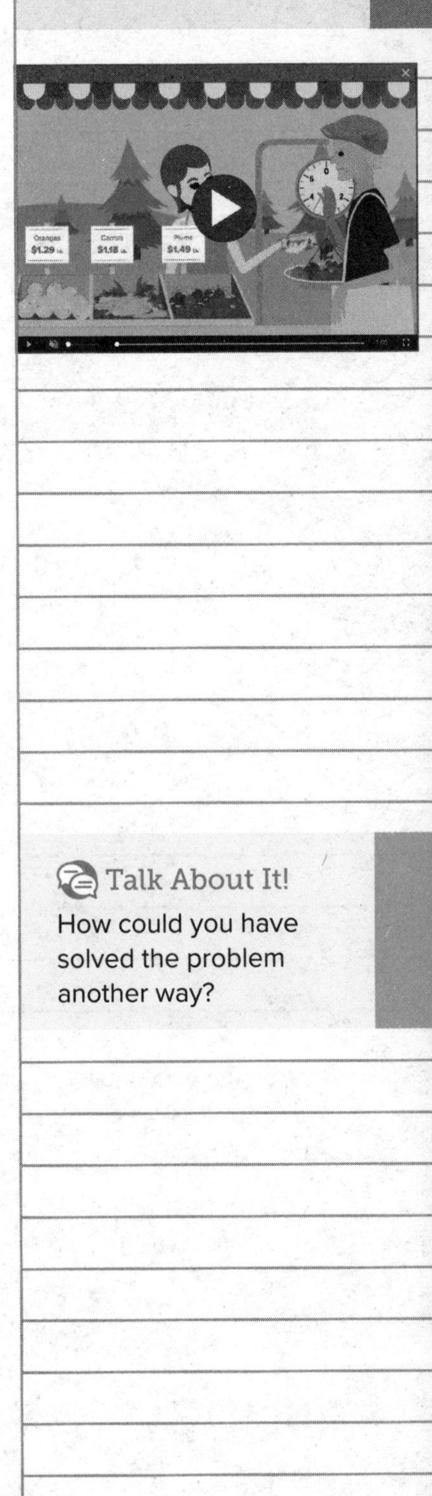

1 What is the task?

Make sure you understand exactly what question to answer or problem to solve. You may want to read the problem three times. Discuss these questions with a partner.

First Time Describe the context of the problem, in your own words.
Second Time What mathematics do you see in the problem?
Third Time What are you wondering about?

2 How can you approach the task? What strategies can you use?

3 What is your solution?

Use your strategy to solve the problem.

4 How can you show your solution is reasonable?

Write About It! Write an argument that can be used to defend your solution.

Check

There are two types of granola being sold at a local grocery store. Jerome wants to buy 1.5 pounds of cranberry granola for $5.99 per pound and 0.9 pound of dark chocolate granola for $7.99 per pound. If Jerome pays for his granola with a $20 bill, how much change will he receive?

Go Online You can complete an Extra Example online.

Pause and Reflect

Where did you encounter struggle in this lesson, and how did you deal with it? Write down any questions you still have.

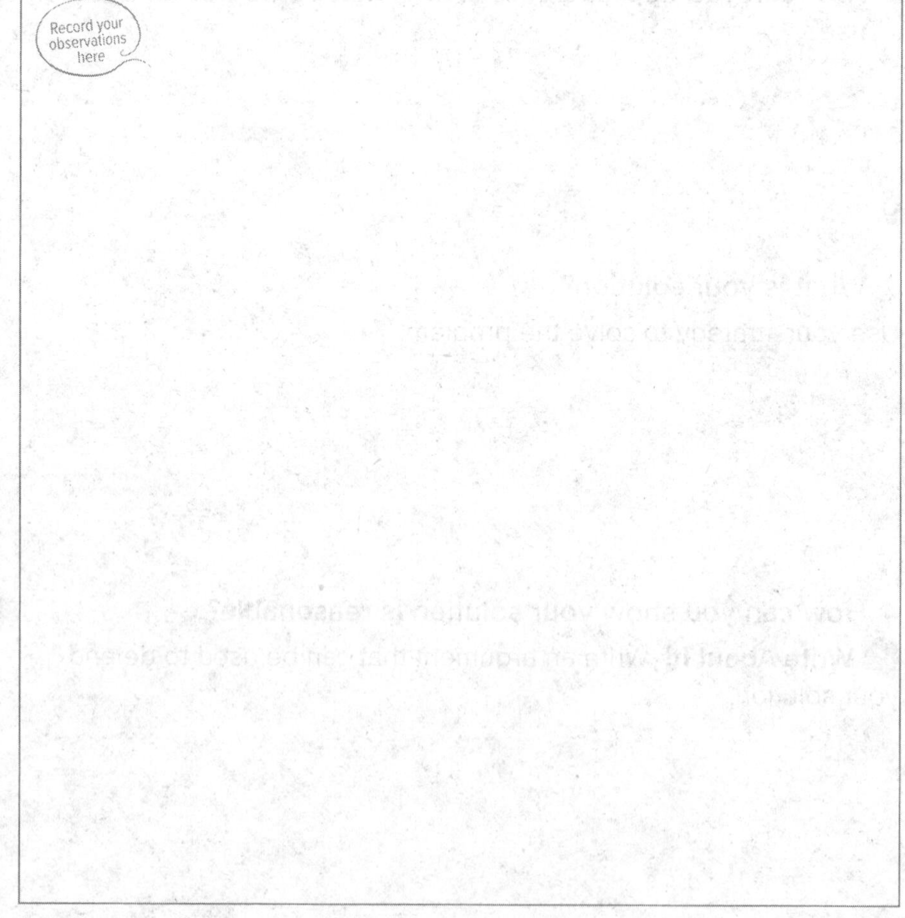

Name ______ Period ______ Date ______

Practice

Go Online You can complete your homework online.

Find each sum. (Example 1)

1. $34.672 + 15.31 =$ ______

2. $152.875 + 35.4 =$ ______

Find each difference. (Examples 2 and 3)

3. $139.65 - 59.623 =$ ______

4. $352.37 - 231.975 =$ ______

Find each product. (Example 4)

5. $0.025 \times 1.24 =$ ______

6. $17.15 \times 1.062 =$ ______

Find each quotient. (Example 5)

7. $32.674 \div 0.016 =$ ______

8. $3.825 \div 0.25 =$ ______

9. The table shows the number of miles Roberto hiked each weekend. How much more did he hike on weekend two than on weekend one?

Weekend	Miles Hiked
One	21.48
Two	30

Test Practice

10. Equation Editor What is the value of the expression 2,965.7 + 5.8?

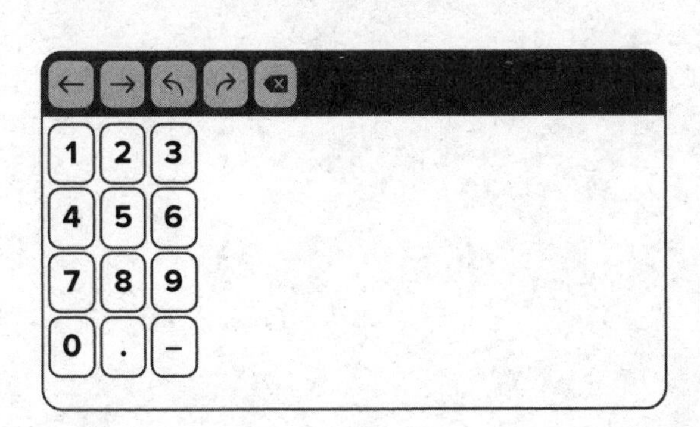

Apply

11. The table shows the cost per pound of food items you can buy in bulk at a grocery store. Mrs. Linden bought 1.25 pounds of dried fruit and 0.5 pound of cereal. If Mrs. Linden paid for her items with a $5 bill, how much change will she receive?

Item	Cost per Pound ($)
Beans	2.86
Cereal	2.38
Dried Fruit	1.84
Rice	0.52

12. Chloe is making hair bows to sell at a craft show. The table shows the cost per yard of different types of ribbon. Chloe bought 5.5 yards of satin ribbon and 3.8 yards of tulle. If Chloe paid with a $20 bill, how much change will she receive?

Ribbon	Cost per Yard ($)
Chiffon	5.88
Satin	1.50
Lace	3.29
Tulle	2.25

13. MP **Construct an Argument** Explain how you can mentally determine if the product of 5.5 and 0.95 is less than, greater than, or equal to 5.5?

14. MP **Persevere with Problems** Brand A dish detergent costs $2.48 for a 21.6-ounce bottle. Brand B costs $1.55 for a 12.6 ounce bottle. Which brand costs less per ounce?

15. Explain how you know that the sum of 26.541 and 14.2 will be greater than 40.

16. MP **Find the Error** A student is multiplying 1.02 × 2.55. Find the student's mistake and correct it.

```
     1.02
   × 2.55
     1510
     5100
 + 20400
   260.10
```

Divide Whole Numbers by Fractions

I Can... apply what I previously learned about multiplication, division, and operations on multi-digit numbers to divide whole numbers by fractions.

Today's Standards
6.NS.A.1
MP1, MP2, MP3, MP5, MP6, MP7, MP8

What Vocabulary Will You Learn?
Inverse Property of Multiplication
multiplicative inverse
reciprocal

Learn. Reciprocals

Two numbers whose product is 1 are called **multiplicative inverses** or **reciprocals**. The **Inverse Property of Multiplication** states that the product of a number and its multiplicative inverse is 1.

Numbers	$\frac{2}{3} \times \frac{3}{2} = 1$
Algebra	For every number $\frac{a}{b}$ where a and $b \neq 0$, there is exactly one number, $\frac{b}{a}$, such that $\frac{a}{b} \times \frac{b}{a} = 1$.

Example 1 Find Reciprocals

Find the reciprocal of $\frac{1}{8}$.

Since $\frac{1}{8} \times \frac{8}{1} = 1$, the reciprocal of $\frac{1}{8}$ is $\frac{8}{1}$ or ________.

So, the reciprocal of $\frac{1}{8}$ is 8.

Check

Find the reciprocal of $\frac{1}{7}$.

Go Online You can complete an Extra Example online.

Your Notes

Example 2 Find Reciprocals of Fractions

What number multiplied by $\frac{3}{4}$ has a product of 1?

$$\frac{3}{4} \times \frac{\square}{\square} = 1$$

So, the reciprocal of $\frac{3}{4}$ is $\frac{4}{3}$.

Check

What number multiplied by $\frac{4}{7}$ has a product of 1?

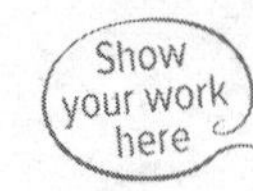

Example 3 Find Reciprocals of Whole Numbers

Find the reciprocal of 5.

The whole number 5 can be written as the fraction ________.

Since $\frac{5}{1} \times \frac{1}{5} = 1$, the reciprocal is ________.

So, the reciprocal of 5 is $\frac{1}{5}$.

Check

Find the reciprocal of 4.

Talk About It!

Can you write any whole number as a fraction? Explain.

Go Online You can complete an Extra Example online.

Explore Divide Whole Numbers by Fractions

Go Online You will use models to divide whole numbers by fractions and make a conjecture about finding the quotient without using a model.

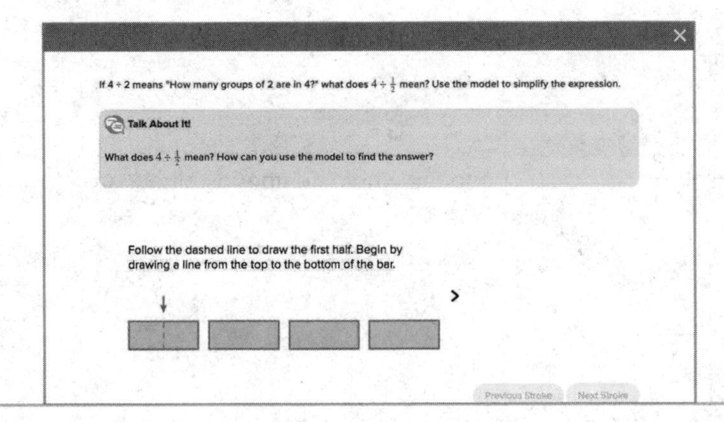

Learn Divide Whole Numbers by Fractions

You can use a visual model to solve division problems involving whole numbers and fractions.

Find $3 \div \frac{3}{4}$.

Draw a model to represent the dividend, 3.

Divide each whole into fourths, because the denominator of the divisor is 4.

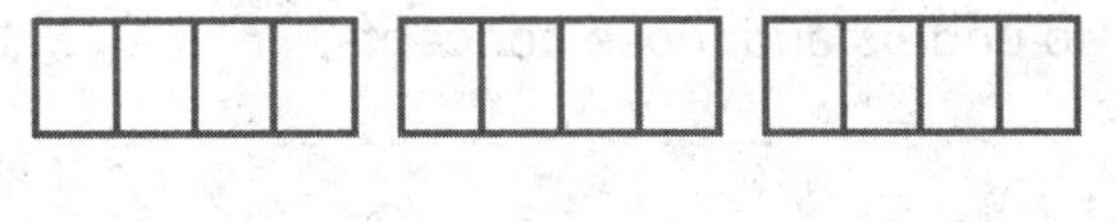

Identify groups of three-fourths. Shade each group of $\frac{3}{4}$.

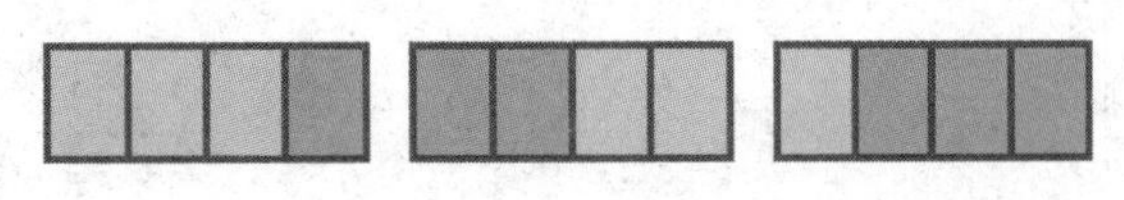

There are four groups of $\frac{3}{4}$ in 3 wholes.

1 2 3 4

So, $3 \div \frac{3}{4}$ is 4.

(continued on next page)

Talk About It!

Why is each whole divided into fourths?

You can also use an equation to solve division problems involving whole numbers and fractions. Recall that multiplication and division are inverse operations, so you can divide a whole number by a fraction by multiplying the whole number by the reciprocal of the fraction.

$3 \div \frac{3}{4} = \square$ — Write the equation.

$3 \div \frac{3}{4} = \frac{\square}{\square} \div \frac{3}{4}$ — Write the whole number as a fraction.

$= \frac{3}{1} \times \frac{\square}{\square}$ — Multiply by the reciprocal of $\frac{3}{4}$, $\frac{4}{3}$.

$= \frac{\overset{1}{\cancel{3}}}{1} \times \frac{4}{\underset{1}{\cancel{3}}}$ — Divide by common factors.

$= \frac{1 \times 4}{1 \times 1}$ — Simplify.

$= \frac{4}{1}$ or 4 — Multiply.

So, $3 \div \frac{3}{4}$ is _______.

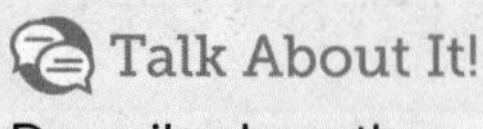
Talk About It!

Describe how the visual model supports the equation.

Pause and Reflect

Did you struggle with any of the concepts in this Learn? How do you feel when you struggle with math concepts? What steps can you take to understand those concepts?

Record your observations here

Example 4 Divide Whole Numbers by Fractions

Find $2 \div \frac{2}{3}$.

Method 1 Use a visual model.

Draw a model to represent the whole-number dividend, 2.

Divide each whole into thirds because the denominator of the divisor is 3.

Determine how many groups of $\frac{2}{3}$ are in 2. Shade each group of $\frac{2}{3}$.

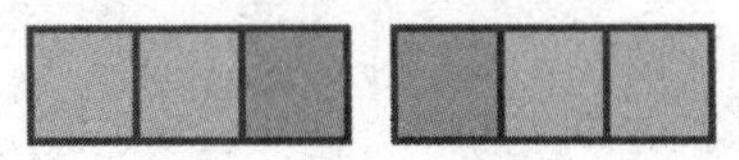

Label the number of groups.

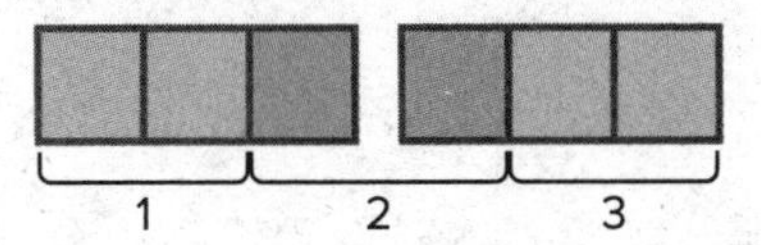

How many whole groups of $\frac{2}{3}$ were labeled? ________

There are ________ sections left over.

So, $2 \div \frac{2}{3}$ is 3.

(continued on next page)

Think About It!

The quotient represents the number of groups of $\frac{2}{3}$ that are in what number?

Method 2 Use an equation.

$2 \div \frac{2}{3} = \square$ Write the equation.

$2 \div \frac{2}{3} = \frac{\square}{\square} \div \frac{2}{3}$ Write the whole number as a fraction.

$= \frac{2}{1} \times \frac{\square}{\square}$ Multiply by the reciprocal of $\frac{2}{3}$, $\frac{3}{2}$.

$= \frac{\overset{1}{\cancel{2}}}{1} \times \frac{3}{\cancel{2}_1}$ Divide by common factors.

$= \frac{1 \times 3}{1 \times 1}$ Simplify.

$= \square$ Multiply.

So, $2 \div \frac{2}{3}$ is 3.

Talk About It!

Compare and contrast the two methods used to find $2 \div \frac{2}{3}$.

Check

Find $4 \div \frac{5}{8}$.

Go Online You can complete an Extra Example online.

Example 5 Divide Whole Numbers by Fractions

At summer camp, the duration of each activity is $\frac{3}{4}$ hour. The camp counselors have set aside 4 hours in the afternoon for activities.

Find $4 \div \frac{3}{4}$. Then interpret the quotient.

Part A Find $4 \div \frac{3}{4}$.

Method 1 Use a model.

Draw a model to represent the dividend, 4.

Divide each whole into fourths.

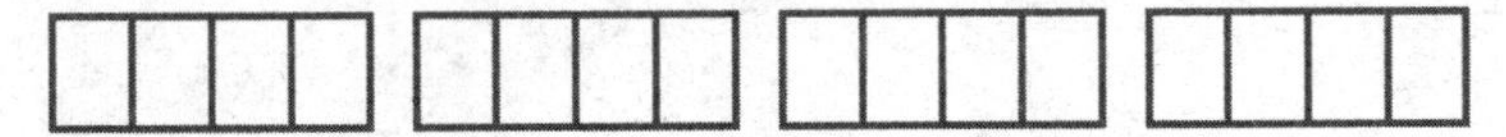

Identify groups of $\frac{3}{4}$.

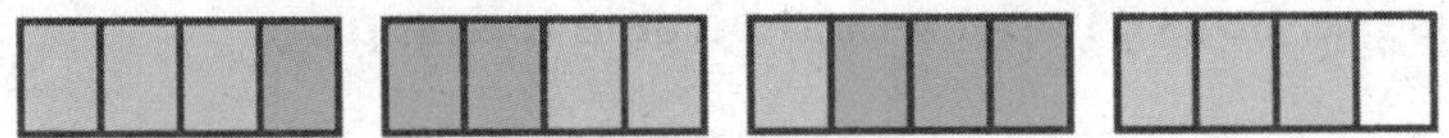

Label the number of groups.

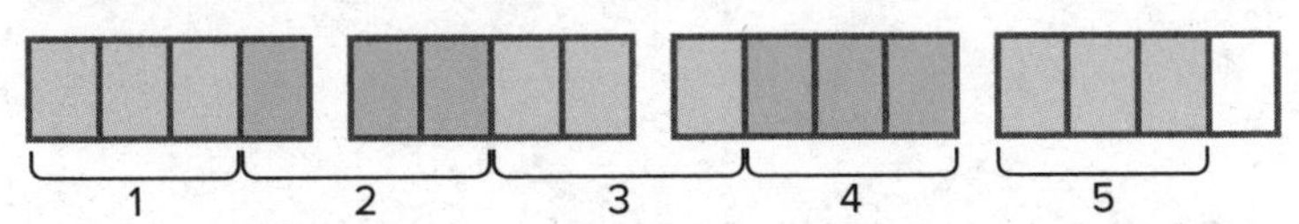

There are ________ whole groups of $\frac{3}{4}$.

There is ________ section left over.

One section is $\frac{1}{3}$ of a group.

So, $4 \div \frac{3}{4}$ is $5\frac{1}{3}$.

Think About It!

The quotient represent the number of $\frac{3}{4}$ that are in what number?

(continued on next page)

Method 2 Use an equation.

$4 \div \frac{3}{4} = \square$ Write the equation.

$4 \div \frac{3}{4} = \frac{\square}{\square} \div \frac{3}{4}$ Write the whole number as a fraction.

$= \frac{4}{1} \times \frac{\square}{\square}$ Multiply by the reciprocal of $\frac{3}{4}$, $\frac{4}{3}$.

$= \frac{4 \times 4}{1 \times 3}$ Multiply the numerators and denominators.

$= \frac{16}{3}$ or 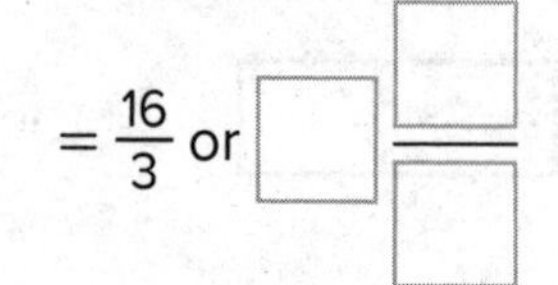 Simplify.

So, $4 \div \frac{3}{4}$ is $5\frac{1}{3}$.

Talk About It!

Compare and contrast the two methods.

Part B Interpret the quotient.

The quotient is $5\frac{1}{3}$. So, a camper would be able to complete

________ activities in 4 hours.

Check

Morgan has a 9-foot-long piece of wood that he wants to cut to build some $\frac{5}{6}$-foot-long shelves for his bedroom. Find $9 \div \frac{5}{6}$. Then interpret the quotient.

Go Online You can complete an Extra Example online.

Apply Cooking

The table shows the ingredients needed to make one batch of salad dressing. A chef has 3 tablespoons (T) of garlic. She made the greatest number of whole batches possible. How much garlic remained?

Ingredient	Amount
Oil	1 c
Vinegar	$\frac{3}{4}$ c
Garlic	$\frac{2}{3}$ T

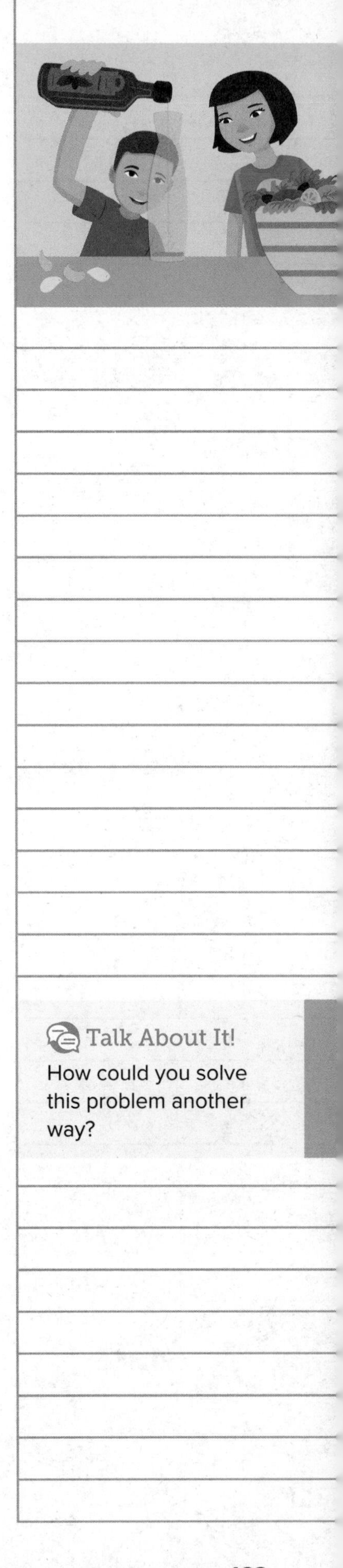

1 What is the task?

Make sure you understand exactly what question to answer or problem to solve. You may want to read the problem three times. Discuss these questions with a partner.

First Time Describe the context of the problem, in your own words.
Second Time What mathematics do you see in the problem?
Third Time What are you wondering about?

2 How can you approach the task? What strategies can you use?

3 What is your solution?

Use your strategy to solve the problem.

4 How can you show your solution is reasonable?

Write About It! Write an argument that can be used to defend your solution.

Talk About It!
How could you solve this problem another way?

Check

The table shows the ingredients needed to make one batch of fudge. A cook has 5 cups of evaporated milk. She made the greatest number of whole batches possible. How much evaporated milk remained?

Ingredient	Amount
Chocolate Chips	$2\frac{1}{2}$ c
Evaporated Milk	$\frac{3}{4}$ c
Butter	$\frac{1}{2}$ c

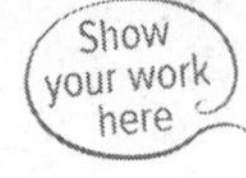

Go Online You can complete an Extra Example online.

Foldables It's time to update your Foldable, located in the Module Review, based on what you learned in this lesson. If you haven't already assembled your Foldable, you can find the instructions on page FL1.

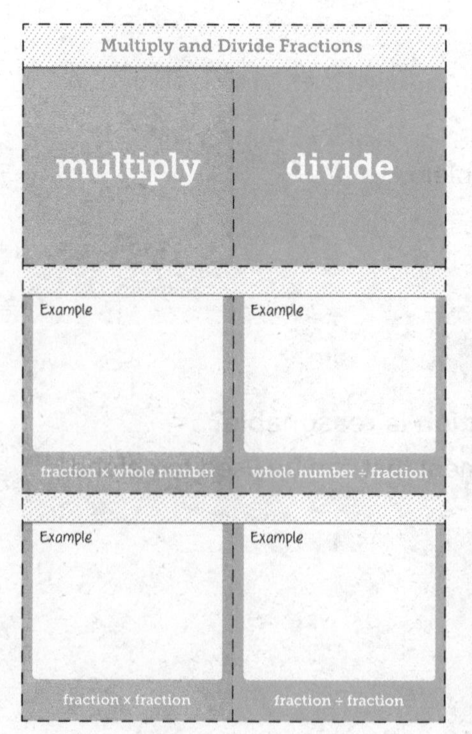

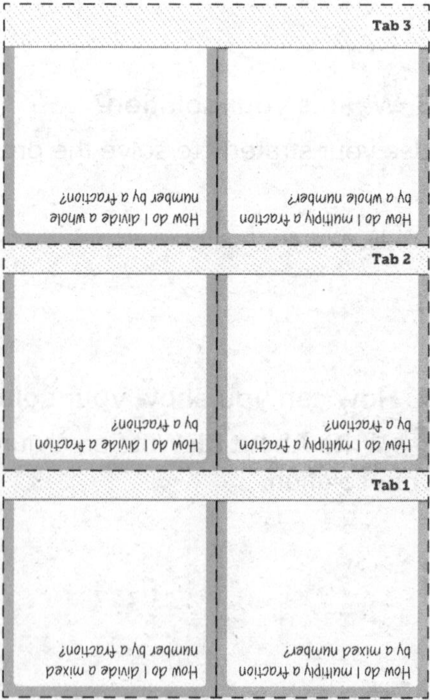

Name ______ Period ______ Date ______

Practice

Go Online You can complete your homework online.

Find the reciprocal of each number. (Example 1 and Example 3)

1. $\frac{1}{2}$

2. $\frac{1}{5}$

3. 8

4. What number multiplied by $\frac{3}{5}$ has a product of 1? (Example 2)

5. What number multiplied by $\frac{7}{10}$ has a product of 1? (Example 2)

Divide. Write in simplest form. (Example 4)

6. $3 \div \frac{1}{4} =$ ______

7. $4 \div \frac{2}{5} =$ ______

8. $6 \div \frac{2}{3} =$ ______

9. Marie is making scarves. She has 7 yards of fabric and each scarf needs $\frac{5}{8}$ yard of fabric. Find $7 \div \frac{5}{8}$. Then interpret the quotient. (Example 5)

Test Practice

10. Roberto is at a tennis day camp. The coach has set aside 2 hours to play mini matches that last $\frac{3}{5}$ hour. Find $2 \div \frac{3}{5}$. Then interpret the quotient.

11. **Equation Editor** What is the value of $15 \div \frac{5}{9}$?

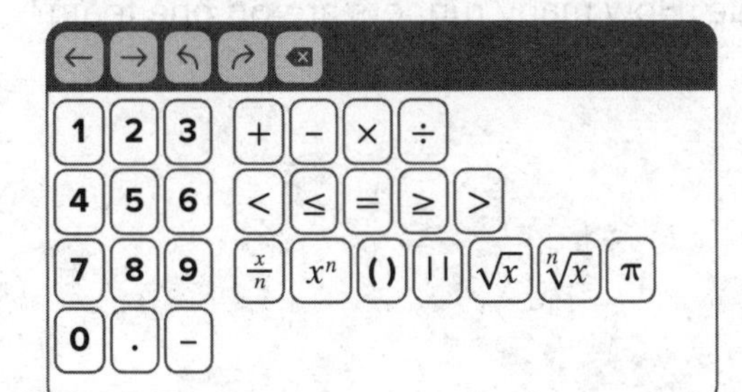

Apply

12. The table shows the amount of each ingredient Jacob is using to make one pizza. If he has 11 cups of mozzarella cheese and makes the greatest number of whole pizzas possible, how much mozzarella cheese remains?

Ingredient	Amount
Mozzarella Cheese	$\frac{3}{4}$ c
Sauce	$\frac{1}{2}$ c

13. The table shows the ingredients for one batch of barbeque sauce. Anne has 9 cups of ketchup and makes the greatest number of whole batches of barbeque sauce possible. How much ketchup remains?

Ingredient	Amount
Brown Sugar	$\frac{1}{4}$ c
Cider Vinegar	$\frac{1}{2}$ c
Ground Cumin	1 tsp
Ketchup	$\frac{2}{3}$ c
Pepper	1 tsp

14. MP **Find the Error** A student is solving $9 \div \frac{3}{4}$. Find the student's mistake and correct it.

$$9 \div \frac{3}{4} = \frac{9}{1} \times \frac{3}{4}$$
$$= \frac{27}{4} \text{ or } 6\frac{3}{4}$$

15. Zach has 20 sub sandwiches for a party. Each sub sandwich is going to be cut into thirds. Zach needs 55 sandwich pieces. Will he have enough sandwich pieces? Justify your answer.

16. MP **Preserve with Problems** In a $\frac{3}{4}$-mile relay race, each runner on one team runs $\frac{3}{16}$ mile. How many runners are on one team?

17. Identify the whole number whose reciprocal has a decimal equivalent between 0.2 and 0.3. Explain.

Divide Fractions by Fractions

I Can... apply what I previously learned about multiplication and division with whole numbers and the division of whole numbers by fractions to divide fractions by fractions.

Today's Standards
6.NS.A.1
MP1, MP2, MP3, MP4, MP5, MP6

Learn Divide Fractions by Fractions

You can use a visual model to solve division problems involving fractions, such as $\frac{1}{2} \div \frac{1}{3}$.

Draw a model to represent the dividend, $\frac{1}{2}$.

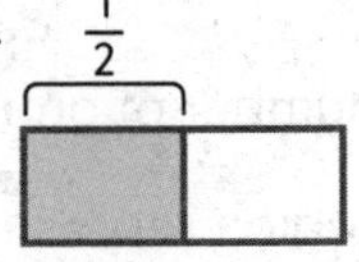

Label groups of $\frac{1}{3}$. Then find the number of groups of $\frac{1}{3}$ that are in the shaded section.

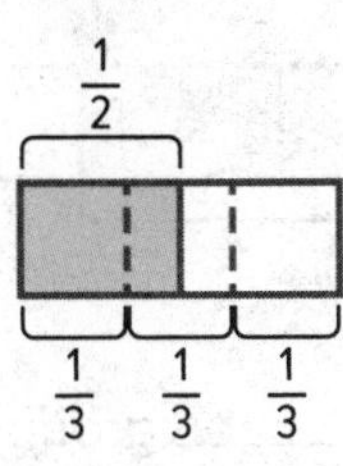

There is one group of $\frac{1}{3}$ and $\frac{1}{2}$ of another third in the shaded section.

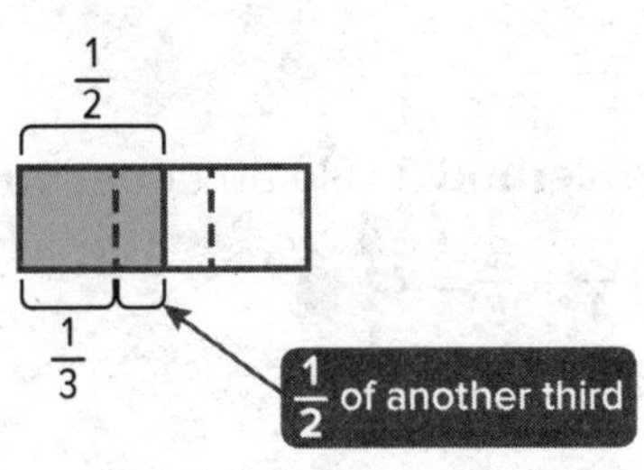

Talk About It!
How does the visual model illustrate the dividend and divisor?

So, there are $1\frac{1}{2}$ groups of $\frac{1}{3}$ in $\frac{1}{2}$. This means that $\frac{1}{2} \div \frac{1}{3} = 1\frac{1}{2}$.

You can also use an equation to solve division problems involving fractions. To divide a fraction by a fraction, multiply the first fraction by the reciprocal of the second fraction, because multiplication and division are inverse operations.

Go Online Watch the animation to see how to find $\frac{1}{3} \div \frac{2}{9}$.

$\frac{1}{3} \div \frac{2}{9} = \frac{1}{\cancel{3}_1} \times \frac{\cancel{9}^3}{2}$ Multiply by the reciprocal. Divide by the common factor, 3.

$= \frac{1 \times 3}{1 \times 2}$ Multiply the numerators and denominators.

$= \frac{3}{2}$ or $1\frac{1}{2}$ Simplify.

Talk About It!
What is the reciprocal of the divisor in the expression $\frac{1}{3} \div \frac{1}{2}$?

Your Notes

Think About It!

The quotient represents the number of groups of $\frac{3}{8}$ that are in what number?

Talk About It!

Compare and contrast the two methods.

Example 1 Divide Fractions by Fractions

Find $\frac{3}{4} \div \frac{3}{8}$.

Method 1 Use a model.

Draw a model to represent the dividend, $\frac{3}{4}$.

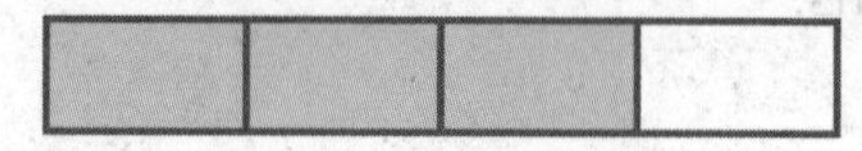

You want to know how many groups of $\frac{3}{8}$ are in $\frac{3}{4}$. Divide the whole into eighths because the denominator of the divisor is 8.

Identify the number of groups of $\frac{3}{8}$ in the shaded section. Remember, the shaded region represents $\frac{3}{4}$.

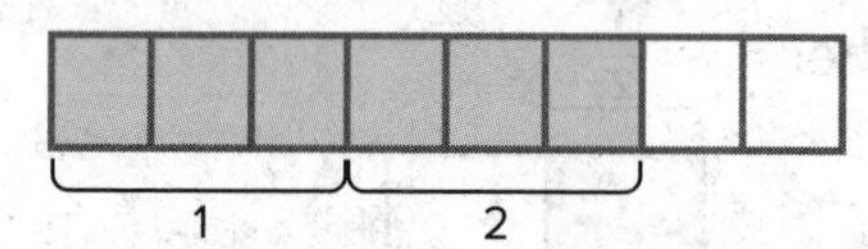

There are ________ groups of $\frac{3}{8}$ in $\frac{3}{4}$.

Method 2 Use an equation.

$\frac{3}{4} \div \frac{3}{8} = \blacksquare$ Write the equation.

$\frac{3}{4} \div \frac{3}{8} = \frac{3}{4} \times \frac{\square}{\square}$ Multiply by the reciprocal of $\frac{3}{8}$, $\frac{8}{3}$.

$= \frac{^{1}3}{_{1}4} \times \frac{8^{2}}{3_{1}}$ Divide by common factors.

$= \frac{1 \times 2}{1 \times 1}$ Simplify.

$= \frac{2}{1}$ or 2 Multiply.

So, $\frac{3}{4} \div \frac{3}{8}$ is ________.

Check

Find $\frac{7}{9} \div \frac{2}{3}$.

Go Online You can complete an Extra Example online.

Example 2 Find and Interpret Quotients

Asahi is making cookies. There is $\frac{5}{6}$ pound of sugar left in the canister. Each batch of cookies requires $\frac{1}{4}$ pound of sugar. He wants to deliver a batch to each of his neighbors. How many neighbors will receive cookies?

Write and solve an equation that models the situation. Then interpret the quotient.

Part A Write and solve an equation.

The expression $\frac{5}{6} \div \frac{1}{4}$ represents the number of batches he can make, since Asahi has $\frac{5}{6}$ pound of sugar left, and each batch of cookies requires $\frac{1}{4}$ pound of sugar.

$\frac{5}{6} \div \frac{1}{4} = \blacksquare$ — Write the equation.

$\frac{5}{6} \div \frac{1}{4} = \frac{5}{6} \times \frac{4}{1}$ — Multiply by the reciprocal of $\frac{1}{4}$, $\frac{4}{1}$.

$= \frac{5}{{}_3\cancel{6}} \times \frac{\cancel{4}^2}{1}$ — Divide by common factors.

$= \frac{5 \times 2}{3 \times 1}$ — Simplify.

$= \frac{10}{3}$ — Multiply.

So, $\frac{5}{6} \div \frac{1}{4}$ is $3\frac{1}{3}$.

Part B Interpret the quotient.

Because Asahi wants to deliver whole batches of cookies, he is only able to make ________ batches of cookies.

Think About It!

What is the divisor? What is the dividend?

Talk About It!

Why do the quotient and the answer to the word problem differ?

Check

Jasmine is mixing paint colors. She has $\frac{3}{4}$ gallon of blue paint. She needs $\frac{1}{6}$ gallon for each new color that she is mixing. Write and solve an equation that models the situation. Then interpret the solution.

Part A

Write and solve an equation.

Part B

Interpret the quotient.

Show your work here

Go Online You can complete an Extra Example online.

Learn Write Story Contexts

You can write a story context, or word problem, to represent any division problem. You can then solve the problem using a model or equation.

For the expression $\frac{4}{5} \div \frac{1}{10}$, you can write a story context by describing each piece of the division problem.

Write the dividend and divisor into the correct location in the story context.

Navid is hanging pictures in his room and has ________ foot of tape to use. He uses ________ foot of tape to hang each photo. How many photos can he hang on the wall?

Example 3 Write Story Contexts

Write a story context for $\frac{2}{3} \div \frac{1}{6}$. Then find the solution.

Part A Write a story context.

To write a story context for the division expression, consider the following situation.

Mimi is very active. She loves to cook, has a couple of hobbies, and has tasks around the house. Choose one the activities shown. Then write a story context using your choice.

cooking dinner	doing laundry	painting
making pasta	feeding birds	swimming

Part B Solve.

$\frac{2}{3} \div \frac{1}{6} = \square$ Write the equation.

$\frac{2}{3} \div \frac{1}{6} = \frac{2}{3} \times \frac{\square}{\square}$ Multiply by the reciprocal of $\frac{1}{6}$, $\frac{6}{1}$.

$= \frac{2}{\cancel{3}_1} \times \frac{\cancel{6}^2}{1}$ Divide by common factors.

$= \frac{2 \times 2}{1 \times 1}$ Simplify.

$= 4$ Multiply.

So, $\frac{2}{3} \div \frac{1}{6}$ is ______.

Think About It!

How would you begin writing the problem?

Talk About It!

If $\frac{2}{3} \div \frac{1}{6} = 4$, what does this mean in the context of the same word problem that you chose?

Check

Write a story context for $\frac{5}{6} \div \frac{1}{12}$. Then find the quotient.

Go Online You can complete an Extra Example online.

Pause and Reflect

Did you make any errors when completing the Check exercise? What can you do to make sure you don't repeat that error in the future?

Apply Food

Alfonso is making snack bags with different types of nuts as shown in the table. Each snack bag contains $\frac{1}{8}$ pound of one type of nut. How many more whole servings of walnuts can he make than peanuts?

Type of Nut	Weight (lb)
Almonds	$\frac{1}{2}$
Cashews	$\frac{1}{4}$
Peanuts	$\frac{2}{5}$
Walnuts	$\frac{3}{4}$

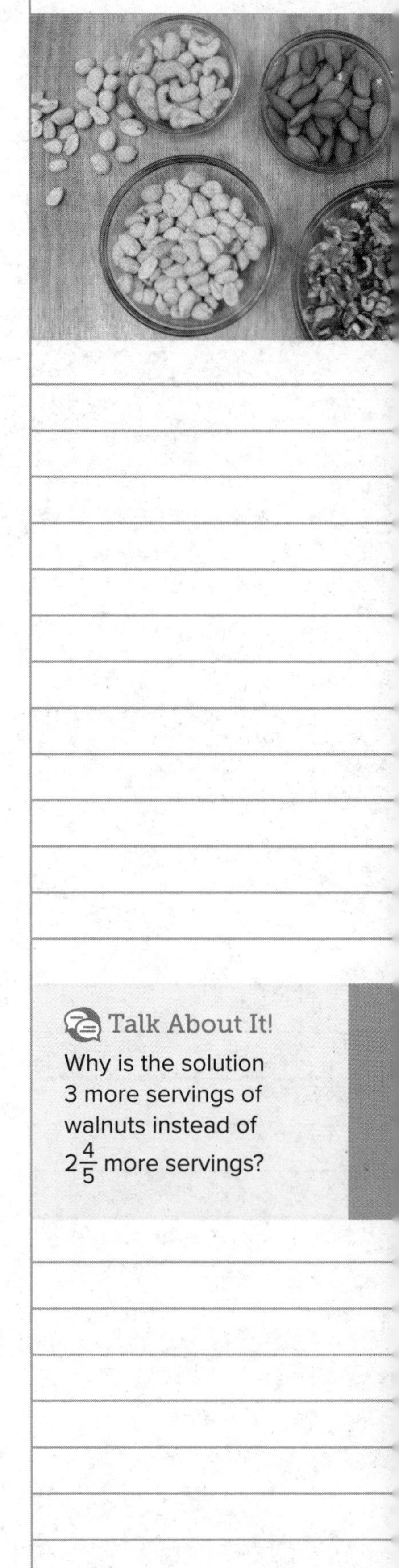

1 What is the task?

Make sure you understand exactly what question to answer or problem to solve. You may want to read the problem three times. Discuss these questions with a partner.

First Time Describe the context of the problem, in your own words.
Second Time What mathematics do you see in the problem?
Third Time What are you wondering about?

2 How can you approach the task? What strategies can you use?

3 What is your solution?

Use your strategy to solve the problem.

4 How can you show your solution is reasonable?

Write About It! Write an argument that can be used to defend your solution.

Talk About It!
Why is the solution 3 more servings of walnuts instead of $2\frac{4}{5}$ more servings?

Check

Stephanie's running schedule is shown in the table. She decides that she wants to do sprint training and will run the total distance by running a series of $\frac{1}{10}$-mile sprints. How many more $\frac{1}{10}$-mile sprints will she have to run on weekdays compared to weekends?

	Total Distance (mile)
Weekdays	$\frac{2}{3}$
Weekends	$\frac{7}{8}$

Go Online You can complete an Extra Example online.

Foldables It's time to update your Foldable, located in the Module Review, based on what you learned in this lesson. If you haven't already assembled your Foldable, you can find the instructions on page FL1.

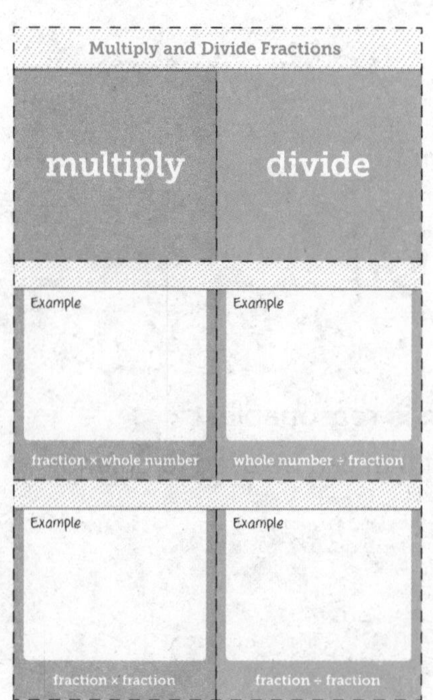

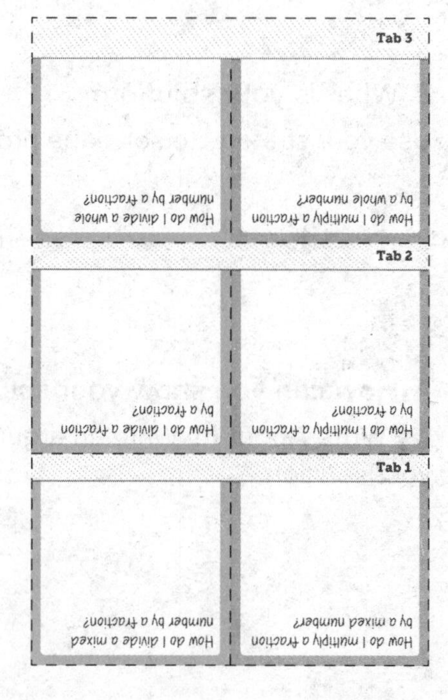

Name ______________________ Period __________ Date __________

Practice

Go Online You can complete your homework online.

Divide. Write in simplest form. **(Example 1)**

1. $\frac{5}{6} \div \frac{5}{12} =$ ________

2. $\frac{1}{3} \div \frac{1}{9} =$ ________

3. $\frac{3}{7} \div \frac{1}{14} =$ ________

4. Romeo had $\frac{3}{4}$ pound of fudge left. He divided the remaining fudge into $\frac{5}{16}$ pound bags. Write and solve an equation that models the situation. Then interpret the solution. **(Example 2)**

5. Chelsea has $\frac{7}{8}$ pound of butter to make icing. Each batch of icing needs $\frac{1}{4}$ pound of butter. Write and solve an equation that models the situation. Then interpret the solution. **(Example 2)**

Test Practice

6. Write a story context for $\frac{2}{3} \div \frac{1}{6}$. Then find the solution. **(Example 3)**

7. Equation Editor What is the value of the expression $\frac{2}{5} \div \frac{1}{6}$?

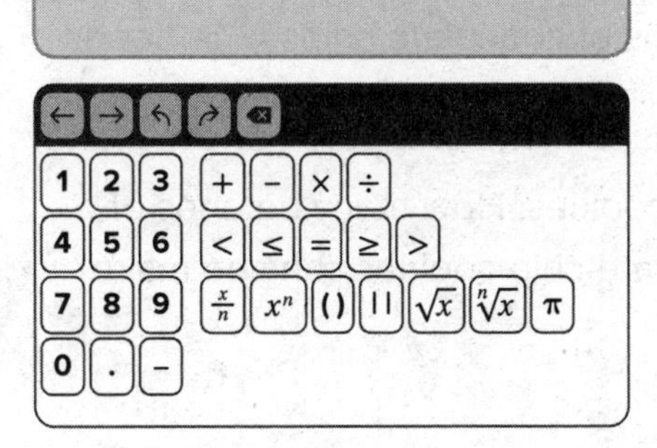

Apply

8. A teacher is making bags of different colors of modeling clay. The table shows the amount of each color she has available. Each color will be divided into $\frac{3}{16}$ pound bags. How many more bags of purple can she make than yellow?

Color	Weight (lb)
Green	$\frac{1}{2}$
Purple	$\frac{15}{16}$
Red	$\frac{2}{3}$
Yellow	$\frac{3}{4}$

9. Mateo is making bookmarks with different colored ribbon. The amount of each color she has is shown in the table. Each bookmark will be $\frac{1}{6}$ yard long. How many more orange bookmarks can she make than aqua bookmarks?

Color	Length (yd)
Aqua	$\frac{3}{4}$
Orange	$\frac{9}{10}$
Yellow	$\frac{15}{16}$

10. MP **Make a Conjecture** Can the quotient of two positive fractions be less than 1? Explain.

11. The length of a race is $\frac{9}{10}$ mile. Andrew wants to place a flag every $\frac{1}{3}$ mile. He has 3 flags. Does he have enough flags? Explain.

12. MP **Persevere with Problems** Lannie has $5\frac{1}{2}$ cups of chocolate chips. She needs $1\frac{3}{4}$ cups to make one batch of chocolate chips cookies. How many batches of chocolate chip cookies can she make?

13. MP **Model with Mathematics** Write a division problem involving the division of two positive fractions whose quotient is equal to 1. Show that your problem is correct.

Divide with Whole and Mixed Numbers

I Can... apply what I previously learned about division and reciprocals to divide fractions by whole and mixed numbers.

Today's Standards
6.NS.A.1
MP1, MP3, MP5, MP6, MP8

Explore Divide Fractions by Whole Numbers

Online Activity You will use Web Sketchpad to divide fractions by whole numbers.

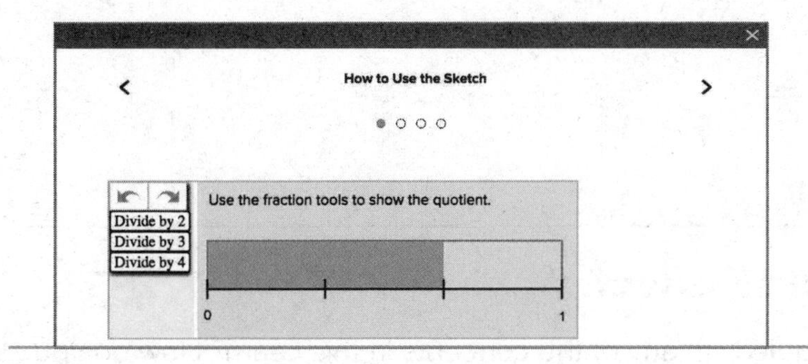

Learn Divide Fractions by Whole Numbers

You can use a visual model to simplify division problems involving whole numbers and fractions.

Find $\frac{3}{5} \div 2$.

Draw a model to represent the dividend, $\frac{3}{5}$.

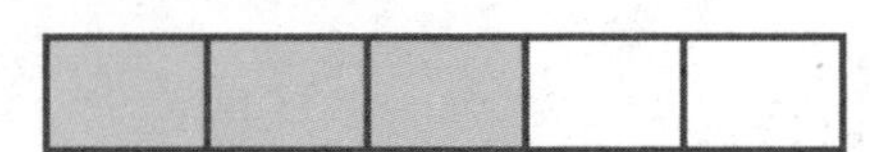

Divide the shaded sections by two, because the divisor is 2.

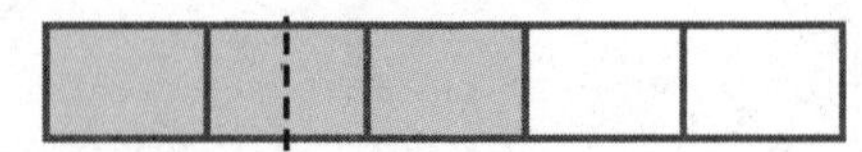

The dotted line divides one of the sections representing $\frac{1}{5}$ into two equal-size sections. Divide each of the remaining fifths into two equal-size sections.

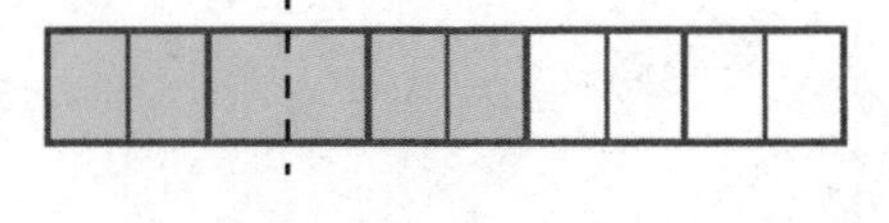

Each of the smaller sections is $\frac{1}{10}$ of the whole. Three fifths divided by two is $\frac{3}{10}$ of the whole.

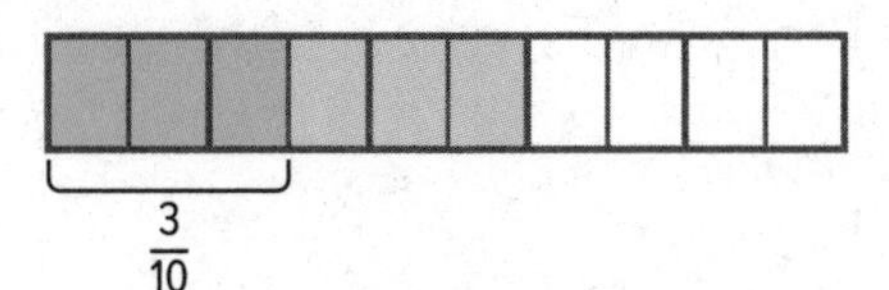

So, $\frac{3}{5} \div 2$ is $\frac{3}{10}$.

(continued on next page)

Your Notes

> **Talk About It!**
> Compare and contrast the two methods.

You can also use an equation to solve division problems involving whole numbers and fractions. To divide a fraction by a whole number, multiply the fraction by the reciprocal of the whole number.

Find $\frac{3}{5} \div 2$.

$\frac{3}{5} \div 2 = \blacksquare$	Write the equation.
$= \frac{3}{5} \div \frac{2}{1}$	Write the whole number as a fraction.
$= \frac{3}{5} \times \frac{1}{2}$	Multiply by the reciprocal of $\frac{2}{1}$, $\frac{1}{2}$.
$= \frac{3 \times 1}{5 \times 2}$	Multiply the numerators and denominators.
$= \frac{3}{10}$	Multiply.

So, $\frac{3}{5} \div 2$ is $\frac{3}{10}$.

Pause and Reflect

Did you struggle with any of the concepts in this Learn? How do you feel when you struggle with math concepts? What steps can you take to understand those concepts?

Example 1 Divide Fractions by Whole Numbers

Faye is making party favors. She is dividing $\frac{3}{4}$ pound of cashews into 12 packages.

How many pounds of cashews are in each package?

Part A Write an equation to model the problem.

Circle the equation that models the problem.

$\frac{3}{4} \div 12 = \blacksquare$ $\qquad$ $12 \div \frac{3}{4} = \blacksquare$

Part B Solve the equation.

Method 1 Use a visual model.

Draw a model to represent the dividend, $\frac{3}{4}$.

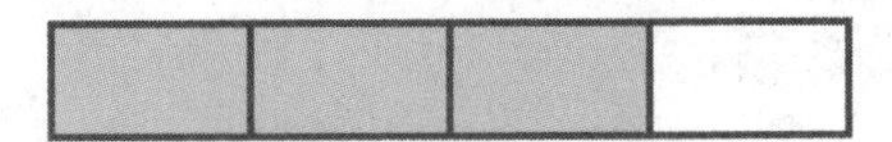

Divide the shaded sections by the divisor, 12.

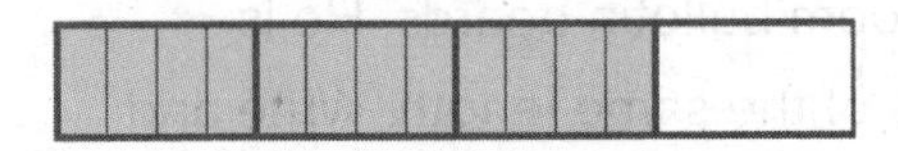

Divide the remaining part of the whole so that each section is of equal size.

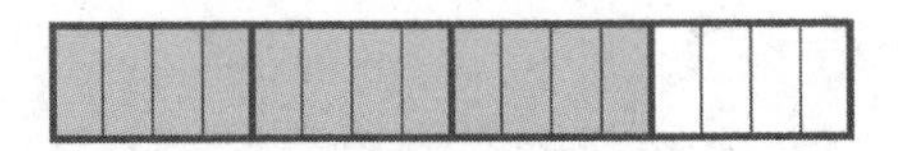

Identify what each of the smallest sections represents. Each section is $\frac{1}{16}$ of the whole.

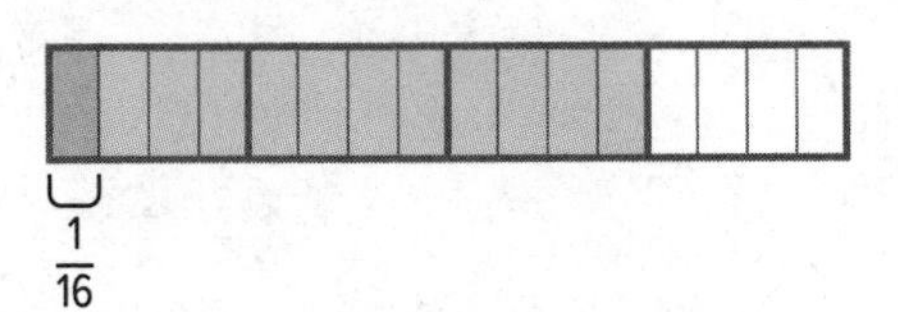

So, $\frac{3}{4} \div 12$ is $\frac{1}{16}$.

(continued on next page)

Think About It!

Will the quotient be less than, greater than, or equal to $\frac{3}{4}$ pound? How do you know?

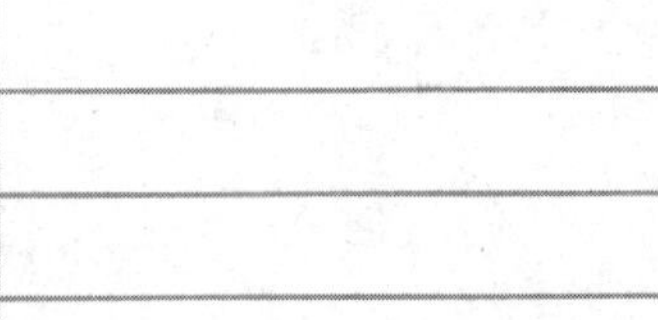

Method 2 Use an equation.

$\frac{3}{4} \div 12 = $ 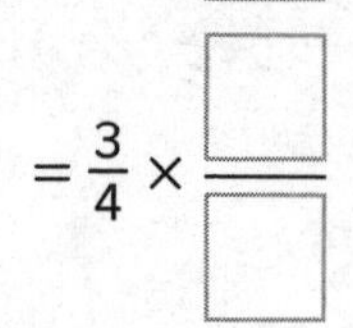 Write the equation.

$\frac{3}{4} \div 12 = \frac{3}{4} \div \frac{\square}{\square}$ Write the whole number as a fraction.

$= \frac{3}{4} \times \frac{\square}{\square}$ Multiply by the reciprocal of $\frac{12}{1}$, $\frac{1}{12}$.

$= \frac{\overset{1}{\cancel{3}}}{4} \times \frac{1}{\cancel{12}_4}$ Divide by common factors.

$= \frac{1 \times 1}{4 \times 4}$ Simplify.

$= \frac{1}{16}$ Multiply.

There are ________ pound(s) of cashews in each package.

Talk About It!

Compare and contrast the two methods.

Check

Ernesto is making designs for classroom bulletin boards. He is cutting $\frac{3}{4}$-yard of fabric into 6 pieces of the same length. Write and solve an equation to find the length of each piece of fabric.

Go Online You can complete an Extra Example online.

Learn Divide Mixed Numbers

Dividing with mixed numbers is similar to dividing with fractions. To divide with mixed numbers, write the mixed number as an improper fraction and then divide as with fractions.

Go Online Watch the animation to learn how to divide with mixed numbers.

Find $2\frac{1}{4} \div \frac{2}{3}$.

$2\frac{1}{4} \div \frac{2}{3} = $ ▇	Write the equation.
$= \frac{9}{4} \div \frac{2}{3}$	Write the mixed number as a fraction.
$= \frac{9}{4} \times \frac{3}{2}$	Multiply by the reciprocal of $\frac{2}{3}$, $\frac{3}{2}$.
$= \frac{9 \times 3}{4 \times 2}$	Multiply the numerators and denominators.
$= \frac{27}{8}$ or $3\frac{3}{8}$	Multiply.
$= \square\frac{\square}{\square}$	

Example 2 Divide Mixed Numbers

Find $3\frac{1}{3} \div 6$.

$3\frac{1}{3} \div 6 = $ ▇	Write the equation.
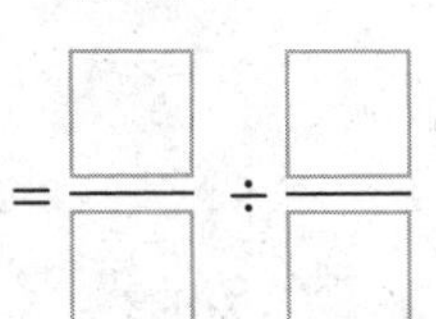	Write the mixed number and the whole number as fractions.
$= \frac{10}{3} \times \frac{\square}{\square}$	Multiply by the reciprocal of $\frac{6}{1}$, $\frac{1}{6}$.
$= \frac{{}^{5}\cancel{10}}{3} \times \frac{1}{\cancel{6}_{3}}$	Divide by common factors.
$= \frac{5 \times 1}{3 \times 3}$	Simplify.
$= \frac{5}{9}$	Multiply.

So, $3\frac{1}{3} \div 6$ is ______.

Check

Find $2\frac{1}{2} \div 3$. Write in simplest form.

Example 3 Divide Mixed Numbers

Find $4\frac{2}{3} \div 1\frac{3}{4}$.

$4\frac{2}{3} \div 1\frac{3}{4} = \square$	Write the equation.
$= \frac{\square}{\square} \div \frac{\square}{\square}$	Write the mixed numbers as fractions.
$= \frac{14}{3} \times \frac{\square}{\square}$	Multiply by the reciprocal of $\frac{7}{4}$, $\frac{4}{7}$.
$= \frac{\overset{2}{\cancel{14}}}{3} \times \frac{4}{\underset{1}{\cancel{7}}}$	Divide by common factors.
$= \frac{2 \times 4}{3 \times 1}$	Simplify.
$= \frac{8}{3}$ or $2\frac{2}{3}$	Multiply.

So, $4\frac{2}{3} \div 1\frac{3}{4}$ is ________.

Check

Find $2\frac{3}{8} \div 1\frac{1}{4}$. Write in simplest form.

Go Online You can complete an Extra Example online.

Apply Decorating

The table shows the side lengths of two square mirrors. How many times greater is the area of mirror A than the area of mirror B?

Mirror	Side Length (ft)
A	$2\frac{1}{2}$
B	$1\frac{3}{4}$

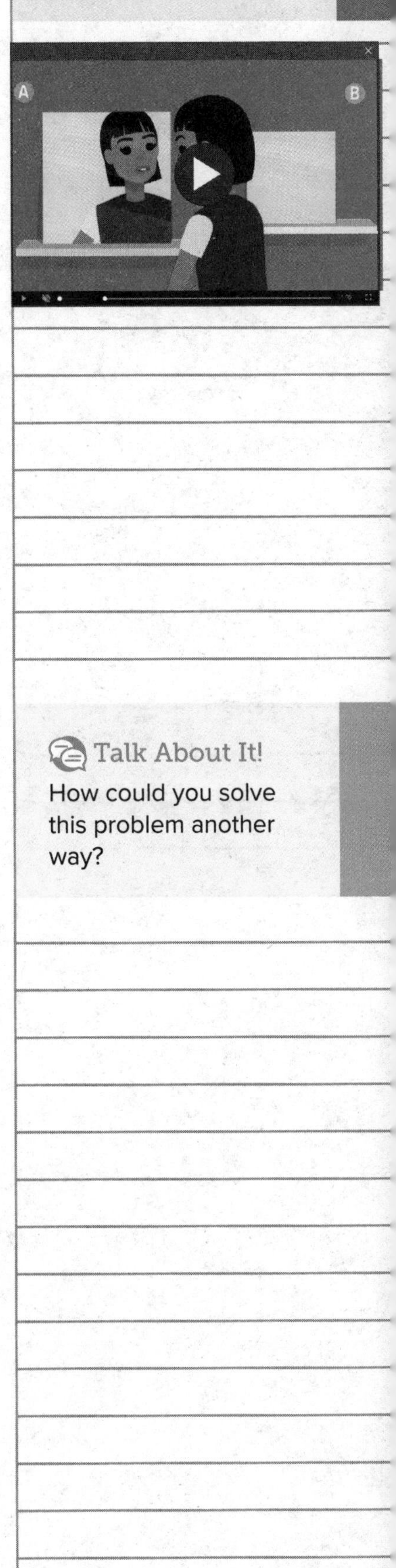

1 What is the task?

Make sure you understand exactly what question to answer or problem to solve. You may want to read the problem three times. Discuss these questions with a partner.

First Time Describe the context of the problem, in your own words.
Second Time What mathematics do you see in the problem?
Third Time What are you wondering about?

2 How can you approach the task? What strategies can you use?

3 What is your solution?

Use your strategy to solve the problem.

4 How can you show your solution is reasonable?

Write About It! Write an argument that can be used to defend your solution.

Check

Mylie has $35\frac{3}{4}$ yards of red ribbon and $30\frac{1}{3}$ yards of green ribbon. She cuts the red ribbon into strips that are each $3\frac{1}{4}$ yards long and the green ribbon into strips that are each $2\frac{1}{6}$ yards long. How many more green strips than red strips does she have?

Go Online You can complete an Extra Example online.

Foldables It's time to update your Foldable, located in the Module Review, based on what you learned in this lesson. If you haven't already assembled your Foldable, you can find the instructions on page FL1.

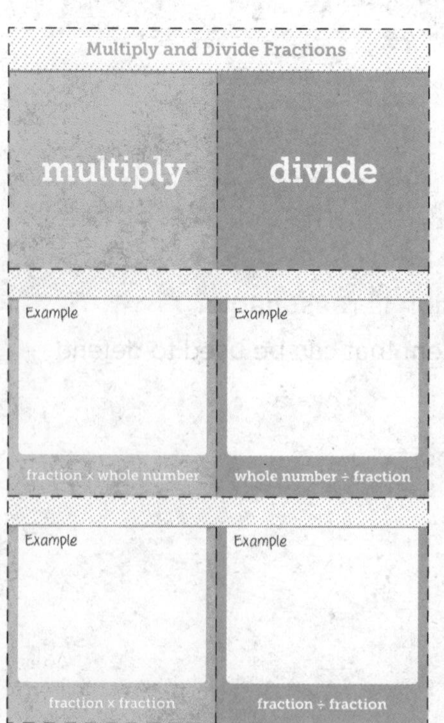

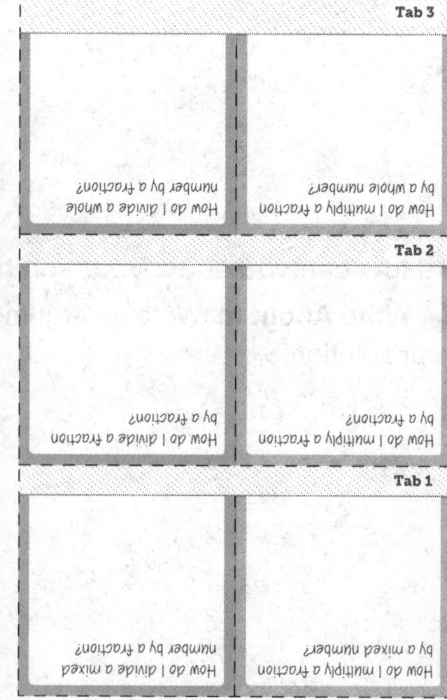

Name ______________________ Period __________ Date __________

Practice

Go Online You can complete your homework online.

1. The drama teacher is making bandanas for costumes. She is cutting $\frac{1}{2}$ yard of fabric into 6 bandanas of the same size. Write and solve an equation to find how much fabric there will be for each bandana. **(Example 1)**

2. A landscape designer has $\frac{4}{5}$ ton of mulch to divide equally among 8 customers. Write and solve an equation to find how much mulch each customer will receive. **(Example 2)**

Divide. Write in simplest form. **(Examples 2 and 3)**

3. $2\frac{4}{5} \div 4 =$ ________

4. $6\frac{2}{3} \div 8 =$ ________

5. $4\frac{2}{3} \div 6 =$ ________

6. $3\frac{3}{5} \div 1\frac{1}{2} =$ ________

7. $3\frac{3}{4} \div 1\frac{2}{3} =$ ________

8. $4\frac{1}{2} \div 2\frac{7}{10} =$ ________

9. Jeanne has $3\frac{7}{8}$ yards of fabric. The table shows the amount of fabric she needs for different items. How many pairs of shorts can she make?

Clothing Item	Fabric Needed (yd)
Shirt	$1\frac{3}{4}$
Shorts	$1\frac{1}{4}$

Test Practice

10. **Equation Editor** What is the value of the expression $5\frac{5}{8} \div 3\frac{3}{4}$?

Apply

11. Kara and Nathan are each painting a poster for the school dance. Their posters have the dimensions shown in the table. How many times greater is the area of Kara's poster than Nathan's?

Student	Poster Length (ft)	Poster Width (ft)
Nathan	$1\frac{1}{2}$	$1\frac{1}{2}$
Kara	$3\frac{3}{4}$	$3\frac{3}{4}$

12. Mrs. Brown is putting different colored sand into cups for her 4 daughters to make sand art bottles. The total amount of each color she has is shown in the table. If each color is divided evenly among the daughters, how much more pink sand will be available for each girl than purple sand?

Sand Color	Weight (lb)
Blue	$\frac{15}{16}$
Pink	$\frac{3}{4}$
Purple	$\frac{1}{2}$
Turquoise	$\frac{7}{8}$

13. Create Write and solve a real-world problem that involves the division of two mixed numbers.

14. Find $2\frac{1}{10} \div 1\frac{1}{5}$. How can you determine if your quotient is reasonable? Explain.

15. MP **Persevere with Problems** Without dividing, explain whether the quotient of $\frac{9}{10} \div 3$ is greater than or less than the quotient of $\frac{9}{10} \div 2$.

16. MP **Reason Inductively** Without computing, which expression is greater $20 \times \frac{1}{2}$ or $20 \div \frac{1}{2}$. Explain your reasoning.

Review

Foldables Use your Foldable to help review the module.

Tab 3	Multiply and Divide Fractions
Tab 2	
Tab 1	
Example	Example
fraction × mixed number	mixed number ÷ fraction

Rate Yourself!   

Complete the chart at the beginning of the module by placing a checkmark in each row that corresponds with how much you know about each topic after completing this module.

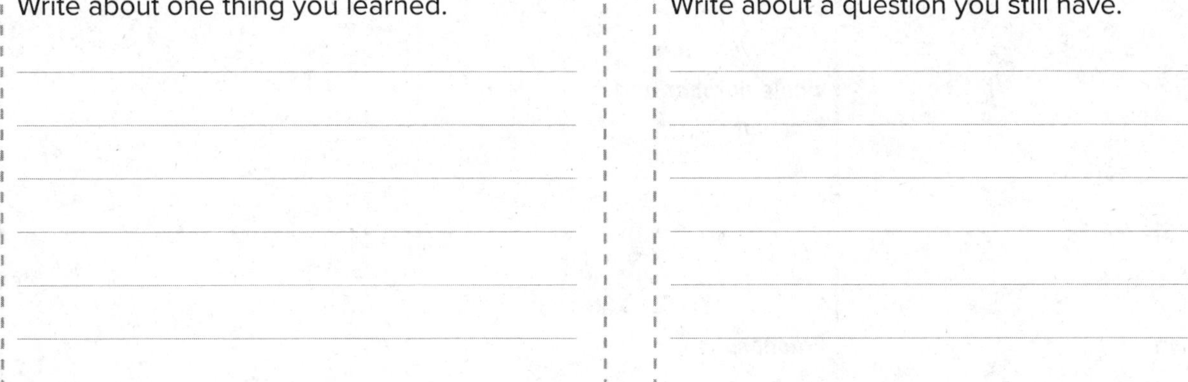

Reflect on the Module

Use what you learned about fractions, decimals, and percents to complete the graphic organizer.

Essential Question

How are operations with fractions and decimals related to operations with whole numbers?

Operation	Numbers Involved	Give an Example
division	whole number and whole number	
addition and subtraction	decimals	
multiplication	decimals	
division	whole number and fraction	
division	fractions	

Name ______________________ Period __________ Date __________

Test Practice

1. Open Response In Jamal's county, there are 60 farms that cover about 8,370 acres of land. If the farms are all approximately the same size, how many acres is each farm? Explain how can you solve the problem. (Lesson 1)

2. Equation Editor At the botanical garden, flower bulbs are planted each spring. The table shows the number of bulbs planted in each color. (Lesson 1)

Color	Number
Yellow	280
Red	245
Purple	393

If each flowerbed can hold 36 bulbs, how many flowerbeds will be completely filled with bulbs?

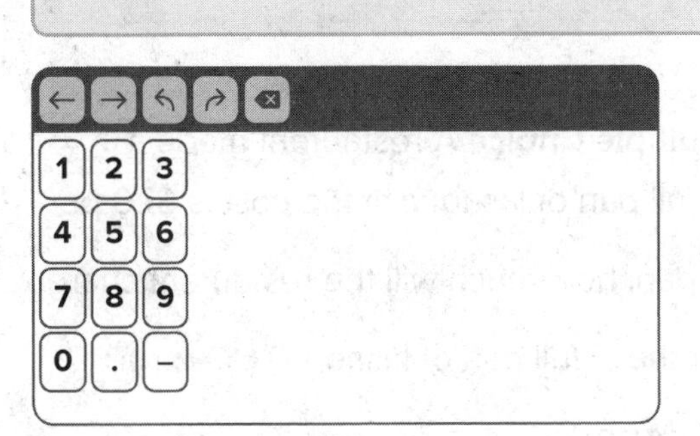

3. Equation Editor Divide 0.008 ÷ 0.25. (Lesson 2)

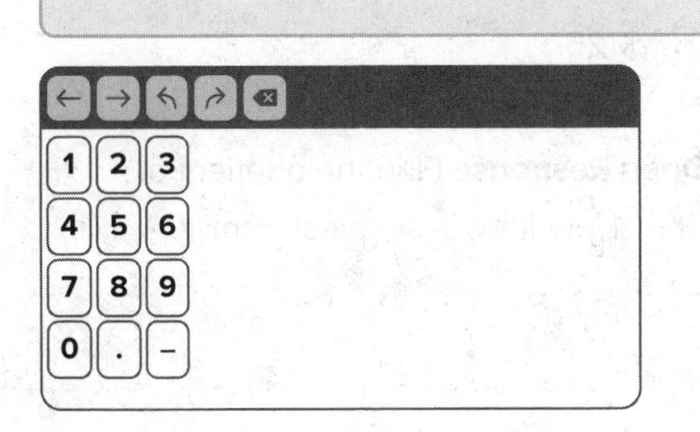

4. Open Response Mariam is making two kinds of paper lanterns. One type of lantern requires 0.75 square foot of construction paper, while the other requires 1.15 square feet. After making 5 of each type of lantern, Mariam has 12.75 square feet of leftover paper. (Lesson 2)

A. How many square feet of paper did Mariam use when making the 10 lanterns? Explain how you found this answer.

B. How many square feet of paper did Mariam begin with? Describe your reasoning.

5. Multiple Choice What number multiplied by $\frac{7}{9}$ has a product of 1? (Lesson 3)

Ⓐ $\frac{7}{9}$

Ⓑ $\frac{9}{9}$

Ⓒ 1

Ⓓ $\frac{9}{7}$

6. Open Response Divide $7 \div \frac{3}{5}$. (Lesson 3)

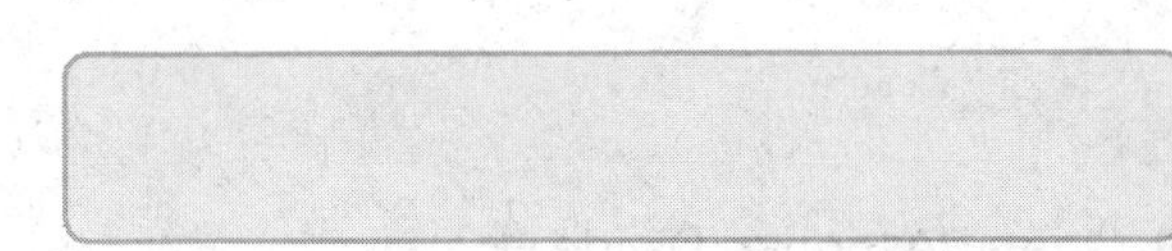

7. Equation Editor The table shows the ingredients needed to make one serving of marinade. Kat has 3 cups of soy sauce. She made the greatest number of servings possible. **(Lesson 3)**

Ingredients	Amount
Ginger	$\frac{1}{8}$ T
Soy sauce	$\frac{5}{6}$ c
Garlic	$\frac{1}{4}$ c

A. How many whole servings of marinade will the 3 cups of soy sauce make?

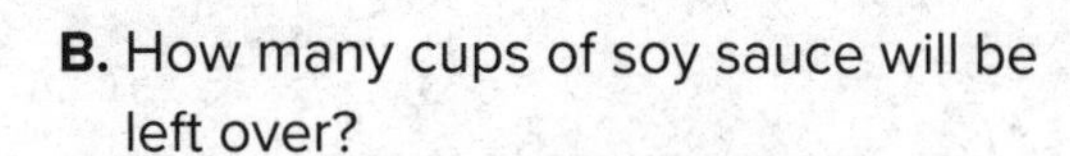

B. How many cups of soy sauce will be left over?

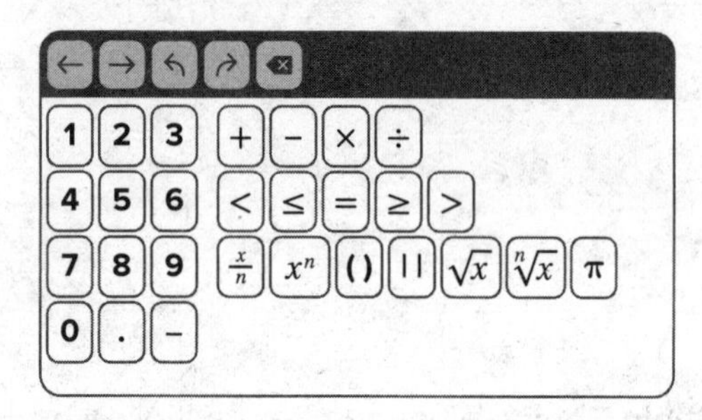

8. Multiple Choice Tony is making chicken enchiladas. He needs $\frac{1}{8}$ jar of sauce for each enchilada. How many chicken enchiladas can Tony make with $\frac{5}{6}$ jar of sauce? **(Lesson 4)**

Ⓐ 5

Ⓑ 6

Ⓒ 7

Ⓓ 8

9. Open Response Divide $\frac{2}{3} \div \frac{3}{4}$. **(Lesson 4)**

10. Multiselect A builder is dividing a hectare (about $2\frac{1}{2}$ acres of land) into $\frac{1}{3}$ acre lots to build houses. Which expression(s) can be used to find how many lots the builder will have to build on? Select all that apply. **(Lesson 5)**

☐ $\frac{5}{2} \div \frac{3}{1}$

☐ $\frac{5}{2} \times \frac{3}{1}$

☐ $\frac{5}{2} \times \frac{1}{3}$

☐ $\frac{5}{2} \div \frac{1}{3}$

11. Open Response Three-fifth pound of pasta is enough to feed 6 people. **(Lesson 5)**

A. Write a division expression to find the number of pounds in each serving.

B. How many pounds are in each serving?

12. Multiple Choice A restaurant made a $\frac{3}{4}$ full pan of lasagna. If the cost is $20 per $\frac{1}{3}$ pan, how much will the restaurant charge for the $\frac{3}{4}$ full pan of lasagna? **(Lesson 5)**

Ⓐ $20

Ⓑ $45

Ⓒ $60

Ⓓ $125

13. Open Response Find the quotient of $13 \div 4\frac{7}{8}$ written in simplest form. **(Lesson 5)**

Module 4

Integers, Rational Numbers, and the Coordinate Plane

Essential Question

How are integers and rational numbers related to the coordinate plane?

6.NS.C.5, 6.NS.C.6, 6.NS.C.6.A, 6.NS.C.6.B, 6.NS.C.6.C, 6.NS.C.7, 6.NS.C.7.A, 6.NS.C.7.B, 6.NS.C.7.C, 6.NS.C.7.D, 6.NS.C.8
Mathematical Practices: MP1, MP2, MP3, MP4, MP5, MP6, MP7, MP8

What Will You Learn?

Place a checkmark (✓) in each row that corresponds with how much you already know about each topic **before** starting this module.

KEY

□ — I don't know. ◇ — I've heard of it. ☆ — I know it!

	Before			After		
	□	◇	☆	□	◇	☆
using integers to represent quantities						
graphing integers on a number line						
finding opposites of integers						
finding absolute values of integers						
comparing and ordering integers						
graphing rational numbers on a number line						
finding absolute values of rational numbers						
comparing and ordering rational numbers						
graphing points in the coordinate plane						
reflecting points in the coordinate plane						
finding distance between points in the coordinate plane						

Foldables Cut out the Foldable and tape it to the Module Review at the end of the module. You can use the Foldable throughout the module as you learn about integers, rational numbers, and the coordinate plane.

What Vocabulary Will You Learn?

Check the box next to each vocabulary term that you may already know.

- ☐ absolute value
- ☐ integer
- ☐ negative integer
- ☐ opposite
- ☐ positive integer
- ☐ quadrants
- ☐ rational number
- ☐ reflection

Are You Ready?

Study the Quick Review to see if you are ready to start this module.
Then complete the Quick Check.

Quick Review

Example 1

Compare decimals.

Fill in the ◯ with $<$, $>$, or $=$ to make a true statement.

$1.6 \bigcirc 1.3$

1.3 1.6

1.0 1.2 1.4 1.6 1.8 2.0

Since 1.6 is to the right of 1.3, $1.6 > 1.3$.

Example 2

Compare fractions.

Fill in the ◯ with $<$, $>$, or $=$ to make a true statement.

$\frac{2}{5} \bigcirc \frac{7}{10}$

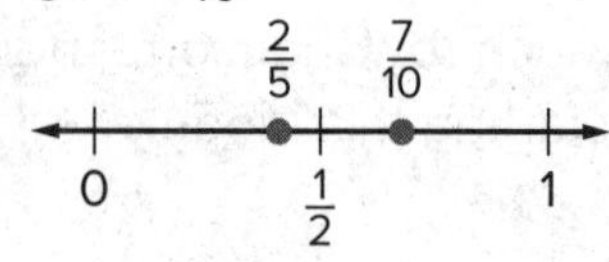

Since $\frac{2}{5}$ is less than $\frac{1}{2}$ and $\frac{7}{10}$ is greater than $\frac{1}{2}$, $\frac{2}{5} < \frac{7}{10}$.

Quick Check

Fill in each ◯ with $<$, $>$, or $=$ to make a true statement.

1. $7.7 \bigcirc 7.5$

2. $4.8 \bigcirc 4.80$

Fill in each ◯ with $<$, $>$, or $=$ to make a true statement.

3. $\frac{4}{11} \bigcirc \frac{9}{10}$

4. $\frac{3}{5} \bigcirc \frac{1}{4}$

How Did You Do?

Which exercises did you answer correctly in the Quick Check? Shade those exercise numbers at the right.

Represent Integers

I Can... use positive and negative numbers, as well as 0, to represent quantities in my everyday life, using a number line or coordinate axes to visually represent the values.

Today's Standards
6.NS.C.5, 6.NS.C.6, 6.NS.C.6.C
MP2, MP3, MP4, MP5

What Vocabulary Will You Learn?
integer
negative integer
positive integer

Explore Represent Integers

Online Activity You will explore how positive and negative values can be represented on a vertical number line.

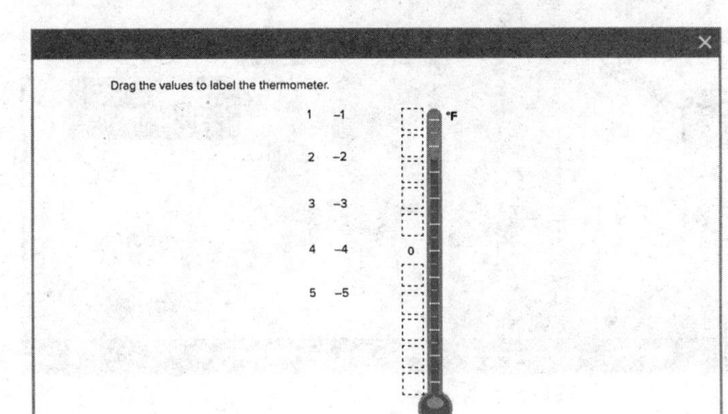

Learn Use Integers to Represent Quantities

An **integer** is any number from the set {..., −3, −2, −1, 0, 1, 2, 3, ...}, where "..." means *continues indefinitely*.

A **negative integer** is an integer less than zero and is written with a − sign. A **positive integer** is an integer greater than zero, and can be written with or without a + sign.

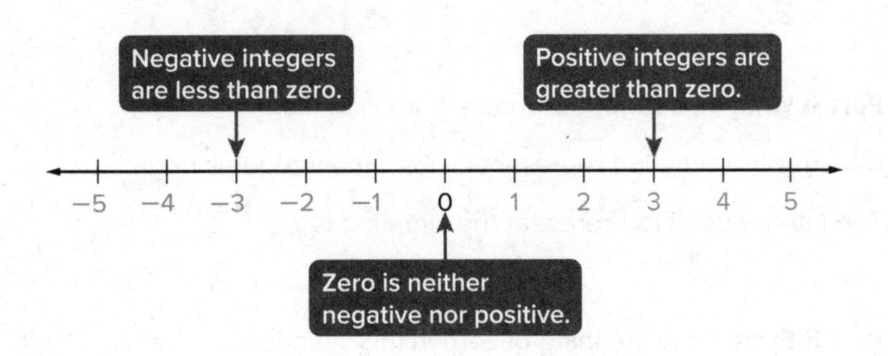

(continued on next page)

Your Notes

Talk About It!

Give another example of when using a vertical number line is useful. Explain your reasoning.

Go Online Watch the animation to see how integers are used in real life.

Suppose Anabeth is traveling to different parts of the country. She logs the temperature in each location. When Anabeth was in Miami, Florida, the temperature was 80 degrees. That same week, she traveled to Caribou, Maine, where it was −10 degrees.

How can Anabeth represent the positive and negative values in her temperature log?

Think About It!

What does the word loss mean?

Example 1 Use Integers to Represent Quantities

A football team has a 10-yard loss in one play.

Write an integer to represent the situation. Explain the meaning of 0 in the situation.

Talk About It!

Describe another real-world situation that can be represented by −10. Explain the meaning of zero in that situation.

Part A Write an integer to represent the situation.

Because the situation represents a loss, the integer is negative.

The integer used to represent the situation is ________.

Part B Explain the meaning of zero in this situation.

In a football play, the integer 0 represents ________ yards gained or lost.

Check

Death Valley National Park contains the lowest elevation in North America, which is 282 feet below sea level.

Part A

Write an integer to represent the situation.

Part B

Explain the meaning of zero in this situation.

Go Online You can complete an Extra Example online.

Learn Graph Integers on a Number Line

Integers and sets of integers can be graphed on a number line. To graph an integer on a number line, place a point on the number line at its location. Positive numbers are graphed to the right of zero on a horizontal number line, or above zero on a vertical number line. Negative numbers are graphed to the left of zero on a horizontal number line, or below zero on a vertical number line.

A set of integers is written using braces, such as {2, −3, 0}.

The set of integers {2, −3, 0} is graphed on each number line.

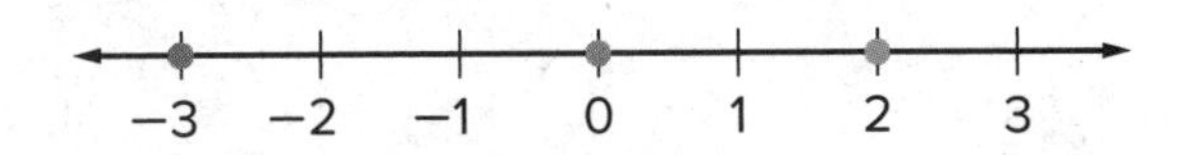

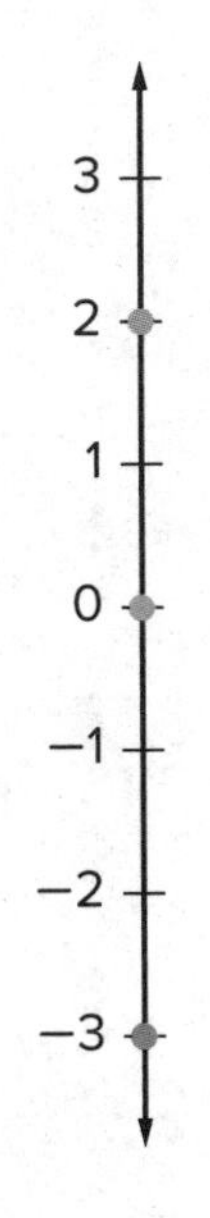

Talk About It!

Compare the horizontal and vertical number lines.

Example 2 Graph Integers on a Number Line

Graph the set of integers {−4, 2, −1} on the number line.

Place a dot at −4, 2, and −1.

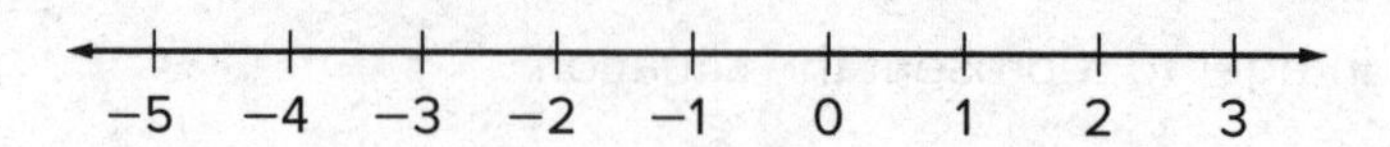

Check

Graph the set of integers {−3, 1, 0} on a number line.

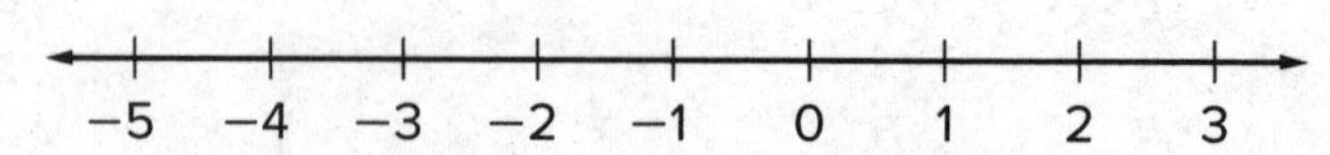

Go Online You can complete an Extra Example online.

Pause and Reflect

How well do you understand the concepts from today's lesson? What questions do you still have? How can you get those questions answered?

Name _______________ Period _______ Date _______

Practice

Go Online You can complete your homework online.

Write an integer for each situation. Explain the meaning of zero in each situation. (Example 1)

1. Since his last vet appointment, a cat lost 2 ounces.

2. On first down, the football team gained 7 yards.

3. Abigail withdrew $15 from her checking account.

4. By noon, the temperature had risen 5 degrees Fahrenheit.

5. For the month of January, the amount of snowfall was 3 inches above average.

6. Carol withdrew $20 from her account.

Graph each set of integers on a number line. (Example 2)

7. {−2, 0, 4}

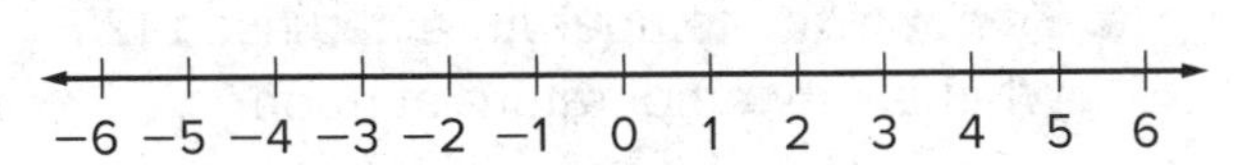

8. {5, −5, −6}

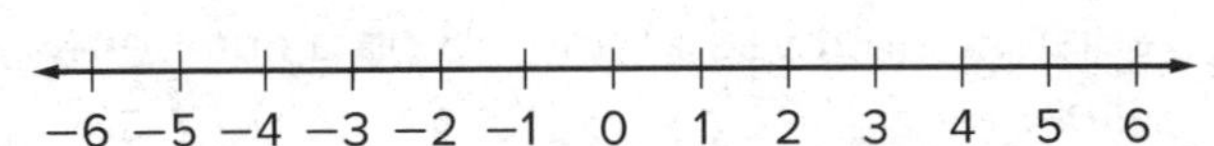

9. {−8, −4, 1}

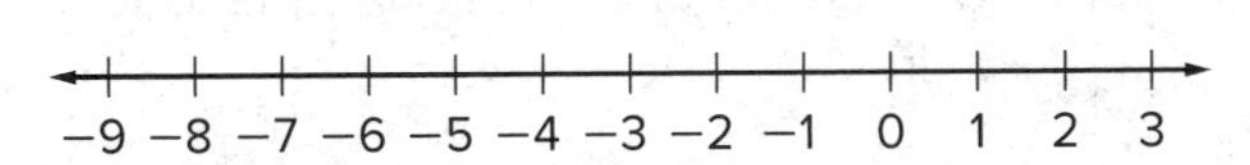

10. {−7, 3, 5}

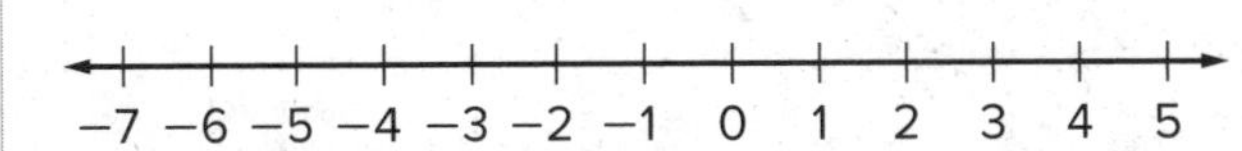

11. {7, −3, −1}

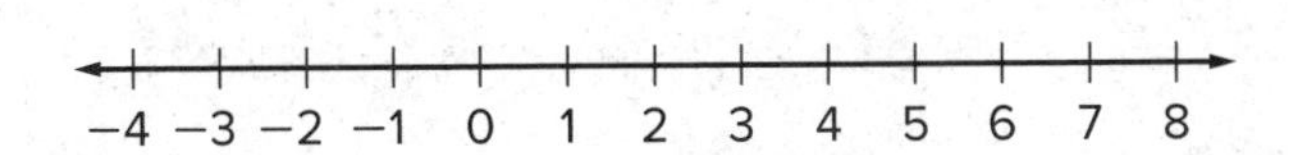

12. {−1, 0, 1}

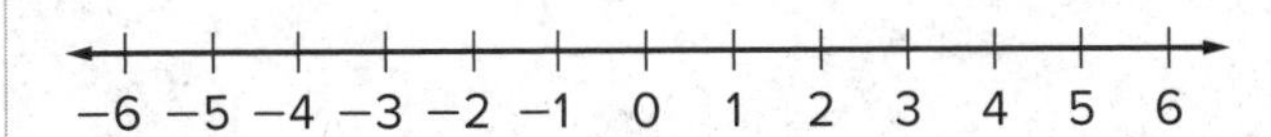

13. The low temperatures for three consecutive days were −5°F, 3°F, and 4°F. Graph this set of integers on a number line.

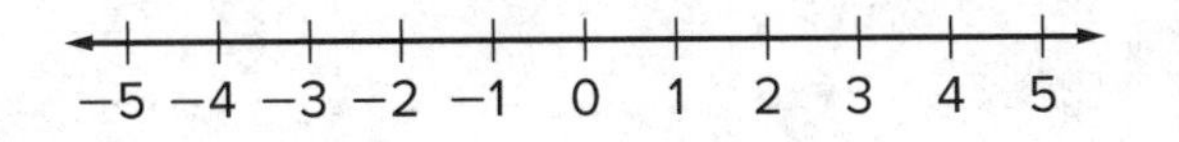

Test Practice

14. Multiple Choice Salton City, California is located 38 meters below sea level. What is a possible elevation for Salton City?

Ⓐ 380 m

Ⓑ 38 m

Ⓒ 0 m

Ⓓ −38 m

Apply

15. Rodney is performing a science experiment. The table shows the temperature of two liquids he is using. Graph the integers that represent the temperatures on a number line. Which beaker's liquid is closer to 0°C? Explain.

Beaker	Temperature
A	−4°C
B	2°C

−5 −4 −3 −2 −1 0 1 2 3

16. Sydney owes her mother \$5 and her brother owes her mother \$7. Graph the integers that represent the amount they owe their mother on a number line. How much more will her brother have to repay their mother than Sydney? Explain using the number line.

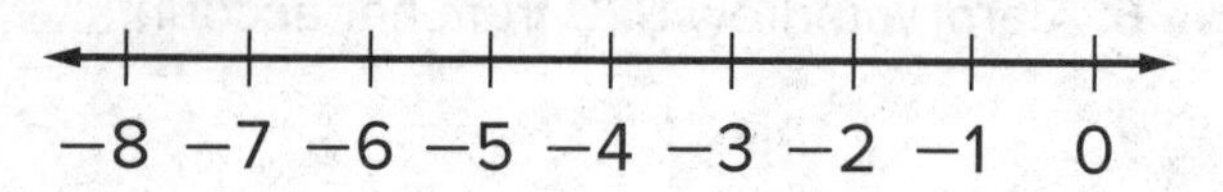

17. MP **Use Math Tools** Explain how to find the distance between 1 and −3 on a number line.

18. At 4 A.M. the outside temperature was 2°F.

a. By noon, the temperature had risen 12°F. What is the temperature at noon?

b. What represents zero in this situation? Explain.

19. Create Describe a real-world situation that can be represented by a negative integer. Then write the integer.

20. MP **Justify Conclusions** Craig has \$28 in his checking account. He makes a withdrawal of \$30. Will his checking account balance be represented by a positive or negative integer? Justify your conclusion.

Opposites and Absolute Value

I Can... understand the absolute value of rational numbers and how to order these numbers.

Explore Opposites and Absolute Value

Online Activity You will use Web Sketchpad to explore opposites and absolute value.

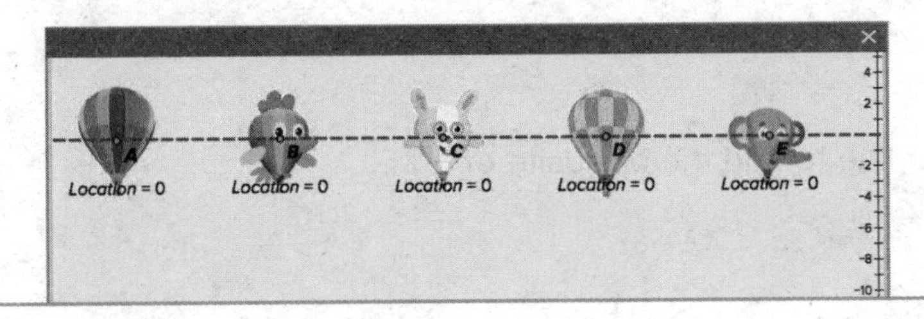

Today's Standards
6.NS.C.5, 6.NS.C.6, 6.NS.C.6.A, 6.NS.C.7, 6.NS.C.7.C
MP1, MP2, MP3, MP4, MP5, MP6, MP8

What Vocabulary Will You Learn?
absolute value
opposites

Learn Find Opposites

Integers are **opposites** when they are the same distance from zero on a number line, in opposite directions. The opposite of a positive integer is indicated by using the notation −(2), which is read *the opposite of two*. The opposite of a negative integer is indicated by using the notation −(−2), which is read *the opposite of negative two*.

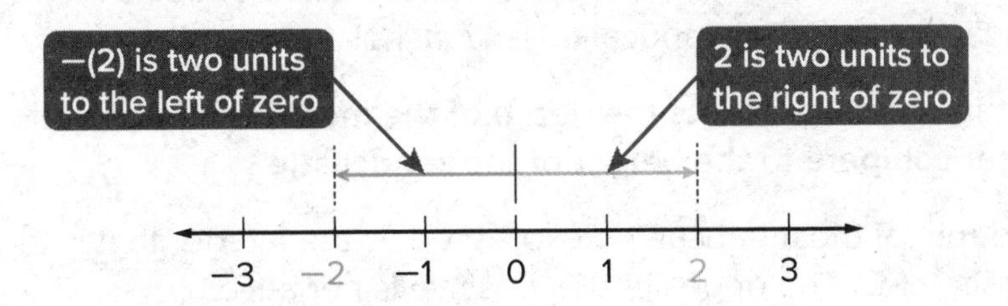

So, −(−2) is 2.

The opposite of the opposite of a number is the number itself.

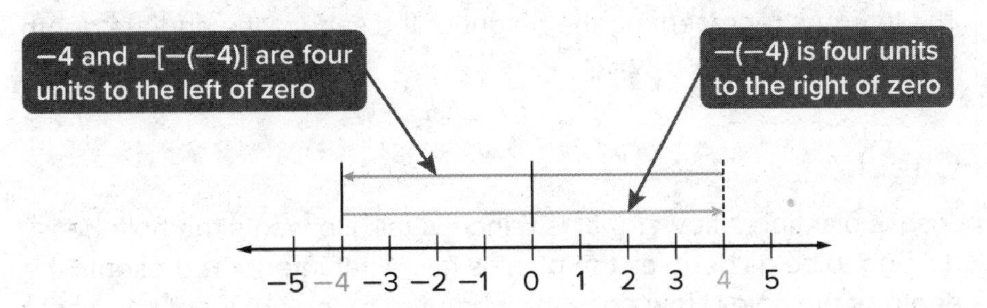

So, −[−(−4)] is −4.

Talk About It!
Explain why 0 is its own opposite.

Your Notes

Example 1 Use a Number Line to Find Opposites of Integers

Find $-(-5)$.

Graph -5 on the number line.

−6 −5 −4 −3 −2 −1 0 1 2 3 4 5

The point graphed at -5 is ________ units to the left of 0. The point that is the same number of units to the right of 0 is 5.

So, the opposite of -5 is ________ .

Check

Use a number line to find the opposite of -21.

Show your work here

Go Online You can complete an Extra Example online.

Example 2 Find Opposites of Integers Using Symbols

Asia and LaToya are building a sandcastle and digging a moat around the sandcastle. They would like the moat to be as deep as the sandcastle is tall. The sandcastle is 17 inches tall.

What integer represents the depth of the moat? How does this integer compare to the height of the sandcastle?

The depth of the moat can be expressed as the integer that is the opposite of 17. The opposite of a positive is negative.

So, the integer that represents the depth of the moat is $-(17)$ or ________.

The integers representing the height of the sandcastle and the depth of the moat are opposites.

Talk About It!

Can all positive integers be written with or without the + sign? Can all negative integers be written with or without the − sign? Explain.

Check

Josh is planting a flower that is 6 inches tall. He wants the hole he is digging to be as deep as the plant is tall. What integer represents the depth of the hole? How does this compare to the height of the plant?

Go Online You can complete an Extra Example online.

Example 3 Find Opposites of Opposites of Integers

Find $-[-(-3)]$.

$$-[-(-3)]$$

3 The opposite of -3 is 3.

-3 The opposite of 3 is -3.

So, the opposite of the opposite of -3 is ______.

Check

Find $-[-(-11)]$.

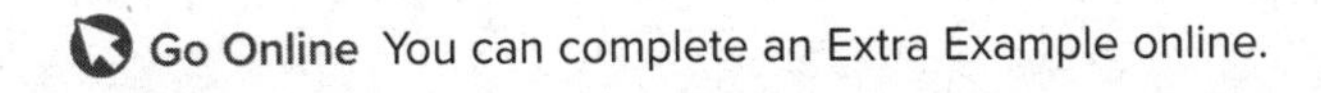

Talk About It!

Compare the opposite of the opposite of a number to the original number.

Learn Absolute Value of Integers

The integers 4 and -4 are each 4 units from 0, even though they are on opposite sides of 0. Numbers that are the same distance from zero on a number line have the same **absolute value**.

Words

The absolute value of a number is the distance between the number and zero on the number line.

Model

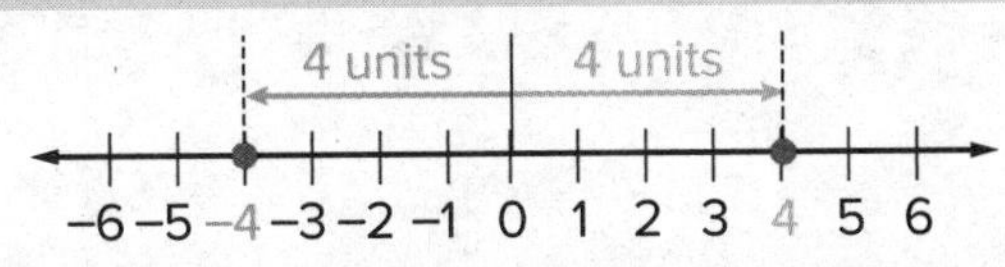

Symbols

The || symbol around a number means the absolute value of that number.

$|4| = 4$ The absolute value of 4 is 4.

$|-4| = 4$ The absolute value of -4 is 4.

Talk About It!

Why is the absolute value of a number never negative?

Example 4 Find the Absolute Value of Integers

A cave explorer started at sea level and descended in a cave. Her location, in relationship to her starting point, can be represented by −150 feet.

How many feet did the cave explorer travel?

To find how many feet the cave explorer traveled, you need to find |−150|.

To find the absolute value, you need to find the distance between the number and zero on the number line.

Graph −150 on the number line.

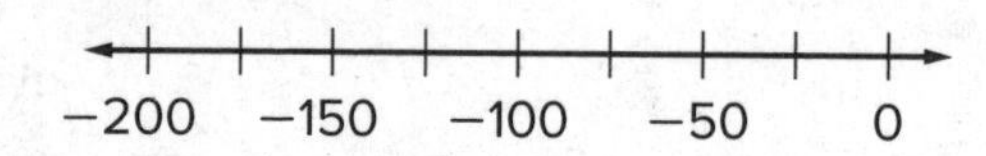

How many units from 0 is −150? ________ units

So, the cave explorer traveled |−150| or 150 feet.

Check

Yixi dropped a coin in a wishing well. The top of the well can be represented by 0 feet. The location of the coin can be represented by −32 feet. How many feet did the coin fall?

 **Go Online** You can complete an Extra Example online.

Pause and Reflect

Did you ask questions about today's lesson? Why or why not?

 **Think About It!**

Is the location represented by a positive or negative integer?

 **Talk About It!**

What other number has the same absolute value as −150? Explain your reasoning.

Name ______ Period ______ Date ______

Practice

Go Online You can complete your homework online.

Find the opposite of each integer. **(Example 1)**

1. −3

2. 2

3. 6

4. Chad is planting a plant that is 4 inches tall. He wants the hole he is digging to be as deep as the plant is tall. What integer represents the location of the bottom of the hole? How does this compare to the height of the plant? **(Example 2)**

5. A hill on a dirt bike course is 5 feet tall. The valley below the hill is as deep as the hill is tall. What integer represents the location of the bottom of the valley? How does this compare to the height of the hill? **(Example 2)**

Find each value. **(Examples 2 and 3)**

6. −(−15) = ______

7. −(−11) = ______

8. −[−(−7)] = ______

9. −[−(−1)] = ______

10. −[−(55)] = ______

11. −[−(100)] = ______

12. A mountain climber started at sea level and descended down a cliff. Her location can be represented by −75 feet. How many feet did the mountain climber travel? **(Example 4)**

13. The temperature was −5°F when Tiffany woke up in the morning. By noon, the temperature was 0°F. How many degrees did the temperature change? **(Example 4)**

Test Practice

14. Multiselect Which of the following represent opposites?

☐ −4 and 4

☐ −1 and 1

☐ −2 and −1

☐ 0 and 1

☐ −7 and −8

☐ 10 and −10

Apply

15. The table shows the minimum and maximum elevations, relative to sea level, of several hiking trails. Which hiking trail has the least change in elevation, related to sea level? Explain how you solved.

Trail	Minimum Elevation (ft)	Maximum Elevation (ft)
Eastern Point	−85	78
Northern Star	−150	34
Southern Moon	−62	48

16. The table shows the lowest and highest record temperatures for three cities. Which city had the greatest change in record temperature? Explain how you solved.

City	Lowest Temperature (°F)	Highest Temperature (°F)
Boston	−30	104
Las Vegas	8	118
Pittsburgh	22	103

17. MP **Reason Inductively** Determine if the following statement is *true* or *false*. Explain your reasoning.

The absolute value of a negative integer is always a negative integer.

18. MP **Find the Error** Judith states that $-|14| = 14$ because the absolute value can never be negative. Find her mistake and correct it.

19. MP **Justify Conclusions** A student states that $-x$ is always equal to a negative integer. Is the student correct? Justify your reasoning.

20. MP **Persevere with Problems** Identify integers for x and y that make the following statement true.

$x > y$ and $|x| < |y|$

Compare and Order Integers

I Can... correctly order rational numbers, including integers and absolute values, and then use a number line to write a statement of inequality.

Today's Standards
6.NS.C.7, 6.NS.C.7.A, 6.NS.C.7.B, 6.NS.C.7.C, 6.NS.C.7.D, *Also addresses 6.NS.C.6, 6.NS.C.6.C*
MP1, MP2, MP3, MP4, MP5, MP6, MP7

Learn Compare Integers

To compare integers, you can compare the signs as well as the magnitude, or size of the numbers. If the signs are different, the positive integer will always be greater than the negative integer.

Different Signs
Compare 2 and −3.
The signs are different, so compare the signs. A positive integer is always greater than a negative integer, so 2 is greater than −3.
$2 > -3$

If the signs of the two integers are the same, you can use a number line to compare them. On a horizontal number line, positive integers are graphed to the right of zero, while negative integers are graphed to the left of zero. The greater numbers will be farther to the right.

On a vertical number line, positive integers are graphed above zero, while negative integers are graphed below zero. The greater numbers are graphed farther above zero.

Same Signs
Compare −2 and −3.
The signs are the same, so use a number line to compare the integers. Because −2 is graphed farther to the right than −3, −2 is greater than −3.
−3 −2 −1 0 1 2 3
$-2 > -3$

Talk About It!
When comparing two negative numbers, like −2 and −3, what do you notice about the absolute value of −2 compared to the absolute value of −3? Does this hold true when comparing other negative numbers?

Your Notes

Think About It!

How can you compare two negative numbers?

Talk About It!

What is another way to write an inequality comparing −3 and −5? Explain why this inequality is also true.

Example 1 Compare Two Integers

Justin has a score of −5 on the Trueville Trivia Game. Desiree's score is −3.

Write an inequality to compare the scores. If a greater score wins the game, explain the meaning of the inequality.

Part A Write an inequality.

Graph the integers on the number line.

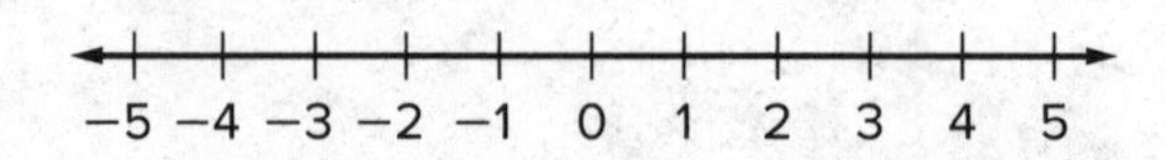

Compare. Which number is farther to the right on the number line? ________

The inequality is $-3 > -5$.

Part B Explain the meaning of the inequality.

Since $-3 > -5$, ________ has a greater score in the trivia game.

So, Desiree has the greater score in the trivia game.

Check

Andrew and his father are hiking near Tackle Box Canyon. Their current elevation, in relation to sea level, is −38 feet. Tackle Box Canyon has an elevation of −83 feet.

Part A Write an inequality to compare the elevations.

Part B Explain the meaning of the inequality.

Go Online You can complete an Extra Example online.

Learn Order Sets of Integers

You can use a number line to order a set of integers from least to greatest or from greatest to least.

Go Online Watch the animation to see how you can use a number line to order a set of integers.

The animation shows how to graph the set of integers {−8, 3, −1, 0, 6} on a number line.

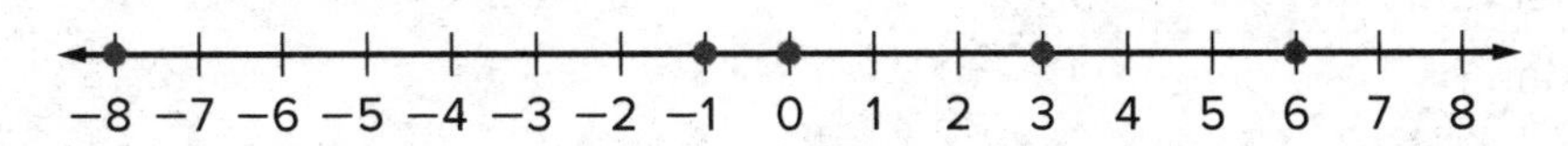

From left to right, the integers from least to greatest are {−8, −1, 0, 3, 6}.

From right to left, the integers from greatest to least are {6, 3, 0, −1, −8}.

Talk About It!
How does a number line help to organize a set of integers?

Example 2 Order Sets of Integers

The table shows the lowest accessible elevations for several continents.

Continent	Lowest Elevation (m)
Antarctica	−50
Australia	−12
North America	−86
South America	−105

Order the continents from least to greatest according to their lowest elevation.

Graph the integers on a number line.

−105 −86 −50 −12

−110 −100 −90 −80 −70 −60 −50 −40 −30 −20 −10 0

Which continent has the least accessible elevation?

____

Which continent has the greatest accessible elevation?

____

So, the contintents written in order from least to greatest elevation are South America, North America, Antarctica, and Australia.

Talk About It!
The lowest elevation in Asia is near the Dead Sea at −423 meters. The lowest elevation in Africa is near Lake Assal at −157 meters. How would adding these points to the data set change the number line and the order of the elevations?

Check

The table shows Kesha's cell phone use over the last four months. Positive values indicate the number of minutes she had remaining, and negative values indicate the number of minutes she went over. Arrange the months from fewest to most minutes remaining for the month.

Month	Number of Minutes Over/Under
February	−156
March	12
April	0
May	−45

Go Online You can complete an Extra Example online.

Pause and Reflect

Did you struggle with any of the concepts in this Check? How do you feel when you struggle with math concepts? What steps can you take to understand those concepts?

Learn Distinguish Absolute Value from Order

You know how to order numbers when you see them on a horizontal number line. The values of the numbers increase as they move to the right, and the values decrease as they move to the left.

What happens to the absolute value, or magnitude, of numbers as the values increase or decrease? Since absolute value is the distance a number is from zero, the absolute value increases the farther the number is from zero.

As a positive value increases, or moves farther from 0, its absolute value also increases.

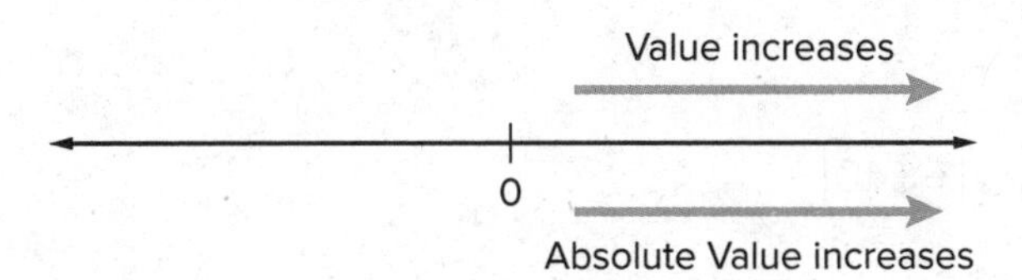

As a negative value decreases, or moves farther from 0, its absolute value increases.

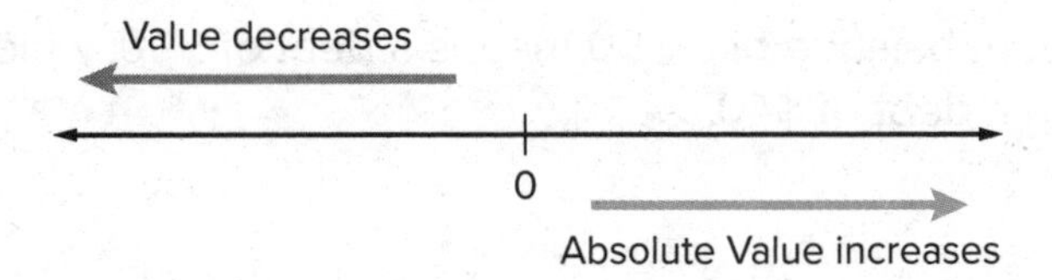

Suppose Kaito and Ember are scuba diving.

Kaito dove to 25 feet below sea level. This can be represented by the integer ________.

Ember dove to 30 feet below sea level. This can be represented by the integer ________. Who reached a greater depth?

You know that $-25 > -30$, but this does not mean that Kaito's depth was greater. When determining who reached a greater depth, you need to consider the magnitude of the numbers, not just their placement on the number line.

The absolute value of a number takes into account the number's magnitude.

What is the absolute value of -30? ________

What is the absolute value of -25? ________

Which absolute value is greater? ________

Since $|-30| > |-25|$, Ember's depth is greater.

 **Talk About It!**

Some words imply a negative value, like depth. What other words imply the sign of the number?

Example 3 Comparisons with Absolute Value

Explain why an account balance less than −$40 represents a debt greater than $40.

Debt is the money owed by one person to another person.

An example of an account balance less than −$40 is −$50.

Write an inequality comparing the two amounts.

−$50 ☐ −$40

Use the absolute value to determine which integer represents a greater debt.

|−$50| ☐ |−$40|

An account balance less than −$50 has a lesser value, but a greater absolute value.

So, an account balance of −$50 means a debt of $50, which is greater than a debt of $40.

Check

Explain why an account balance less than −$5 represents a debt greater than $5.

Show your work here

Go Online You can complete an Extra Example online.

Apply Chemistry

The table shows the freezing points in degrees Celsius for six substances. Nitric acid freezes at −42°C. Between which two substances is the freezing point of nitric acid?

Substance	Freezing Point (°Celsius)
Aniline	−6
Acetic Acid	17
Acetone	−95
Water	0
Carbon Dioxide	−78
Sea Water	−2

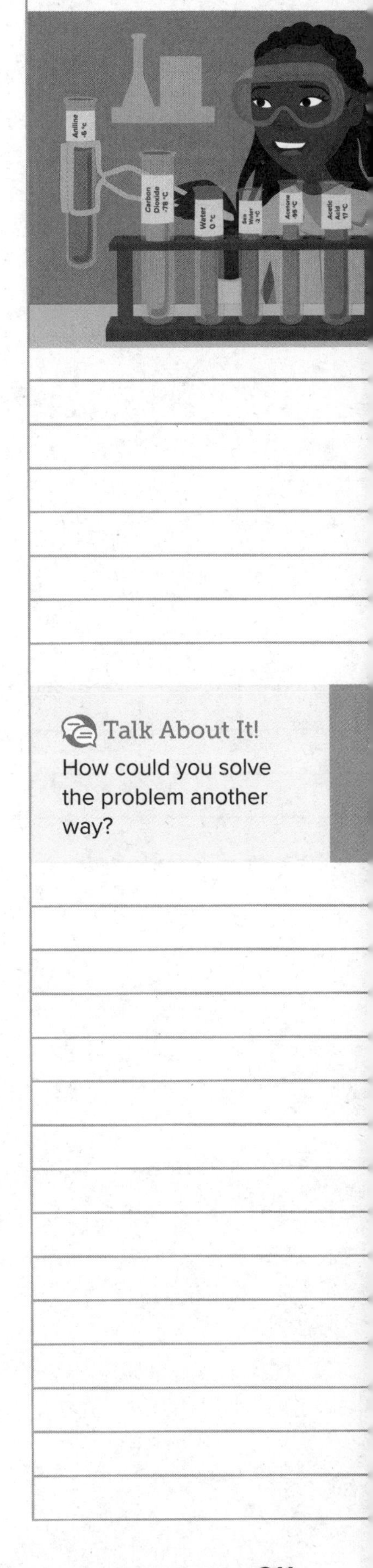

1 What is the task?

Make sure you understand exactly what question to answer or problem to solve. You may want to read the problem three times. Discuss these questions with a partner.

First Time Describe the context of the problem, in your own words.
Second Time What mathematics do you see in the problem?
Third Time What are you wondering about?

Talk About It!
How could you solve the problem another way?

2 How can you approach the task? What strategies can you use?

3 What is your solution?

Use your strategy to solve the problem.

4 How can you show your solution is reasonable?

Write About It! Write an argument that can be used to defend your solution.

Check

When a football player causes a penalty during a game, the team can lose 5, 10, or 15 yards on the play. The table shows the players, by jersey number, and the number of penalty yards the team lost based on each player's penalties. How many players caused more penalty yards than the player with jersey number 10?

Player	Penalty Yards
Chung	15
Davis	25
Diggs	30
Hernandez	10
Lopez	20
Smith	5

Go Online You can complete an Extra Example online.

Foldables It's time to update your Foldable, located in the Module Review, based on what you learned in this lesson. If you haven't already assembled your Foldable, you can find the instructions on page FL1.

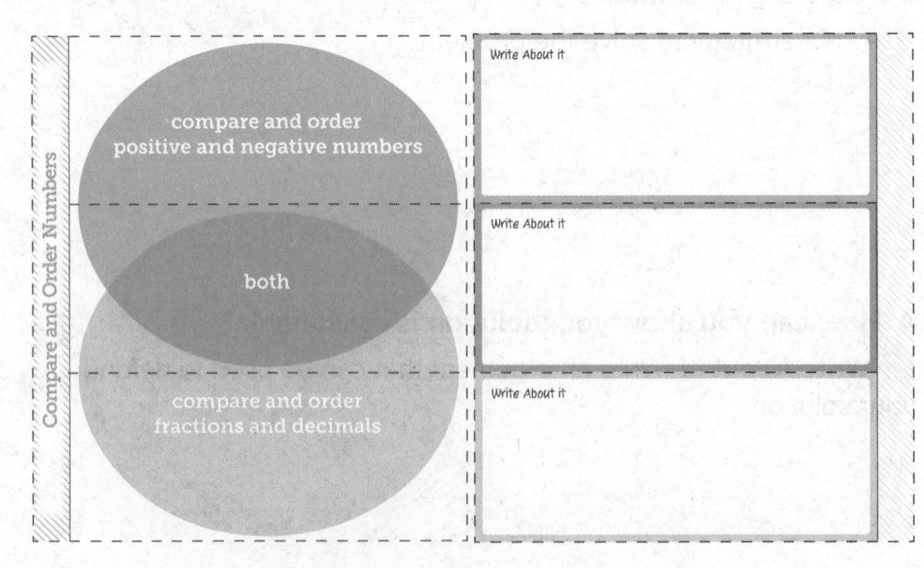

Name ______________________ Period ____________ Date ______________

Practice

Go Online You can complete your homework online.

1. After playing 18 holes of golf, John's score was −4 and Terry's score was −1. Write an inequality to compare the scores. If the person with the lesser score wins the game, explain the meaning of the inequality. **(Example 1)**

2. The record low temperature for Buffalo, New York is −20°F. The record low temperature for Chicago, Illinois is −27°F. Write an inequality to compare the record low temperatures. Explain the meaning of the inequality. **(Example 1)**

3. The table shows the freezing points for gases. Order the gases from least to greatest according to their freezing points. **(Example 2)**

Gas	Freezing Points (°C)
Argon	−189
Carbon Monoxide	−205
Ethane	−297
Helium	−272
Oxygen	−219
Sulfur Dioxide	−72

4. The table shows the scores for players in a trivia game after the first round. Order the players from least to greatest according to their scores. **(Example 2)**

Player	Score
Ace	−11
Diana	3
Jace	−3
Oneida	−7
Nolan	5
Rachel	1

5. Explain why an elevation less than −5 feet represents a distance from sea level greater than 5 feet. **(Example 3)**

6. Explain why a balance of less than −$10 represents a debt greater than $10. **(Example 3)**

7. In a golf match, Jesse scored 5 over par, Neil scored 3 under par, Felipe scored 2 over par, and Dawson scored an even par. Order the players from least to greatest score.

Test Practice

8. **Table Item** Order the integers from least to greatest.

9, −8, −2, 4, −9

least **greatest**

Apply

9. The table shows the lowest elevations for several countries. The lowest elevation in the United States is −86 meters. Between which two countries is the elevation for the United States?

Country	Elevation (m)
Argentina	−105
China	−154
Egypt	−133
Ethiopia	−125
Libya	−47
Morocco	−55

10. A group of students participated in a small business challenge. The table shows results for the students' budgets. The student with the greatest amount under budget wins the challenge. In what place did Dave finish?

Student	Budget
Casey	$2 under
Dave	even
Lily	$5 over
Luke	$4 over
Mike	$1 under
Tyrone	$6 under

11. Create Write a real-world situation that compares two negative integers. Then represent the situation with an inequality.

12. MP **Justify Conclusions** A student said −5 is less than −4 and |−5| is less than |−4|. Is the student correct? Justify your reasoning.

13. Order {−2.5, 4, 23, −1, 5, −3, 0.66} from least to greatest.

14. MP **Identify Structure** Suppose $y = 2$. Identify all the integers for x that make $|x| < |y|$ a true statement.

Rational Numbers

I Can... order rational numbers and understand that the absolute value of rational numbers shows their distance from 0.

Learn Rational Numbers

Recall that natural numbers are from the set {1, 2, 3, 4, ...} where ... means *continues without end.*

The set of whole numbers includes the set of natural numbers and 0.

Integers are any numbers from the set {... −3, −2, −1, 0, 1, 2, 3, ...} where ... means *continues without end.*

Any number that can be written as a fraction $\frac{a}{b}$, where a and b are integers, and $b \neq 0$, is a **rational number**. A rational number can always be represented as a point on the number line.

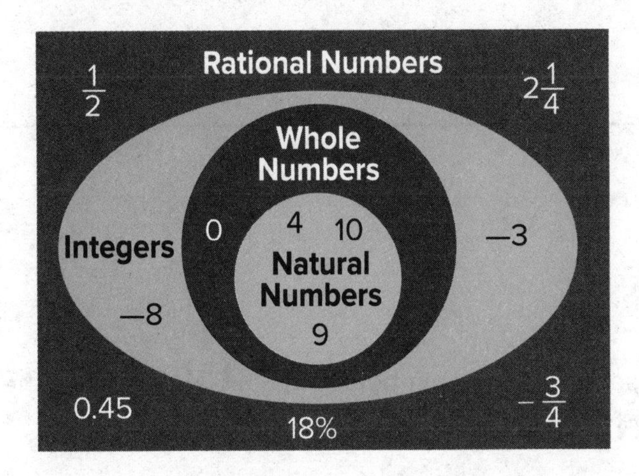

Learn Graph Rational Numbers on a Number Line

Rational numbers and sets of rational numbers can be graphed on a horizontal or vertical number line. A set of rational numbers is written using braces, such as $\left\{2.25, -1\frac{3}{4}, 0\right\}$. To graph a rational number on the number line, place a dot at its location.

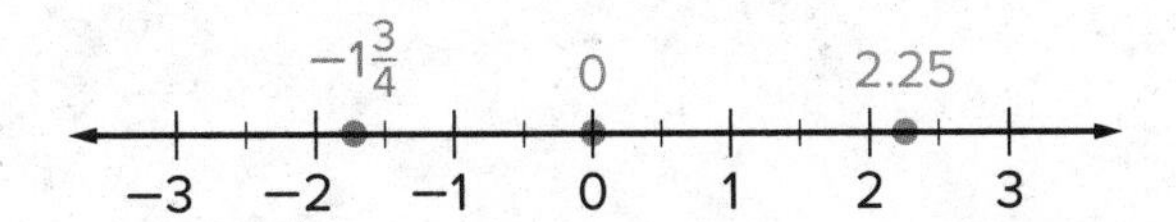

Today's Standards
6.NS.C.6, 6.NS.C.6.C, 6.NS.C.7, 6.NS.C.7.A, 6.NS.C.7.C,
Also addresses 6.NS.C.7.B
MP1, MP2, MP3, MP5, MP6

What Vocabulary Will You Learn?
rational number

Talk About It!

Is −3.77 a rational number? Explain your reasoning.

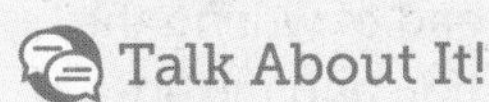

Talk About It!

Suppose the same numbers are graphed on a vertical number line. Compare and contrast the locations of the numbers on the horizontal and vertical number lines.

Your Notes

Think About It!

What do you know about the location of positive rational numbers on a number line? negative numbers?

Talk About It!

Instead of writing the fraction and mixed number as decimals, you can write the decimals as fractions. Compare the two methods.

Example 1 Graph Sets of Rational Numbers

Graph the set of rational numbers $\left\{-\frac{1}{5}, -0.7, 2\frac{3}{5}, -1.8\right\}$ on the number line.

Step 1 Find the integer boundaries of the set.

The values in the set lie between the integers ________ and ________.

Step 2 Graph the rational numbers.

To graph the set, it may be helpful to rewrite the fraction and mixed number as decimals in order to find the locations on the number line.

$-\frac{1}{5} = \square$ $\qquad$ $2\frac{3}{5} = \square$

Then graph each value on the number line. Label each point with the value in its original form.

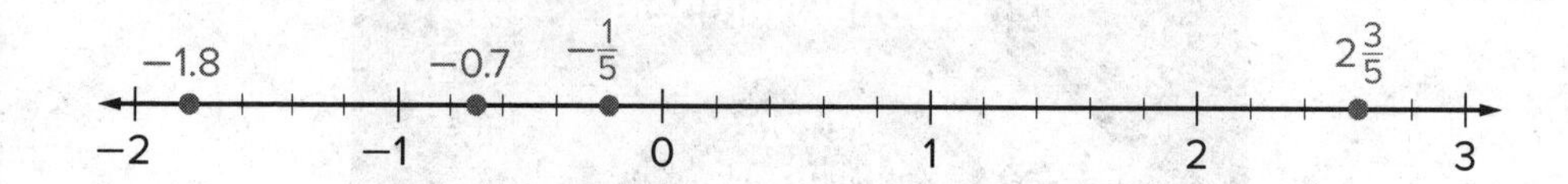

Check

Graph the set of rational numbers $\left\{-1\frac{7}{10}, 1.5, \frac{2}{5}, -0.6\right\}$ on the number line.

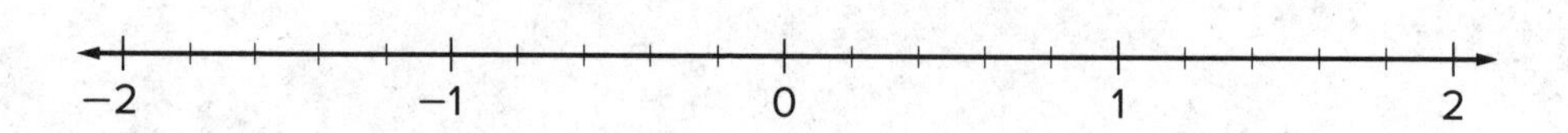

Go Online You can complete an Extra Example online.

Learn Absolute Value of Rational Numbers

The rational numbers 2.5 and −2.5 are each 2.5 units from 0, even though they are on opposite sides of 0. Numbers that are the same distance from zero on a number line have the same absolute value.

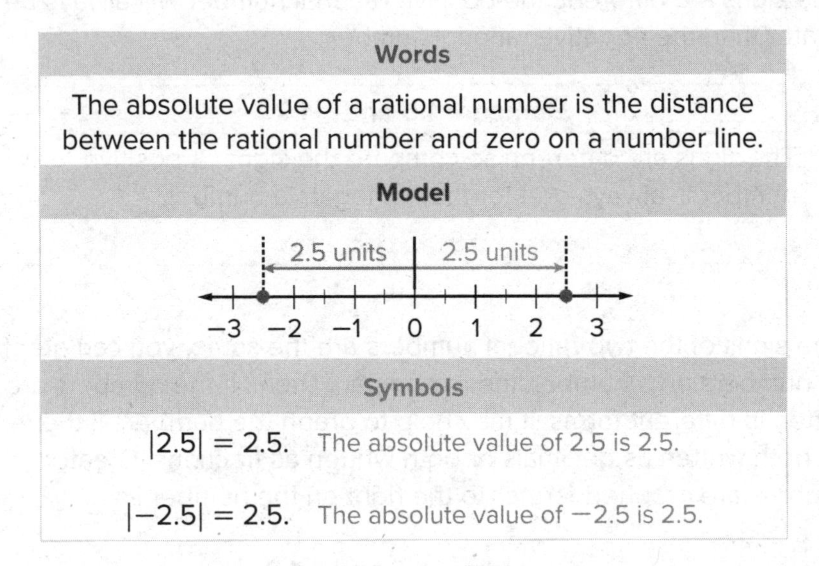

Words

The absolute value of a rational number is the distance between the rational number and zero on a number line.

Model

Symbols

$|2.5| = 2.5.$ The absolute value of 2.5 is 2.5.

$|-2.5| = 2.5.$ The absolute value of −2.5 is 2.5.

Talk About It!

Why is the absolute value of a number not the same as the opposite of a number?

Example 2 Find Absolute Value of Rational Numbers

The lowest point in a certain cave has an elevation of −53.4 meters. **If the cave entrance has an elevation of 0 meters, evaluate $|-53.4|$ to determine the number of meters a hiker would descend to reach the lowest point.**

Graph −53.4 on a number line.

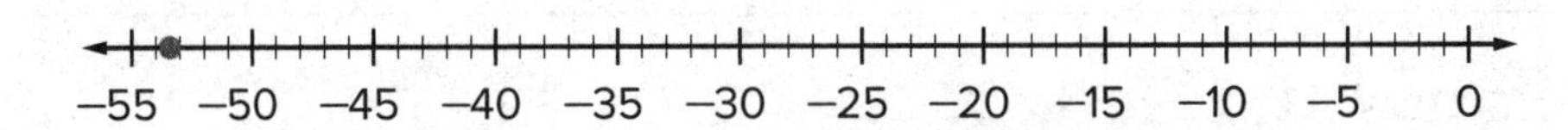

How many units from 0 is −53.4? ________

So, the hiker descended ________ meters.

Check

The Miller family is having an inground pool installed. The deepest point will be −9.75 feet below ground. If the ground has an elevation of 0 feet, evaluate $|-9.75|$ to determine the depth of the pool.

Go Online You can complete an Extra Example online.

Learn Compare Rational Numbers

To compare two rational numbers, you can compare the signs as well as the magnitude, or size of the numbers.

If the signs are different, the positive rational number will always be greater than the negative rational number.

Compare 1.5 and −1.2.
The signs are different, so compare the signs. A positive number is always greater than a negative number.
1.5 > −1.2

If the signs of the two rational numbers are the same, you can graph the numbers on a number line to compare them. If the numbers are written in different forms, it may help to graph the numbers if they are both written as decimals or both written as fractions. Greater numbers are graphed farther to the right on the number line.

Compare −1.5 and −1.2.
The signs are the same, so compare the numbers using a number line.
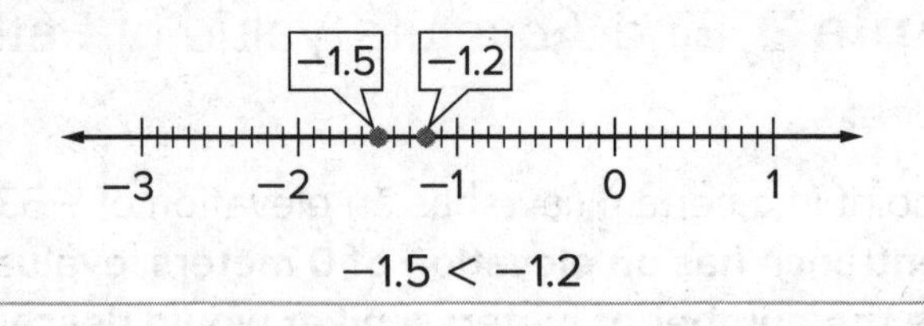
−1.5 < −1.2

Talk About It!

How can you use what you know about the signs of the rational numbers to quickly compare them?

Pause and Reflect

Are you ready to move on to the next Example? If yes, what have you learned that you think will help you? If no, what questions do you still have? How can you get those questions answered?

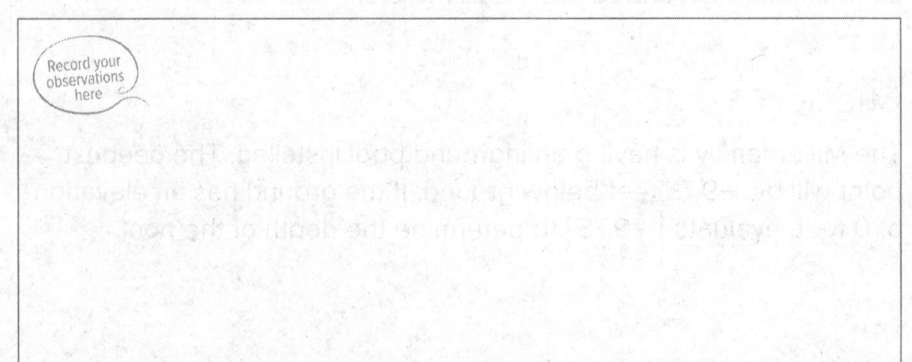

Example 3 Compare Rational Numbers

Compare -0.51 and $-\frac{12}{25}$.

Step 1 Write the fraction as a decimal.

$-\frac{12}{25} = \square$

Rewrite the fraction as a decimal so that the values are in the same form.

Step 2 Graph the values on the number line.

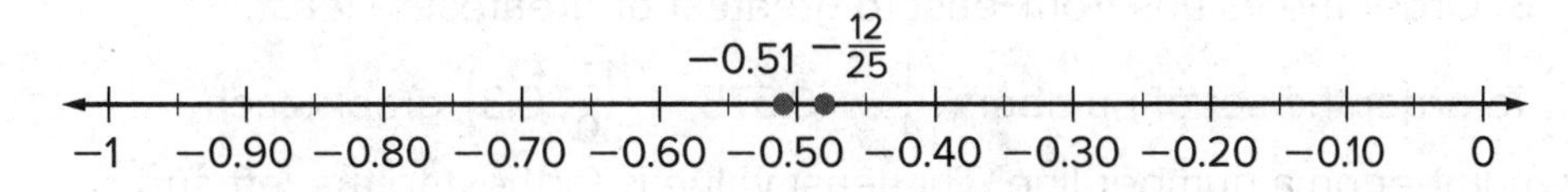

The number -0.51 is farther to the left on the number line.

So, $-0.51 < -\frac{12}{25}$.

Think About It!

How can you compare rational numbers when they are written in different forms?

Talk About It!

How can you compare the numbers without graphing them on a number line?

Check

Compare $-\frac{3}{8}$ and -0.413.

Go Online You can complete an Extra Example online.

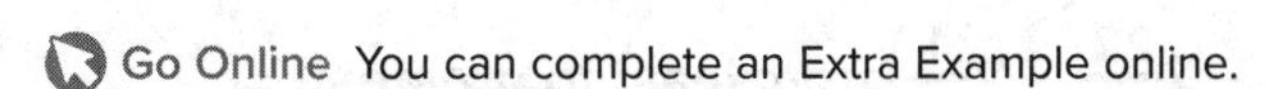

Pause and Reflect

Did you struggle with any of the concepts in this Example and Check? How do you feel when you struggle with math concepts? What steps can you take to understand those concepts?

Learn Order Rational Numbers

Talk About It!

How does place value help you order the set of numbers $\left\{\frac{1}{4}, -0.375, -\frac{17}{50}, 0.3\right\}$?

To order rational numbers, follow these steps:

1. Write each number in the same form. Since there may be different denominators in the fractions, it may be easier to write all of the numbers as decimals.
2. Use the signs of the numbers, place value, or a number line to compare the numbers.
3. Order the values from least to greatest or greatest to least.

To order the set of numbers $\left\{\frac{1}{4}, -0.375, -\frac{17}{50}, 0.3\right\}$, graph each number on a number line. The least value is farthest to the left and the greatest value is farthest to the right.

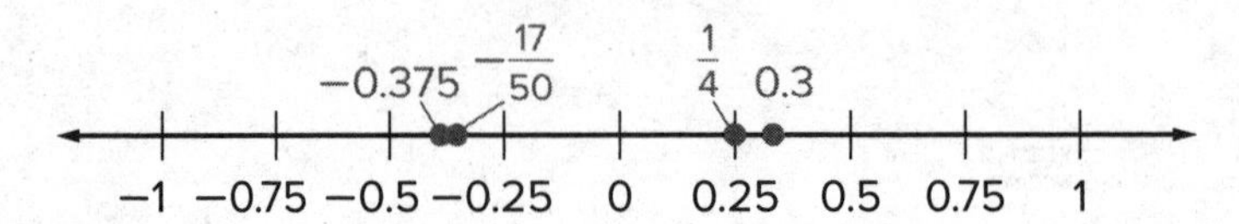

So, the set of numbers in order from least to greatest is

____________________ and from greatest to least is

____________________.

Think About It!

How can you order rational numbers when they are written in different forms?

Example 4 Order Sets of Rational Numbers

Talk About It!

How does a number line help you visualize the order of rational numbers?

Order the set $\left\{-2.46, -2\frac{22}{25}, -2\frac{1}{10}\right\}$ from least to greatest.

Step 1 Write the mixed numbers as decimals.

$-2.46 = -2.46$ $-2\frac{22}{25} = $ ☐ $-2\frac{1}{10} = $ ☐

Step 2 Graph the numbers on a number line.

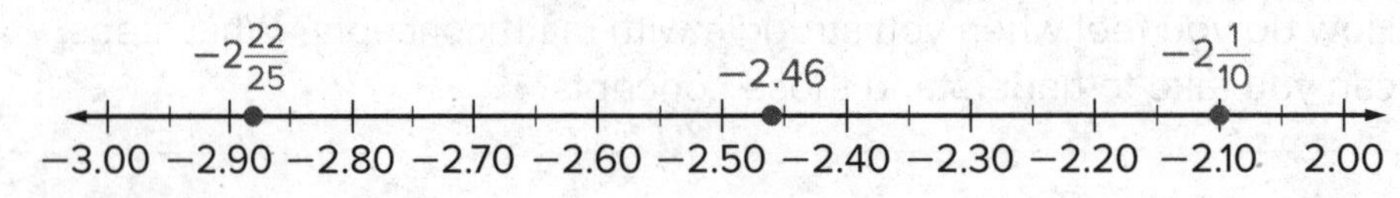

So, the set of numbers in order from least to greatest is $-2\frac{22}{25}$, -2.46, $-2\frac{1}{10}$.

Check

Order the set $\left\{2.12, -2.1, 2\frac{1}{10}, -2\frac{1}{5}\right\}$ from least to greatest.

Go Online You can complete an Extra Example online.

Apply Gardening

Mr. Plumb's agriculture class is growing pumpkins under different conditions. The table shows the change in weight for each student's pumpkin in relation to the weight of the pumpkin with the current class record. Which student's pumpkin(s) broke the record? Which student's pumpkin was closest to the record?

Student	Change
Ricky	$\frac{1}{5}$ lb
Debbie	−0.18 lb
Suni	$3\frac{1}{4}$ oz
Leonora	−3 oz

Go Online watch the animation.

1 What is the task?

Make sure you understand exactly what question to answer or problem to solve. You may want to read the problem three times. Discuss these questions with a partner.

First Time Describe the context of the problem, in your own words.
Second Time What mathematics do you see in the problem?
Third Time What are you wondering about?

2 How can you approach the task? What strategies can you use?

3 What is your solution?

Use your strategy to solve the problem.

4 How can you show your solution is reasonable?

Write About It! Write an argument that can be used to defend your solution.

Talk About It!
Why was it important to notice the units were different?

Check

The table shows the change in the actual amounts of rainfall, in inches, that a city received over four weeks and the average amount that it usually receives during those weeks. In which week was the rainfall closest to the average?

Week	Change (in.)
1	$\frac{1}{2}$
2	−1.6
3	0.3
4	$-1\frac{1}{2}$

Go Online You can complete an Extra Example online.

Foldables It's time to update your Foldable, located in the Module Review, based on what you learned in this lesson. If you haven't already assembled your Foldable, you can find the instructions on page FL1.

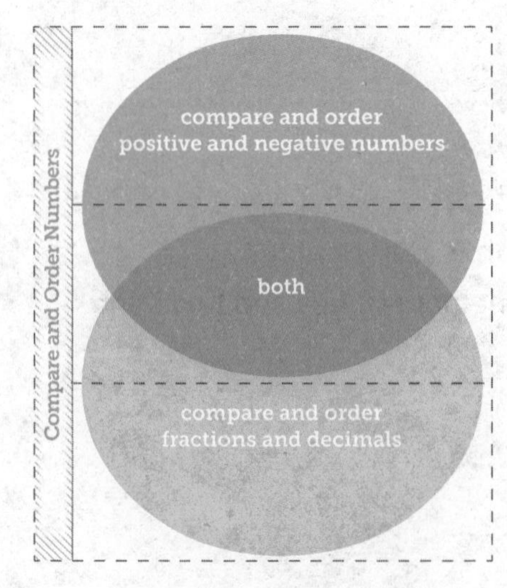

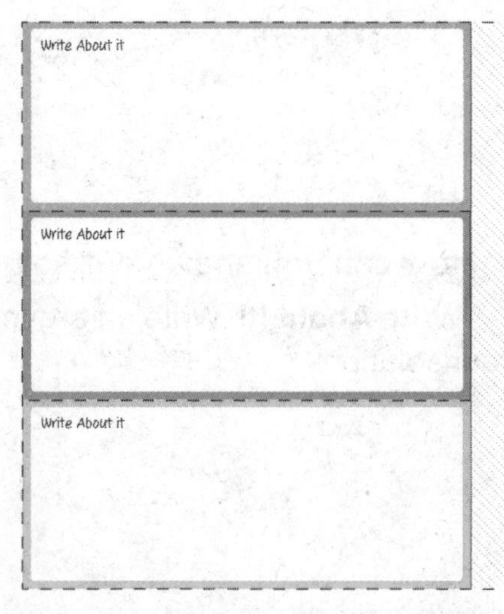

Name ______________________ Period __________ Date __________

Practice

Go Online You can complete your homework online.

Graph each set of rational numbers on a number line. (Example 1)

1. $\left\{-0.9, -2\frac{1}{2}, 0.25, -\frac{3}{4}\right\}$

2. $\left\{-\frac{1}{4}, -1.4, -1\frac{4}{5}, -0.15\right\}$

3. Mammoth Cave in Kentucky has a minimum elevation of −124.1 meters. Suppose a hiker traveled to the bottom of the cave. How many meters did the hiker travel? (Example 2)

4. A scuba diver was at a depth of $-80\frac{1}{2}$ feet. How many feet did the scuba diver travel if the diver traveled to the surface of the ocean? (Example 2)

Fill in the ◯ with <, >, or = to make a true statement. (Example 3)

5. -0.24 ◯ $-\frac{3}{16}$

6. $-\frac{5}{8}$ ◯ -0.76

7. $-4\frac{4}{25}$ ◯ -4.16

8. -5.52 ◯ $-5\frac{7}{15}$

Order each set of rational numbers from least to greatest. (Example 4)

9. $\left\{-4.25, -4\frac{7}{10}, -4\frac{3}{20}\right\}$

10. $\left\{-1.55, -1\frac{11}{100}, -1\frac{23}{25}\right\}$

11. The change in runners' goals and their actual times is shown in the table. Order the changes from least to greatest.

Runner	Change (min)
Sean	−3.2
Lacy	$1\frac{2}{5}$
Maura	1.43
Amos	$-2\frac{1}{5}$

Test Practice

12. **Table Item** Order the numbers from least to greatest.

−1.75, 2, 1.25, −2, 0

least				greatest

Apply

13. Saeng wants to run the 100-meter-dash in a certain number of seconds. The table shows the change in times from her goal and her actual times for 5 races. Between which two race numbers is Saeng's third race?

Race	Change in Time from Goal (s)
1	-1.2
2	$+1\frac{1}{10}$
3	$-1\frac{1}{4}$
4	-1.4
5	$+1\frac{1}{2}$

14. In science class, students are growing plants. The table shows the changes in the heights between the heights of some students' plants and the height of last year's tallest plant. Order the changes from least to greatest.

Student	Change
Ellen	$-2\frac{3}{4}$ in.
Juan	$\frac{1}{4}$ ft
Patty	3.1 in.
Sonny	$-\frac{1}{5}$ ft

15. Create Write about a real-world situation in which you compare two negative rational numbers. Then write an inequality comparing the two negative rational numbers.

16. MP **Justify Conclusions** A student said $-2\frac{1}{4}$ is less than -2.2 and $\left|-2\frac{1}{4}\right|$ is less than $|-2.2|$. Is the student correct? Justify your reasoning.

17. MP **Reason Inductively** Determine whether the following statement is *always, sometimes*, or *never true*. Justify your reasoning. Use an example.

If x and y are both less than 0 and x < y, then –x > –y.

The Coordinate Plane

I Can... recognize rational numbers and graph them in the coordinate plane.

Today's Standards
6.NS.C.6, 6.NS.C.6.B, 6.NS.C.6.C, 6.NS.C.8
MP1, MP2, MP3, MP5, MP6, MP7

What Vocabulary Will You Learn?
quadrants

Explore The Coordinate Plan

Online Activity You will use Web Sketchpad to explore the coordinate plane.

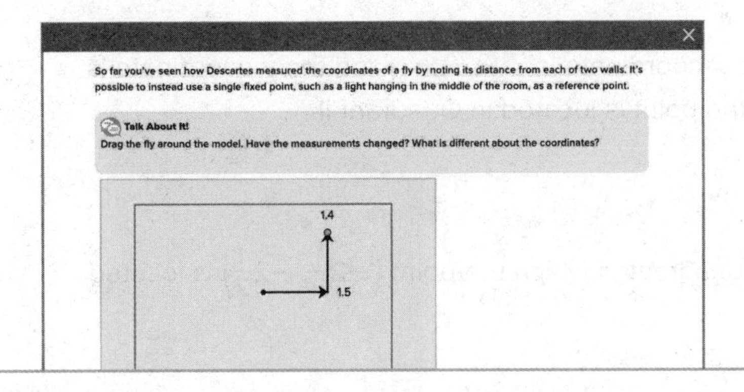

Learn The Coordinate Plane

The coordinate plane is formed by the intersection of two number lines, or axes, that meet at right angles at their zero points. The intersection of these number lines separates the coordinate plane into four **quadrants**: Quadrants I, II, III, and IV.

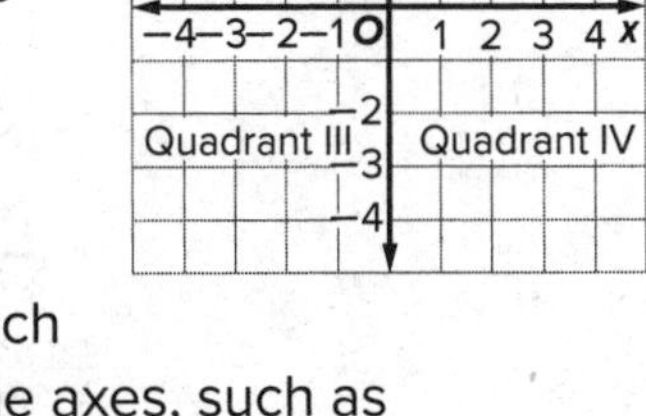

You can use the *x*-coordinates and *y*-coordinates to identify the quadrant in which a point is located. The axes and points on the axes, such as (−3, 0) and (0, 0.5), are not located in any of the quadrants.

Use what you know about the coordinate plane to complete the table.

Quadrant	*x*-coordinate	*y*-coordinate
I	positive	
II		positive
III	negative	
IV	positive	

Axis	*x*-coordinate	*y*-coordinate
x	positive	
y	0	
	negative	0
	0	negative

Talk About It!
How can you tell in which quadrant the point $\left(\frac{2}{3}, -7\right)$ lies?

Your Notes

Example 1 Identify the Quadrant

Identify the quadrant in which the point $\left(-\frac{3}{4}, 1\frac{1}{2}\right)$ is located.

You can use the signs of the *x*- and *y*-coordinates to identify the quadrant.

Because the *x*-coordinate is ________, and the *y*-coordinate is ________, the point is located in Quadrant II.

Check

Identify the quadrant in which the point $\left(-2\frac{1}{2}, -2\frac{1}{2}\right)$ is located.

Example 2 Identify the Axis

Identify the axis on which the point $\left(0, \frac{2}{5}\right)$ is located.

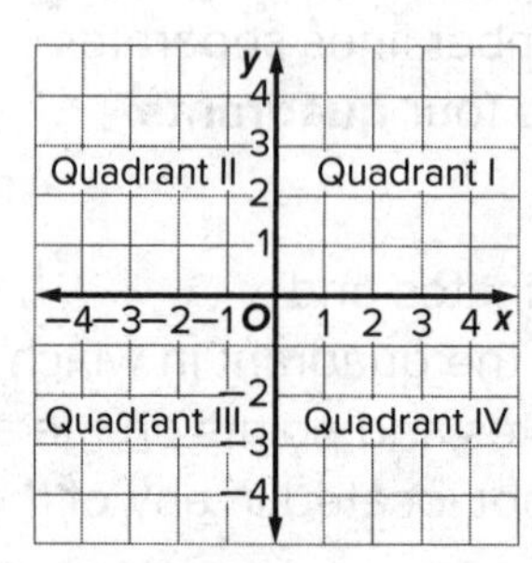

Look at which coordinate has the nonzero value.

The ________ -coordinate has the nonzero value.

So, the point lies on the *y*-axis.

Check

Identify the axis on which the point (0.25, 0) is located.

Go Online You can complete an Extra Example online.

Learn Identify Ordered Pairs

Go Online Watch the animation to learn how to identify ordered pairs of points graphed in the coordinate plane.

To identify an ordered pair graphed on the coordinate plane, start at the origin.

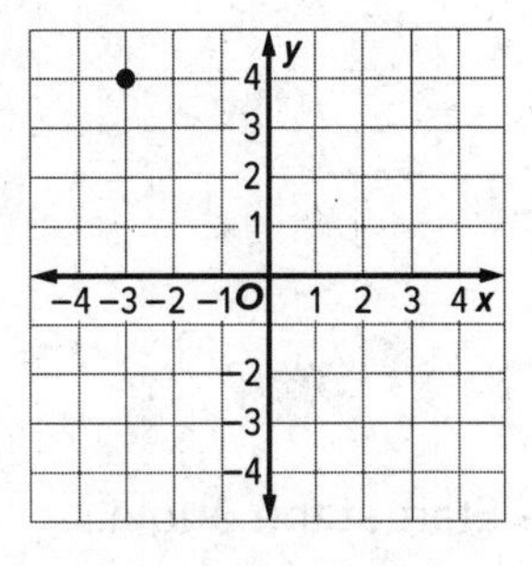

First, move horizontally along the *x*-axis, counting the units.

The *x*-coordinate of the point is −3.

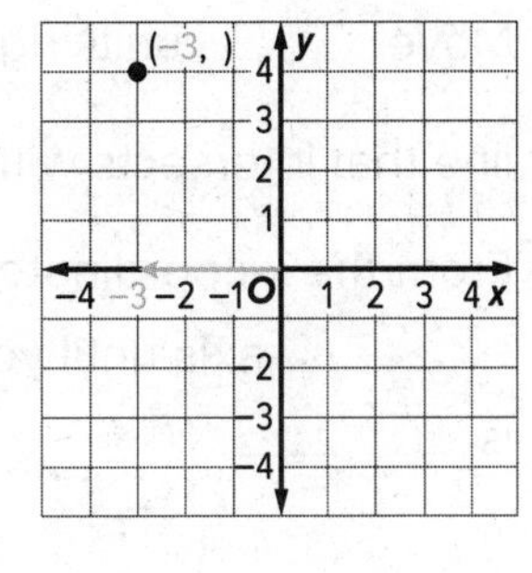

Next, move vertically toward the point, counting the units.

The *y*-coordinate of the point is 4.

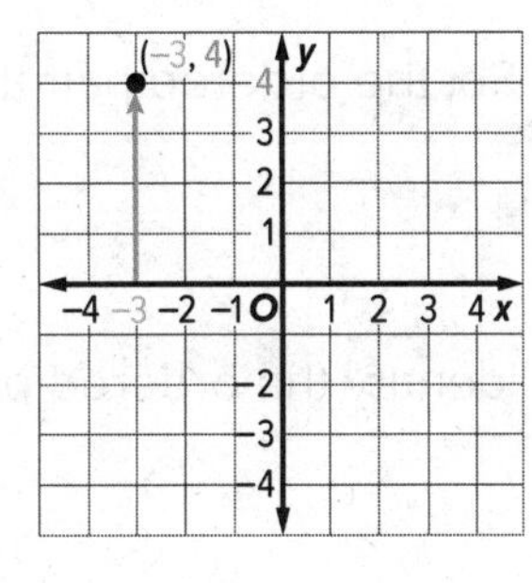

So, the ordered pair for the point is (______, ______).

Talk About It!

When identifying an ordered pair that represents a graphed point, why is it important to count the *horizontal* movement from the origin to that point first?

Pause and Reflect

Are you ready to move on to the next Example? If yes, what have you learned that you think will help you? If no, what questions do you still have? How can you get those questions answered?

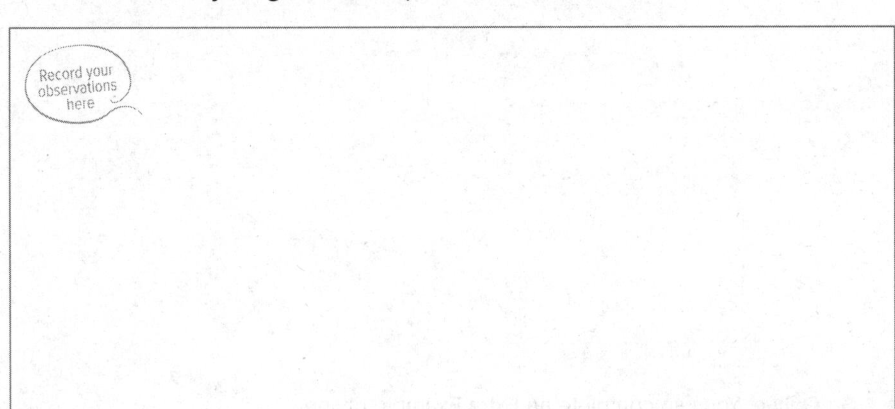

Example 3 Identify Ordered Pairs

Identify the ordered pair that names point *D*.

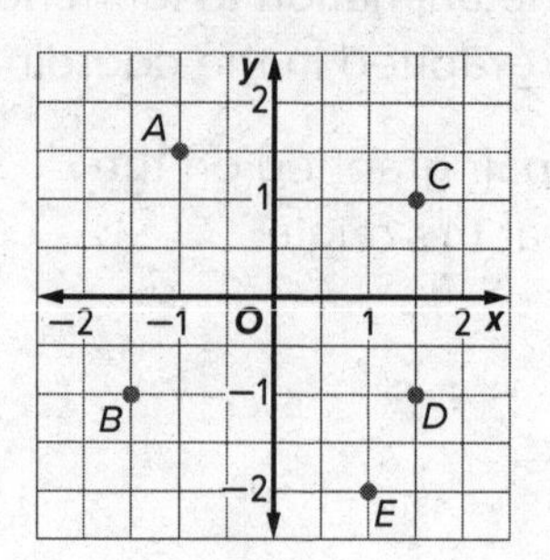

Start at the origin.

Move ______ units right on the ______-axis until you reach the vertical line that intersects with point *D*. The *x*-coordinate of point *D* is ______.

From the *x*-coordinate, $1\frac{1}{2}$, move ________ unit down on the ________-axis until you reach point *D*. The *y*-coordinate of point *D* is ________.

So, the ordered pair that names point *D* is $\left(1\frac{1}{2}, -1\right)$.

Check

Identify the ordered pair that names point *B*.

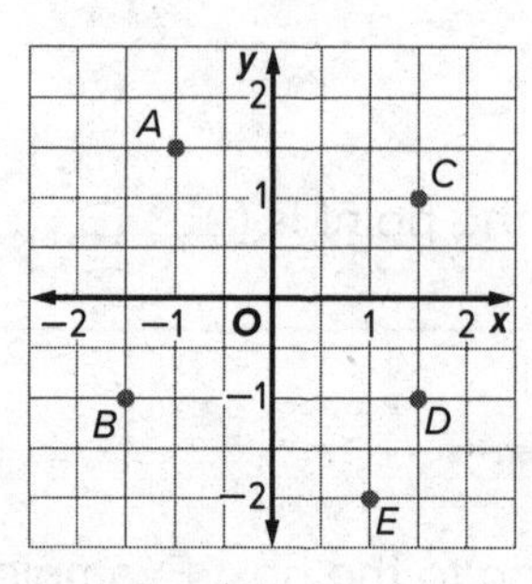

Go Online You can complete an Extra Example online.

Think About It!

In which quadrant does point *D* lie?

Talk About It!

Why is the ordered pair $\left(-1, 1\frac{1}{2}\right)$ incorrect for naming point *D*?

Learn Identify Points

Go Online Watch the animation to learn how to identify points graphed in the coordinate plane, given the ordered pair.

The animation explains that you can identify a point graphed on the coordinate plane using the *x*- and *y*-coordinates. The *x*-coordinate indicates how far left or right to move from the origin. The *y*-coordinate indicates how far up or down to move from the origin.

Identify the point graphed at (−2, 4).

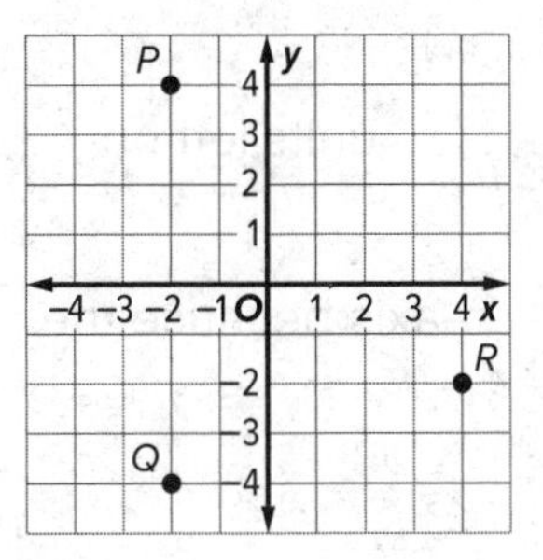

Because the *x*-coordinate is negative, move left two units on the *x*-axis.

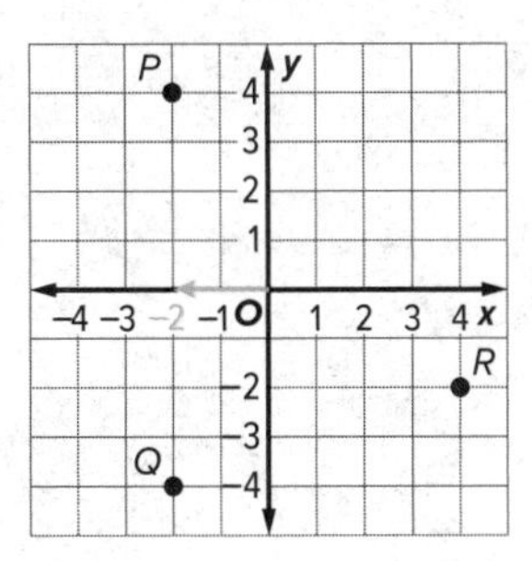

Because the *y*-coordinate is positive, move up four units.

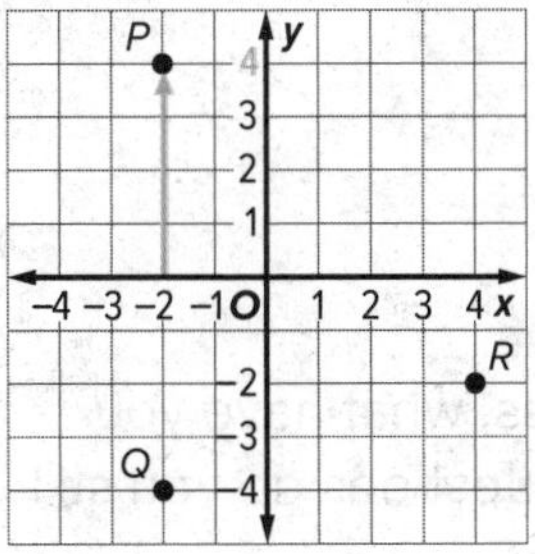

Point ________ is located at (−2, 4).

Talk About It!

How can you use what you know about the signs of the coordinates in each quadrant to quickly identify the point?

Example 4 Identify Points

Identify the point located at $\left(-2, \frac{1}{2}\right)$.

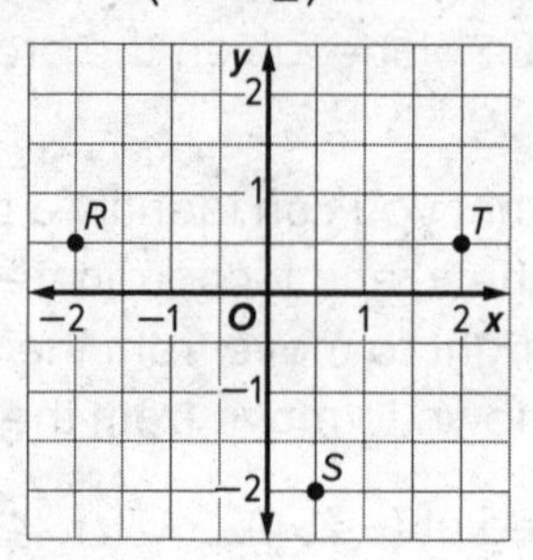

Start at the origin.

Because the *x*-coordinate is negative, move ________ units left on the ________ -axis.

From −2, move ________ unit up on the ________ -axis because the *y*-coordinate

is positive.

So, point *R* is located at $\left(-2, \frac{1}{2}\right)$.

Check

Identify the point located at $\left(\frac{1}{2}, -2\right)$.

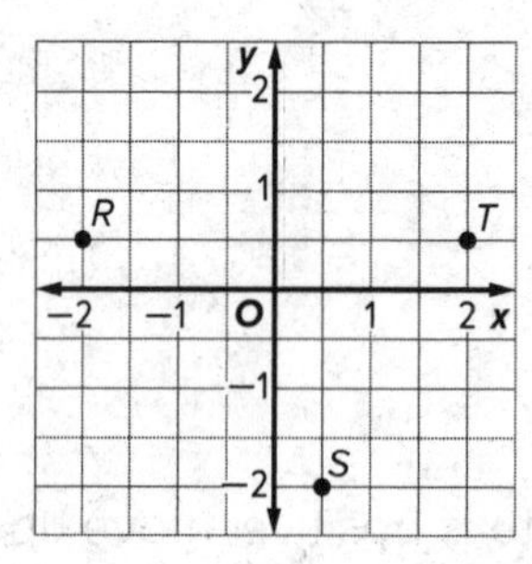

Go Online You can complete an Extra Example online.

Pause and Reflect

Are you ready to move on to the next Learn? If yes, what have you learned that you think will help you? If no, what questions do you still have? How can you get those questions answered.

Record your observations here

Learn Graph Ordered Pairs

To graph an ordered pair, place a dot at the point that corresponds to the coordinates.

Go Online Watch the animation to see how to graph ordered pairs.

The animation explains that you can graph a point on the coordinate plane using the *x*- and *y*-coordinates.

Graph *A*(−4, 3). The *x*-coordinate is −4. The *y*-coordinate is 3.

Because the *x*-coordinate is negative, move left four units on the *x*-axis from the origin.

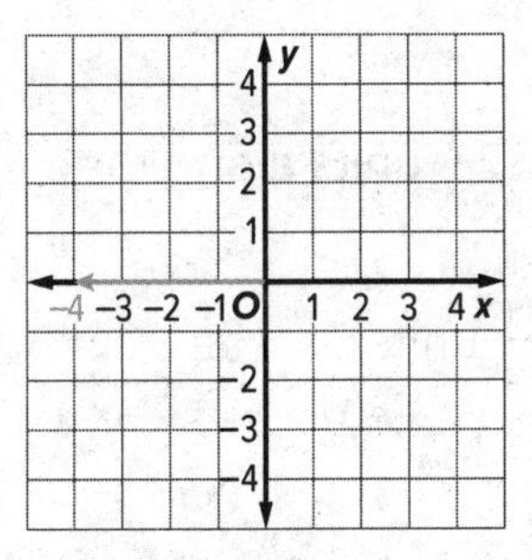

Because the *y*-coordinate is positive, move up three units.

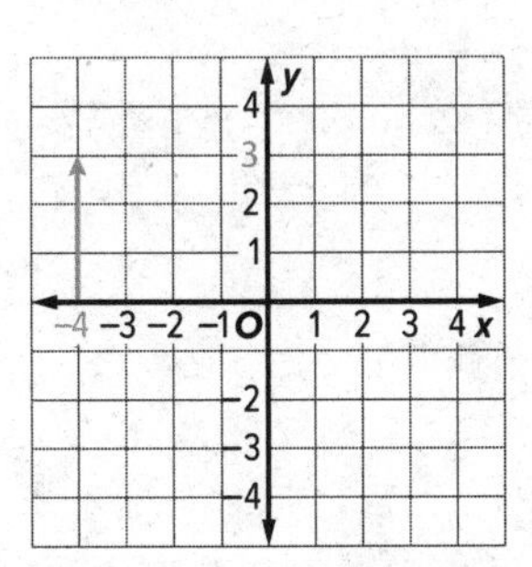

Graph point *A* by placing a dot at (−4, 3).

Example 5 Graph Ordered Pairs

Graph point $N\left(-2\frac{1}{2}, -3\frac{1}{2}\right)$.

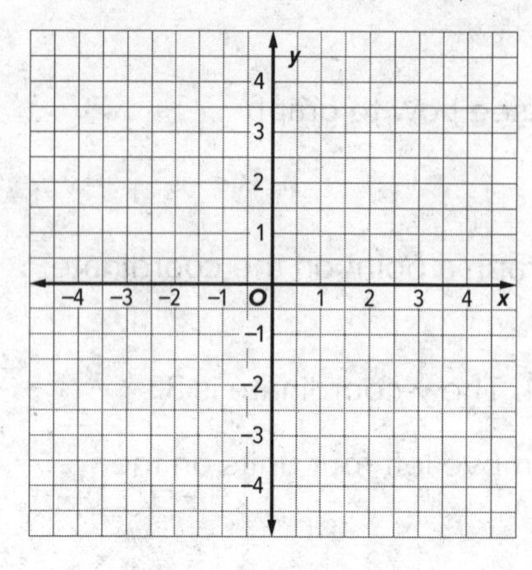

Start at the origin.

The x-coordinate is negative, so move ________ $2\frac{1}{2}$ units along the x-axis.

Next, since the y-coordinate is negative, move $3\frac{1}{2}$ units ________ . Place a dot at this location.

Check

Graph point $M(4.5, -1)$.

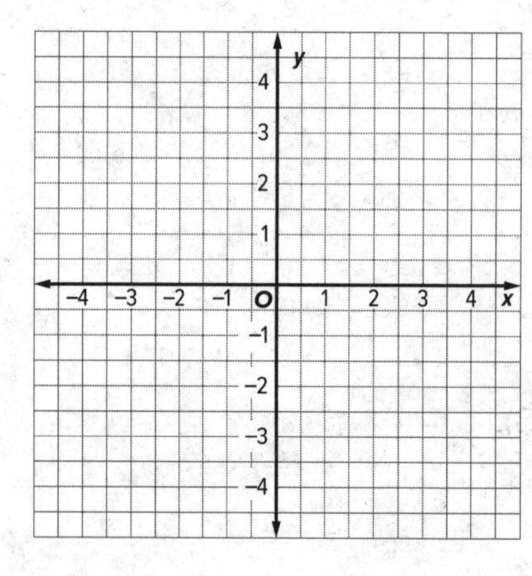

Go Online You can complete an Extra Example online.

Apply Maps

The table shows the locations for several different places around town. The grid shows a map of the town, and each square on the grid represents one city block. Ben needs to go to the dry cleaner, which is 3 blocks west and 5 blocks north of the library. Where on the grid should he go?

Place	Location
Bank	$\left(1\frac{1}{4}, -1\right)$
Grocery	$\left(-\frac{3}{4}, 0\right)$
Library	$\left(0, -\frac{3}{4}\right)$
Post Office	$\left(-1, 1\frac{1}{4}\right)$

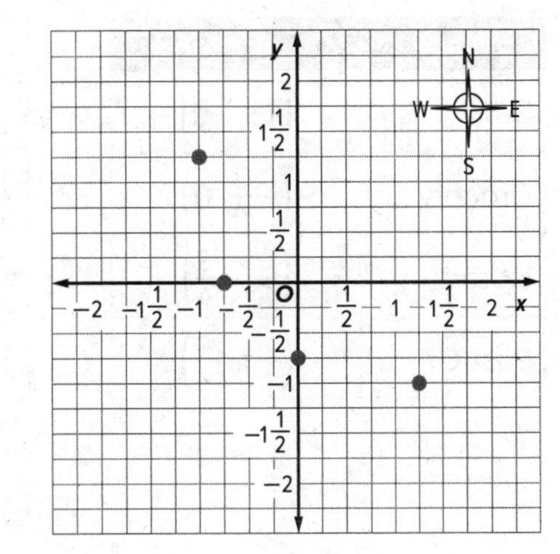

1 What is the task?

Make sure you understand exactly what question to answer or problem to solve. You may want to read the problem three times. Discuss these questions with a partner.

First Time Describe the context of the problem, in your own words.
Second Time What mathematics do you see in the problem?
Third Time What are you wondering about?

2 How can you approach the task? What strategies can you use?

3 What is your solution?

Use your strategy to solve the problem.

4 How can you show your solution is reasonable?

Write About It! Write an argument that can be used to defend your solution.

Talk About It!
Why was the location of the library important?

Check

The table shows the locations for several different places around town. The grid shows a map of the town, and each square on the grid represents one city block. Yamenah needs to go to the farmer's market, which is 6 blocks east and 2 blocks south of the post office. Where on the grid should she go?

Place	Location
Bank	$\left(1\frac{1}{4}, -1\right)$
Grocery	$\left(-\frac{3}{4}, 0\right)$
Library	$\left(0, -\frac{3}{4}\right)$
Post Office	$\left(-1, 1\frac{1}{4}\right)$

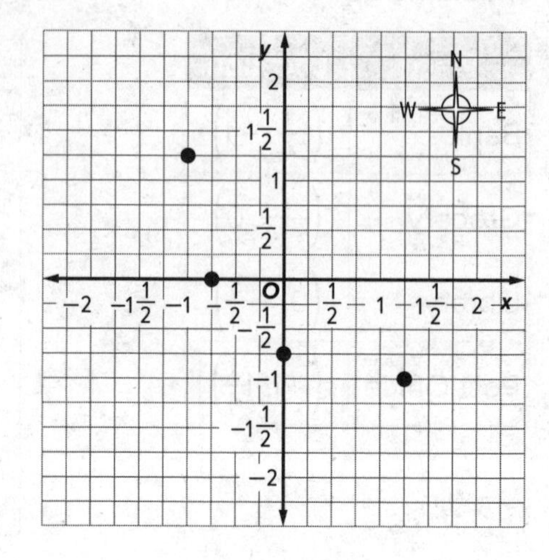

Go Online You can complete an Extra Example online.

Pause and Reflect

Create a graphic organizer that will help you study the concepts you learned today in class.

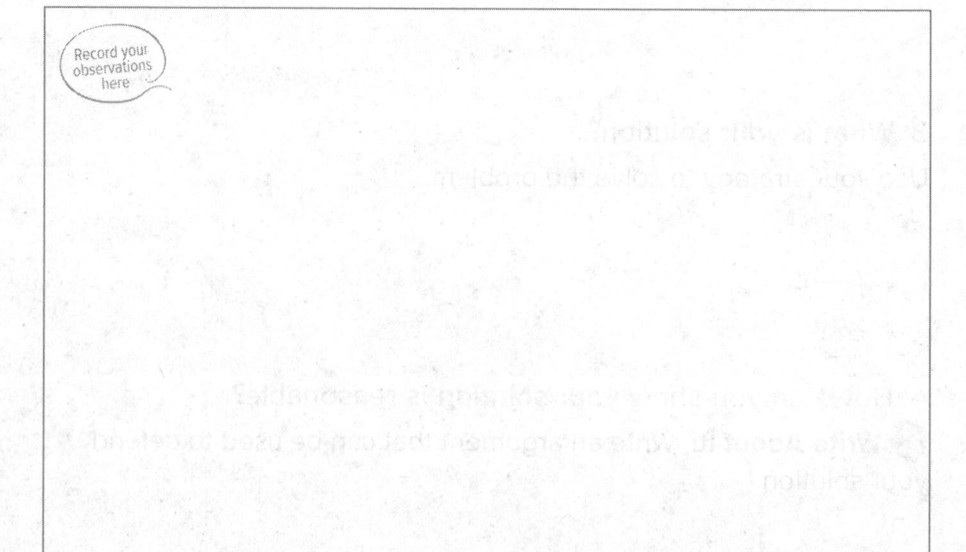

Name ______________________ Period ____________ Date ____________

Practice

Go Online You can complete your homework online.

Identify the quadrant in which each point is located. **(Example 1)**

1. $\left(-1\frac{1}{2}, -2\frac{1}{4}\right)$ ____________

2. $\left(5\frac{3}{4}, -6\frac{1}{5}\right)$ ____________

3. $\left(\frac{4}{5}, 3\frac{3}{4}\right)$ ____________

4. $\left(-3\frac{1}{2}, 2\frac{4}{5}\right)$ ____________

5. Identify the axis on which the point $\left(-\frac{2}{3}, 0\right)$ is located. **(Example 2)**

____

6. Identify the axis on which the point $\left(0, 6\frac{3}{5}\right)$ is located. **(Example 2)**

____

Use the coordinate plane. Identify the ordered pair that names each point. **(Example 3)**

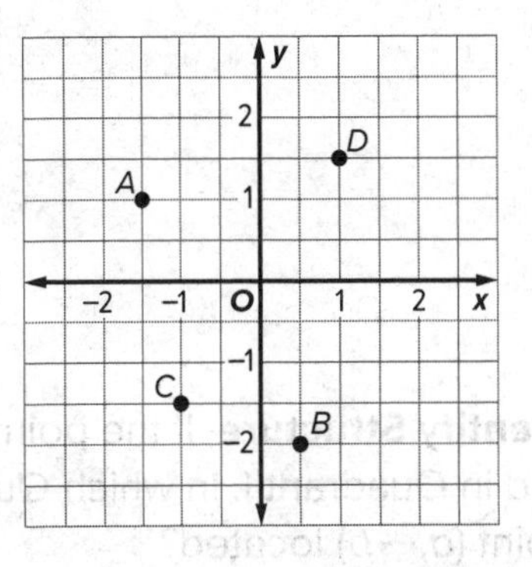

7. *A* ____________

8. *B* ____________

9. *C* ____________

Use the coordinate plane. Identify the point for each ordered pair. **(Example 4)**

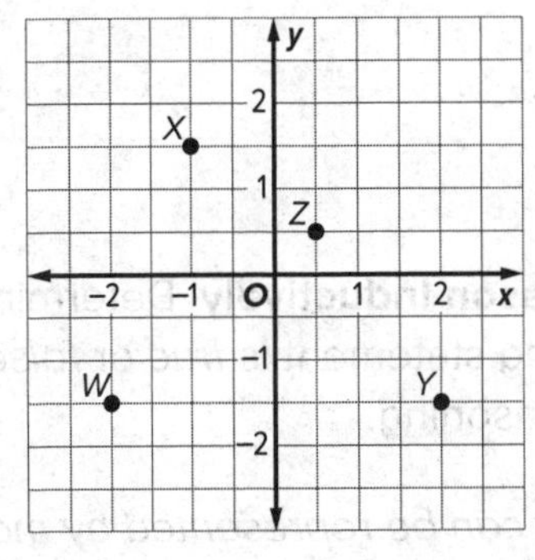

10. $\left(\frac{1}{2}, \frac{1}{2}\right)$ ____________

11. $\left(-1, 1\frac{1}{2}\right)$ ____________

12. $\left(-2, -1\frac{1}{2}\right)$ ____________

13. Graph point $A\left(\frac{1}{2}, 1\right)$. **(Example 5)**

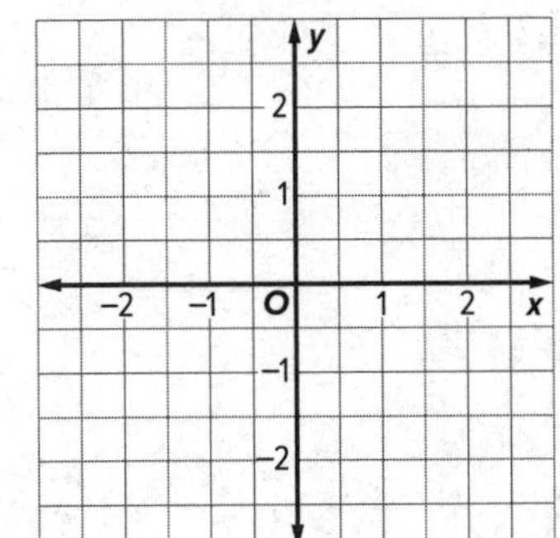

Test Practice

14. Grid Graph point $X\left(-1\frac{1}{2}, 2\right)$.

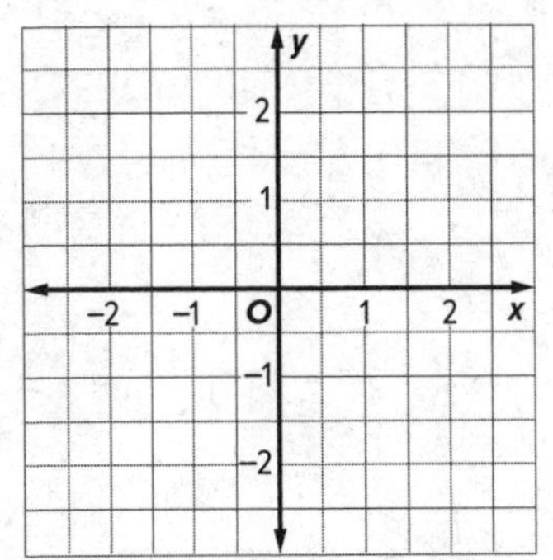

Apply

15. The table shows the locations for several different places around a small city. The grid shows a map of the city, and each square on the grid represents one city block. Shannon needs to go to the library that is 2 blocks east and 3 blocks south of the bakery. Where on the grid should she go?

Place	Location
Bakery	$\left(-\frac{3}{4}, -\frac{1}{2}\right)$
Courthouse	$\left(0, \frac{1}{2}\right)$
Restaurant	(1, −1)
Town Hall	$\left(-1\frac{1}{4}, \frac{3}{4}\right)$

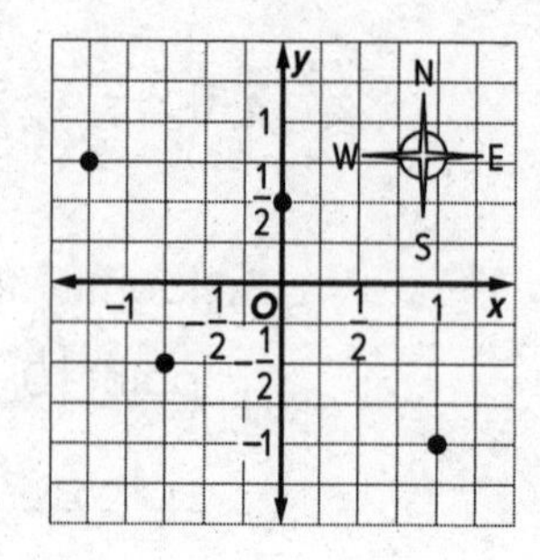

16. MP **Identify Structure** If the point (a, b) is located in Quadrant I, in which Quadrant is the point $(a, -b)$ located?

17. MP **Identify Structure** If the point $(-m, n)$ is located in Quadrant I, what must be true about the value of m? the value of n?

18. MP **Reason Inductively** Determine if the following statement is *true* or *false*. Explain your reasoning.

A point can be represented by more than one ordered pair.

19. MP **Find the Error** A student stated that if the point $(-a, b)$ is located in Quadrant I, then the point (a, b) is located in Quadrant IV. Find the student's mistake and correct it.

Graph Reflections of Points

I Can... recognize that the coordinates of points reflected across either axis differ only by the sign of one of the coordinates.

Explore Reflect a Point

Online Activity You will use Web Sketchpad to explore reflections of points.

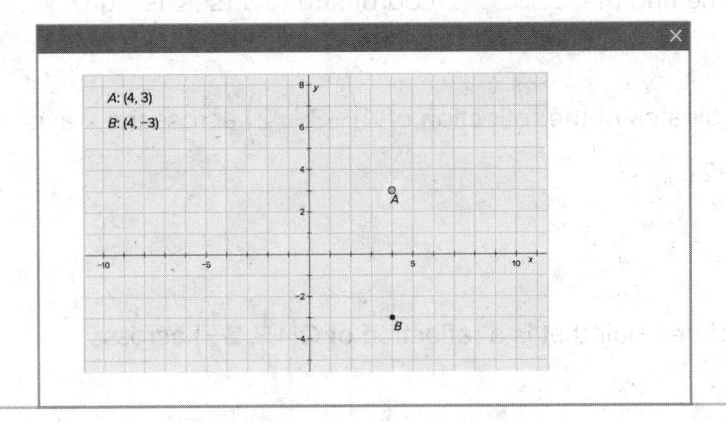

Today's Standards
6.NS.C.6, 6.NS.C.6.B, 6.NS.C.6.C, 6.NS.C.8, *Also addresses* *6.NS.C.6.A*
MP1, MP2, MP3, MP4, MP5, MP6, MP7, MP8

What Vocabulary Will You Learn?
reflection

Talk About It!
What do you notice about the *x*- and *y*-coordinates of points *A* and *A'*?

Learn Reflections of Points

The number line shows that −4 and 4 are opposites. They are the same distance from 0 in opposite directions.

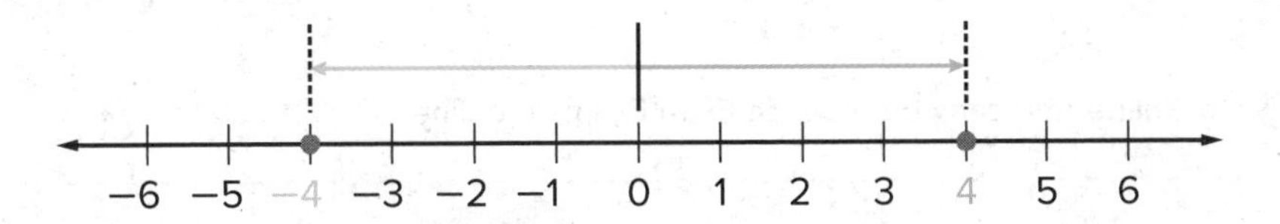

In a coordinate plane, the points (−4, 0) and (4, 0) are the same distance from the origin in opposite directions. These points are reflections across the *y*-axis. Notice that the *x*-coordinates are opposite integers and the *y*-coordinates are the same.

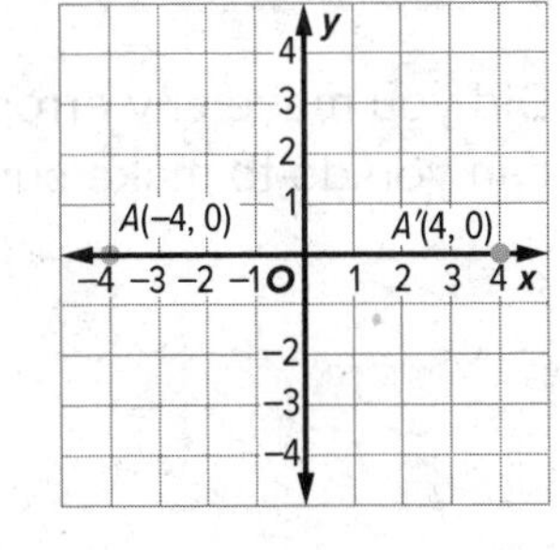

A **reflection** is the mirror image produced by flipping a figure across a line.

In the coordinate plane, when you reflect a point across a line, you name the reflected point using prime notation. In the figure, the reflection of *A*(−4, 0) across the *y*-axis is *A'*(4, 0).

Talk About It!
You can also reflect a point across the *x*-axis. Point *P* is graphed at (3, 2). How can you find the coordinates of *P'* after a reflection across the *x*-axis?

Your Notes

Think About It!
In what quadrant is point A located? In what quadrant will the reflection of point A across the x-axis be located?

Talk About It!
How do you know, without graphing, that the point $A'\left(-3\frac{1}{2}, -2\right)$ is the reflection of the point $A\left(-3\frac{1}{2}, 2\right)$ across the x-axis?

Example 1 Identify Reflections of Points Across the x-axis

Write the ordered pair that is a reflection of $A\left(-3\frac{1}{2}, 2\right)$ across the x-axis.

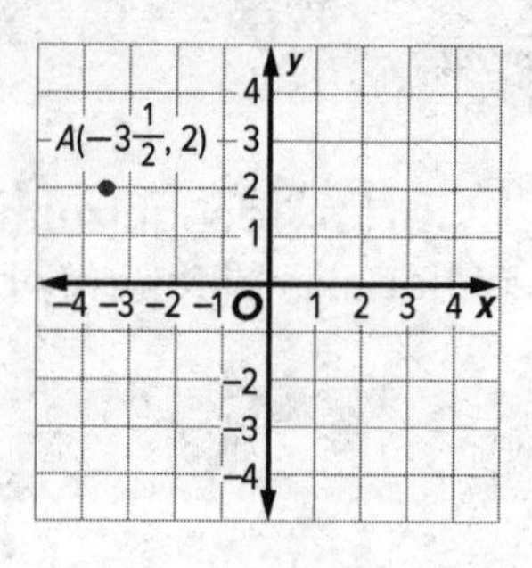

Find the point on the coordinate plane that is the same distance away from the x-axis as the original point. Graph the point on the coordinate plane above and label it.

When a point is reflected across the x-axis, the ________-coordinate stays the same and the ________-coordinate reverses its sign.

So, the coordinates of the reflection of $A\left(-3\frac{1}{2}, 2\right)$ across the x-axis are $\left(-3\frac{1}{2}, -2\right)$.

Check

Write the ordered pair that is a reflection of $Q\left(1\frac{1}{2}, 2\frac{1}{4}\right)$ across the x-axis.

Go Online You can complete an Extra Example online.

Pause and Reflect

Did you make any errors when completing the Check exercise? What can you do to make sure you don't repeat that error in the future?

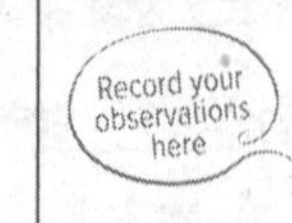

Example 2 Identify Reflections of Points Across the y-axis

Kendall is building a square fence. She places fence posts at the locations indicated on the grid.

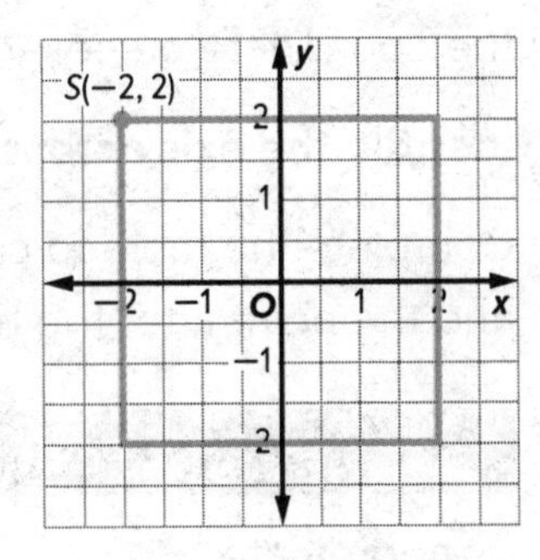

What is the location of the post that reflects $S(-2, 2)$ across the y-axis?

Find the point on the grid that is the same distance away from the y-axis as the original point. Graph the point and label it.

When a point is reflected across the y-axis, the ________-coordinate stays the same and the ________-coordinate reverses its sign.

So, the coordinates of the reflection of $S(-2, 2)$ across the y-axis are (2, 2).

Check

Rico is building a garden fence in the shape of a square. He placed a corner post of the fence at (10.2, −5.3). What is the location of the corner that reflects that corner post across the y-axis?

Go Online You can complete an Extra Example online.

Pause and Reflect

Did you struggle with any of the concepts in this Example? How do you feel when you struggle with math concepts? What steps can you take to understand those concepts?

Example 3 Identify the Axis of Reflection

The point $A'\left(-2\frac{3}{4}, -4\right)$ is the result of reflecting $A\left(2\frac{3}{4}, -4\right)$ in the coordinate plane.

Identify the axis across which the point was reflected.

Complete the table to compare the coordinates of the original point and the point after the reflection.

	Point	Reflected Point
x-coordinate		
y-coordinate		

The *x*-coordinates are __________ and the *y*-coordinates are __________.

So, point *A* was reflected across the *y*-axis.

Check

The point $M'\left(2\frac{1}{3}, -1\right)$ is the result of reflecting $M\left(-2\frac{1}{3}, -1\right)$ in the coordinate plane. Identify the axis across which the point was reflected.

Go Online You can complete an Extra Example online.

Pause and Reflect

Are you ready to move on? If yes, what have you learned in this lesson that you think will help you? If no, what questions do you still have? How can you get those questions answered?

Apply Geography

Samantha drew a map of the park in her neighborhood. She graphed the point $P(-3.5, -3.5)$ for the playground. The fountain is located at P', a reflection of P across the y-axis. The picnic tables are located at P'', a reflection of P' across the x-axis. Identify the ordered pair that describes the location of the picnic tables.

Go Online watch the animation.

1 What is the task?

Make sure you understand exactly what question to answer or problem to solve. You may want to read the problem three times. Discuss these questions with a partner.

First Time Describe the context of the problem, in your own words.
Second Time What mathematics do you see in the problem?
Third Time What are you wondering about?

2 How can you approach the task? What strategies can you use?

3 What is your solution?

Use your strategy to solve the problem.

4 How can you show your solution is reasonable?

Write About It! Write an argument that can be used to defend your solution.

Talk About It!
Where would the picnic tables be located if the playground was located at $(-1, -2)$?

Check

Michele drew a map of the route she walks every day after school. She starts at the front entrance of the school, which she graphed at point $S(-3.5, -2.5)$. She walks to the bird feeder, located at S', a reflection of S across the x-axis. Then she walks to where her mother picks her up, at S'', a reflection across the y-axis. Identify the ordered pair that describes the location where her mother picks her up.

Go Online You can complete an Extra Example online.

Pause and Reflect

What was your most positive experience with math in this module? Why was it positive?

Name ______________________ Period __________ Date __________

Practice

Go Online You can complete your homework online.

Write the ordered pair that is a reflection of each point across the *x*-axis. (Example 1)

1. $A\left(-2\frac{3}{4}, 1\right)$

2. $B\left(1\frac{1}{4}, -\frac{1}{2}\right)$

3. $C\left(-4, -2\frac{1}{2}\right)$

4. $D\left(\frac{3}{4}, 3\right)$

5. Aika is building a square garden. She places a garden post at (3.5, 3.5). What is the location of the corner that reflects (3.5, 3.5) across the *y*-axis? (Example 2)

6. A farmer is installing a chicken pen in the shape of a square. He placed a corner of the enclosure at (−5.25, −5.25). What is the location of the corner that reflects (−5.25, −5.25) across the *y*-axis? (Example 2)

7. The point $C'(-4, -2)$ is the result of reflecting $C(4, -2)$ in the coordinate plane. Identify the axis across which the point was reflected. (Example 3)

8. The point $B'\left(-5\frac{1}{4}, -3\frac{1}{2}\right)$ is the result of reflecting $B\left(-5\frac{1}{4}, 3\frac{1}{2}\right)$ in the coordinate plane. Identify the axis across which the point was reflected. (Example 3)

9. Graph point $Z(-4, -2.5)$ on the coordinate plane. Then graph its reflection across the *y*-axis.

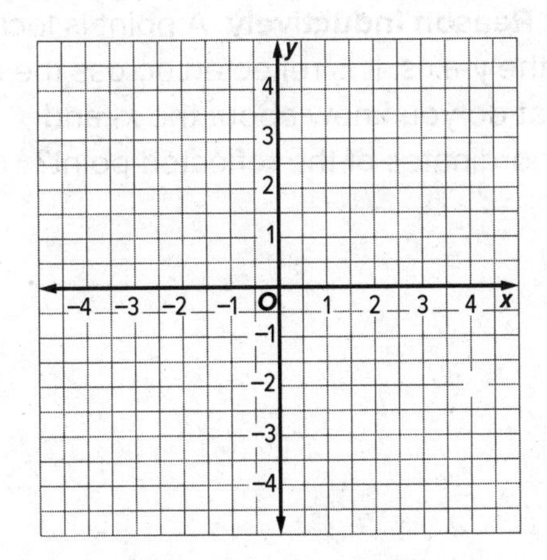

Test Practice

10. Multiple Choice Which ordered pair represents a reflection of point $Y\left(1\frac{3}{4}, -4\right)$ across the *x*-axis?

Ⓐ $\left(-4, 1\frac{3}{4}\right)$

Ⓑ $\left(1\frac{3}{4}, -4\right)$

Ⓒ $\left(1\frac{3}{4}, 4\right)$

Ⓓ $\left(-1\frac{3}{4}, 4\right)$

Apply

11. Trey drew a map of the summer camp he is staying at this summer. He graphed the point $D(-4.5, 4.5)$ for the dining hall. The flag pole is located at D', a reflection of D across the y-axis. The campfire is located at D'', a reflection of D' across the x-axis. Identify the ordered pair that describes the location of the campfire.

12. Liv drew a map of her favorite park. She graphed the point $S\left(2\frac{1}{2}, -2\right)$ for the swings. The picnic tables are located at S', a reflection of S across the x-axis. The lake is located at S'', a reflection of S' across the y-axis. Identify the ordered pair that describes the location of the lake.

13. MP **Find the Error** A student was finding the ordered pair for point $Y(1.5, -2)$ after its reflection across the x-axis. Find the student's mistake and correct it.

$Y(1.5, -2) \rightarrow Y'(-1.5, -2)$

14. MP **Persevere with Problems** Determine whether the statement is always, sometimes, or never true. Justify your response.

When a point is reflected across the x-axis, the new point has a negative y-coordinate.

15. Identify the coordinates of a point located in Quadrant III. Reflect the point across the y-axis. Then give the coordinates of the reflected point.

16. MP **Reason Inductively** A point is located on the y-axis. It is reflected across the x-axis. What do you know about the x- and y-coordinates of the reflected point?

Absolute Value and Distance

I Can... solve real-world problems by graphing points in a coordinate plane and then finding the distances between points using absolute values.

Today's Standards
6.NS.C.8,
Also addresses
6.NS.C.6, 6.NS.C.7.C
MP1, MP2, MP3, MP4, MP5, MP6, MP7

Explore Distance on the Coordinate Plane

Online Activity You will use Web Sketchpad to explore distance on the coordinate plane.

Two points are graphed in the coordinate plane.

The points are graphed in [ ] quadrant(s).

Reset Check Answer

What is the distance in units? Press the *Show Distance* button to check your answer.

Talk About It!
Without using a graph, how could you find the distance using only the coordinates?

Show Distance

E: (3.50, 1.50)

Learn Find Horizontal Distance

You can find the horizontal distance between two points with the same *y*-coordinate on the coordinate plane using the absolute values of the *x*-coordinates.

Go Online Watch the animation to learn how to find horizontal distance in the coordinate plane.

When two points are in the same quadrant and they have the same *y*-coordinate, subtract the absolute values of the *x*-coordinates to find the distance between the two points.

Consider the points (−5, −4) and (−1, −4). They have the same *y*-coordinates, so find the absolute value of each *x*-coordinate.

$|-1| =$ ☐ $|-5| =$ ☐

Subtract the absolute values.

$5 - 1 =$ ☐

The distance between the two points is 4 units.

(−5, −4) (−1, −4)

(continued on next page)

Talk About It!
If both points are in Quadrant III, will the distance be a negative number? Explain why or why not.

Your Notes

When two points are in different quadrants and they have the same y-coordinate, add the absolute values of the x-coordinates to find the distance between the two points.

Consider the points $(-4, 2)$ and $(1, 2)$. They have the same y-coordinates, so find the absolute value of each x-coordinate.

$|-4| =$ ☐ $\quad |1| =$ ☐

Add the absolute values. $4 + 1 =$ ☐

The distance between the two points is 5 units.

Example 1 Find Horizontal Distance in the Same Quadrant

Find the horizontal distance between the two points.

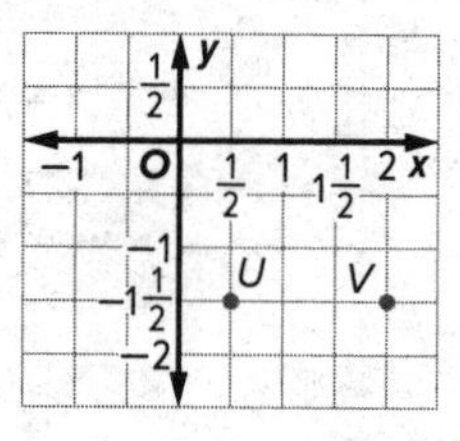

To find the horizontal distance between the two points, consider the scale on each axis. The scale of the axes is in $\frac{1}{2}$-unit increments.

Identify the ordered pair for each point.

U: (☐) $\quad V$: (☐)

Since the y-coordinates are the same, find the absolute value of each x-coordinate.

U: $\left|\frac{1}{2}\right| =$ ☐ $\quad V$: $|2| =$ ☐

Because the points are in the same quadrant, subtract the absolute values of the x-coordinates to find the distance between the points.

$2 - \frac{1}{2} = 1\frac{1}{2}$

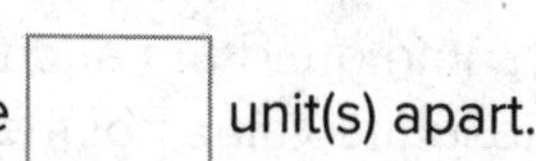

So, points U and V are ☐ unit(s) apart.

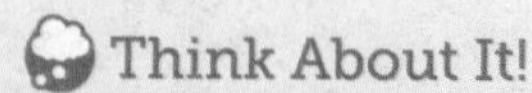
Think About It!
Are the x-coordinates the same or different? Are the y-coordinates the same or different?

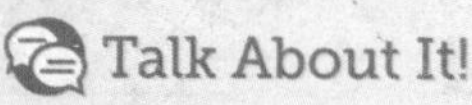

Talk About It!
How can you check your solution? Explain the process you would use.

Check

Find the horizontal distance between the two points.

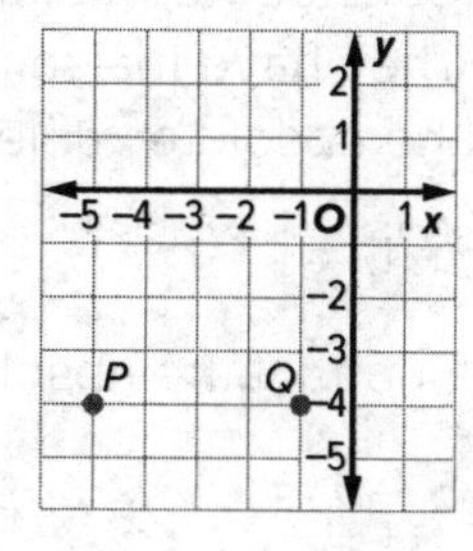

Go Online You can complete an Extra Example online.

Example 2 Find Horizontal Distance in Different Quadrants

Find the horizontal distance between the two points.

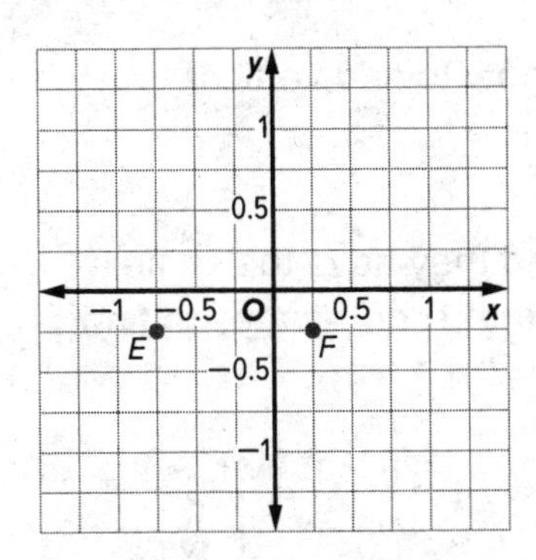

To find the horizontal distance between the two points, consider the scale on each axis. The scale of the axes are in 0.25-unit increments.

Identify the coordinates for each point.

E: ( ) F: ( )

Since the y-coordinates are the same. Find the absolute value of each x-coordinate.

E: $|-0.75| =$ ☐ F: $|0.25| =$ ☐

Because the points are in different quadrants, add the absolute values of the x-coordinates to find the distance between points.

$0.75 + 0.25 = 1$

So, points U and V are ☐ unit(s) apart.

Check

Find the horizontal distance between the two points.

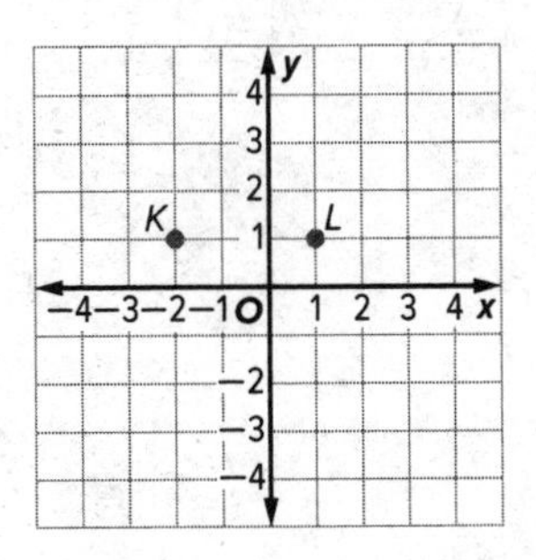

Show your work here

Go Online You can complete an Extra Example online.

Think About It!

Are the points in the same quadrant? How will that affect how you find the distance?

Talk About It!

Use the graph to explain why the absolute values of the x-coordinates are added when the points are in different quadrants.

Learn Find Vertical Distance

You can find vertical distance between two points on the coordinate plane with the same *x*-coordinates.

Go Online Watch the animation to learn how to find vertical distance on the coordinate plane.

When two points are in the same quadrant and they have the same *x*-coordinate, subtract the absolute values of the *y*-coordinates to find the distance between the two points.

Consider the points (3, −1) and (3, −5). They have the same *x*-coordinates, so find the absolute value of each *y*-coordinate.

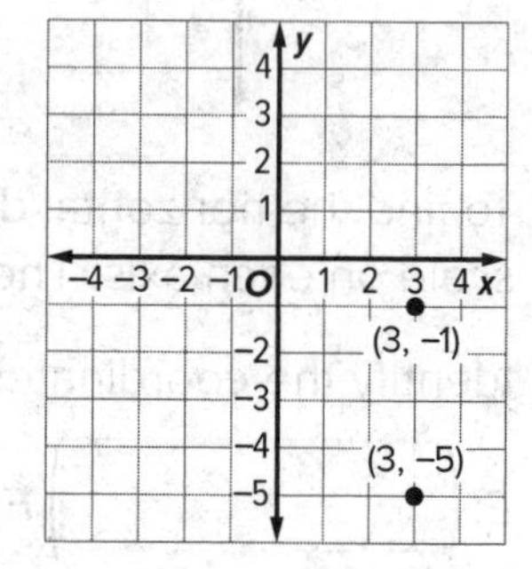

$|-1| = \square$ $\quad |-5| = \square$

Subtract the absolute values.

$5 - 1 = \square$

The distance between the two points is 4 units.

Talk About It!

How can you find the distance between two points with the same x-coordinates, but different y-coordinates, if you are only given the coordinates, and not the graph?

When two points are in different quadrants and they have the same *x*-coordinate, add the absolute values of the *y*-coordinates to find the distance between the two points.

Consider the points (2, 1) and (2, −4). They have the same *x*-coordinates, so find the absolute value of each *y*-coordinate.

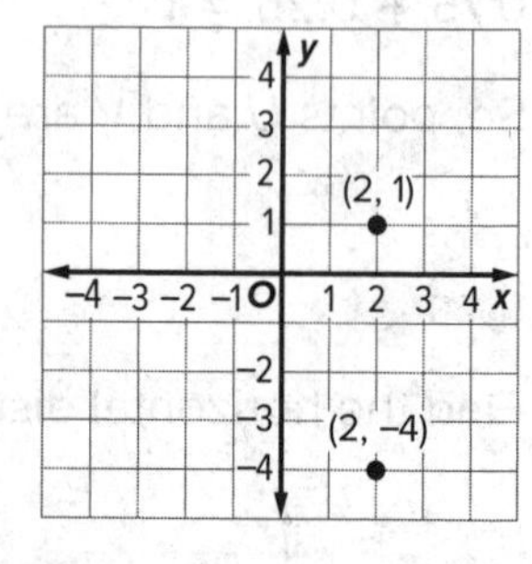

$|-4| = \square$ $\quad |1| = \square$

Add the absolute values.

$4 + 1 = \square$

The distance between the two points is 5 units.

Example 3 Find Vertical Distance in the Same Quadrant

Find the vertical distance between the points $D\left(-1, -\frac{1}{2}\right)$ and $C(-1, -2)$.

The x-coordinates are ________ and the y-coordinates are ________.

This means the points are in Quadrant ________.

The x-coordinates are the same. To find the distance each point is from the x-axis, find the absolute value of each y-coordinate.

C: $|-2| = \square$ D: $\left|-\frac{1}{2}\right| = \square$

Because the points are in the same quadrant, subtract the absolute values of the y-coordinates to find the distance between points.

$2 - \frac{1}{2} = 1\frac{1}{2}$

So, points C and D are ________ unit(s) apart.

Think About It!

Are the x-coordinates the same or different? Are the y-coordinates the same or different?

Talk About It!

How can you check your solution? Explain a process you could use.

Check

Find the vertical distance between the points $A\left(-\frac{1}{3}, -\frac{2}{3}\right)$ and $B\left(-\frac{1}{3}, -1\frac{1}{3}\right)$.

Go Online You can complete an Extra Example online.

Pause and Reflect

Did you struggle with any of the concepts in this Example and Check? How do you feel when you struggle with math concepts? What steps can you take to understand those concepts?

Think About It!
Are the points in the same quadrant? How will that affect how you find the distance?

Talk About It!
Compare and contrast finding vertical and horizontal distance between two points in the coordinate plane.

Example 4 Find Vertical Distances in Different Quadrants

Find the vertical distance between points $S(1, 0.5)$ and $T(1, -0.5)$.

The x-coordinates have the ________ signs.

The y-coordinates have ________ signs.

This means the points are in ________________.

The x-coordinates are the same. To find the distance each point is from the x-axis, find the absolute value of each y-coordinate.

$S: |0.5| = \square$ $T: |-0.5| = \square$

Because the points are in different quadrants, add the absolute values of the y-coordinates to find the distance between points.

$0.5 + 0.5 = 1$

So, points S and T are ________ unit(s) apart.

Check

Find the vertical distance between points $E(0.5, 1.5)$ and $F(0.5, -2)$.

Go Online You can complete an Extra Example online.

Pause and Reflect

Are you ready to move on? If yes, what have you learned in this lesson that you think will help you? If no, what questions do you still have? How can you get those questions answered?

Apply Distance

Fritz and Manolo like to skateboard together at a nearby park. They want to determine who has to walk the farther distance to get to the park, so they graph the locations on a coordinate plane, with the city's main square at the origin. The coordinates for each location are shown in the table. Each unit represents a city block. Who has to walk the farther distance to get to the park?

Location	Coordinates
Fritz's house	$\left(-2\frac{1}{2}, 2\right)$
Manolo's house	$\left(3, -3\frac{3}{4}\right)$
Park	(3, 2)

Go Online watch the animation.

1 What is the task?

Make sure you understand exactly what question to answer or problem to solve. You may want to read the problem three times. Discuss these questions with a partner.

First Time Describe the context of the problem, in your own words.
Second Time What mathematics do you see in the problem?
Third Time What are you wondering about?

2 How can you approach the task? What strategies can you use?

3 What is your solution?

Use your strategy to solve the problem.

4 How can you show your solution is reasonable?

Write About It! Write an argument that can be used to defend your solution.

Talk About It!
Who would have the farther distance to get to the park, if the park was located at (4, 4)?

Check

Fernando has a dog-walking job and will walk the dogs from his house to one of the two parks shown. He wants to go to the park that will give the dogs a longer walk. To which park should he go?

Location	Coordinates
Fernando's house	$\left(-2\frac{1}{2}, 2\frac{1}{4}\right)$
Cobblestone Dog Park	$\left(1\frac{3}{4}, 2\frac{1}{4}\right)$
Blue Limestone Park	$\left(-2\frac{1}{2}, -1\frac{1}{2}\right)$

Go Online You can complete an Extra Example online.

Name ______________________ Period __________ Date __________

Practice

Go Online You can complete your homework online.

For Exercises 1–8, find the horizontal or vertical distance between the two points. (Examples 1–4)

1. 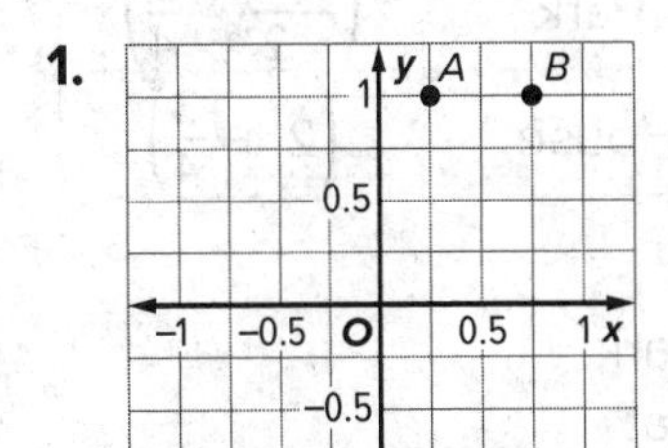

______

2. 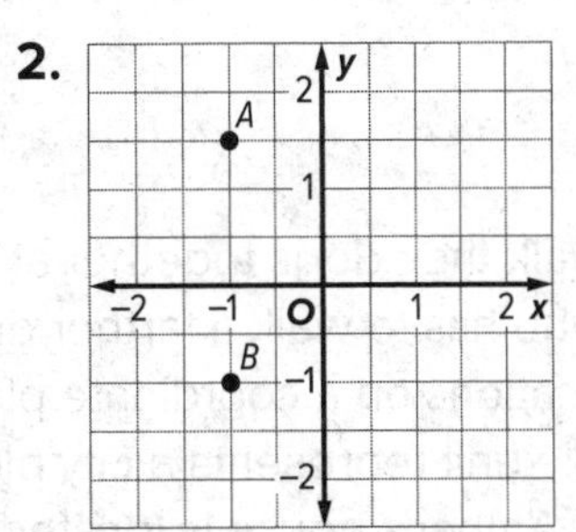

______

3. 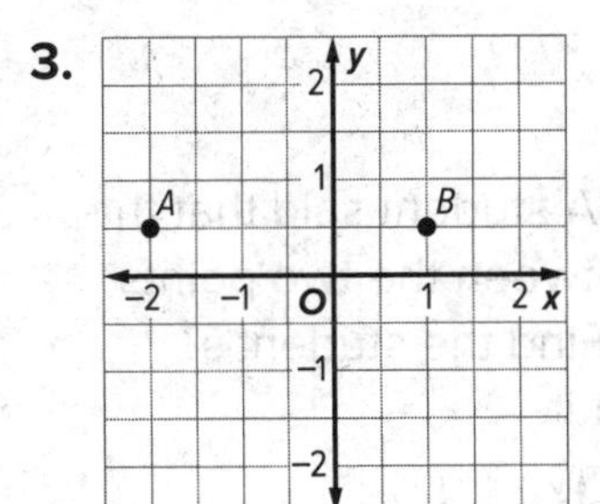

______

4. 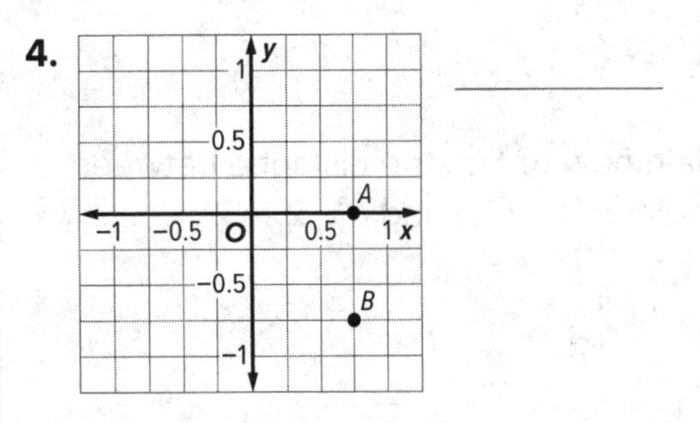

______

5. $X(-2, 3)$ and $Y\left(-2, 1\frac{1}{4}\right)$

______

6. $Y\left(1, -\frac{3}{4}\right)$ and $Z\left(-1, -\frac{3}{4}\right)$

______

7. $A(-1, 1.5)$ and $B(-1, -1.5)$

______

8. $C(3.5, -0.25)$ and $D(0.5, -0.25)$

______

Test Practice

9. Multiple Choice What is the vertical distance between the points $C(2, -0.8)$ and $D(2, 1.2)$?

Ⓐ 0 units

Ⓑ 0.4 unit

Ⓒ 1 unit

Ⓓ 2 units

Apply

10. There are two parks near Kennedy's house. She wants to go the park closer to her house. To which park should Kennedy go?

Location	Coordinate
Maple Avenue Park	$\left(2, 1\frac{1}{2}\right)$
Oak Woods Park	$\left(-\frac{1}{2}, -\frac{3}{4}\right)$
Kennedy's House	$\left(2, -\frac{3}{4}\right)$

11. James and Amber walk their dogs together at a nearby dog park. They want to determine who has to walk a farther distance to get to the dog park, so they graph the locations on a coordinate plane, with the town square at the origin. Each whole unit represents a city block. James's house is located at the point (−1.5, 4). Amber's house is located at the point (2, 0.25). The dog park is located at the point (2, 4). Who has to walk the farther distance to get to the dog park?

12. Explain how to find the distance between the points $A(-2, 2)$ and $B(-2, -2)$.

13. MP **Find the Error** A student said that the vertical distance between the two points graphed is 3 units. Find the student's mistake and correct it.

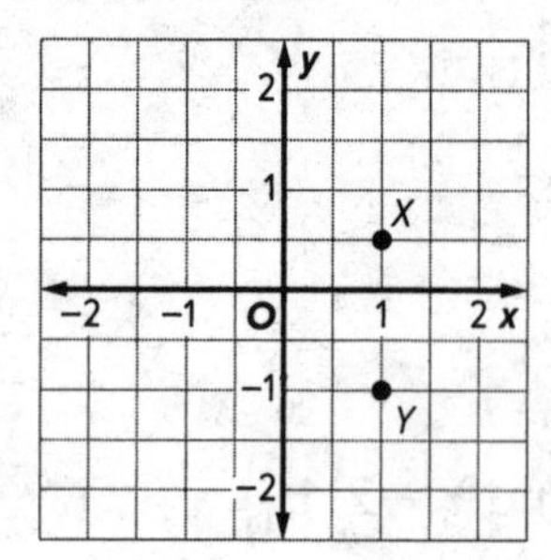

14. Give the coordinates for two points that have a vertical distance of 1.5 units.

15. Yara said that the vertical distance between two points was −1.5 units. How do you know that Yara's answer is incorrect?

Review

Foldables Use your Foldable to help review the module.

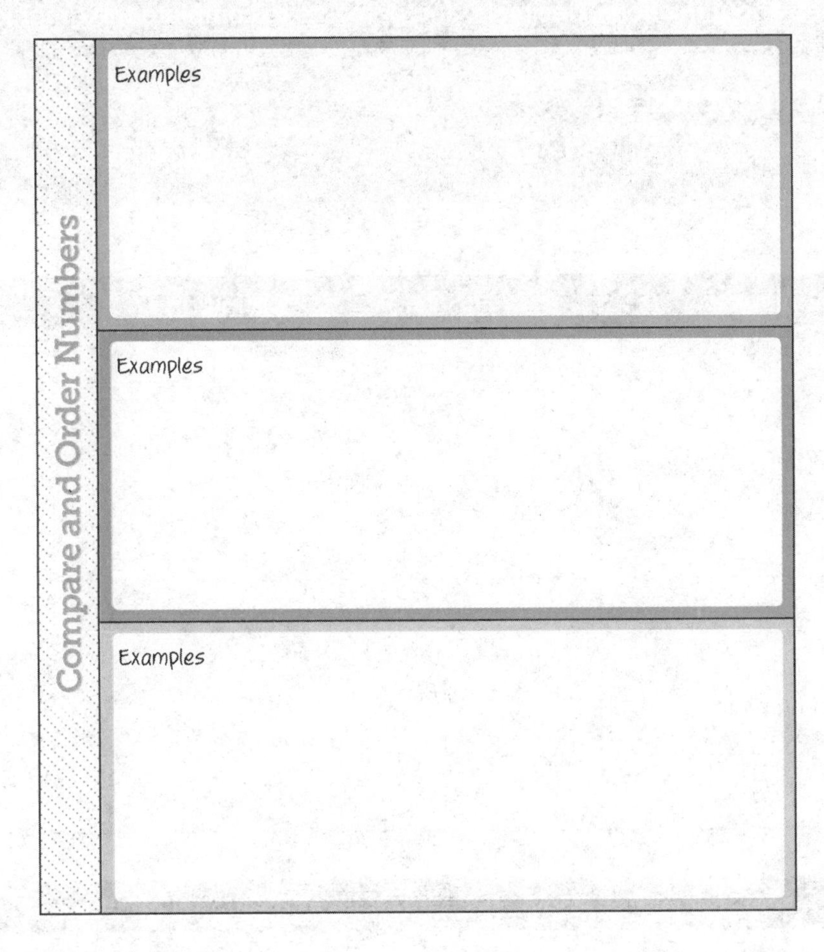

Rate Yourself!   

Complete the chart at the beginning of the module by placing a checkmark in each row that corresponds with how much you know about each topic after completing this module.

Write about one thing you learned.	Write about a question you still have.

Reflect on the Module

Use what you learned about fractions, decimals, and percents to complete the graphic organizer.

Essential Question

How are integers and rational numbers related to the coordinate plane?

Vocabulary	Definition
integer	
rational number	

Describe what you know about integers and rational numbers.

Describe what you know about the coordinate plane.

Name _______________ Period _______ Date _______

Test Practice

1. Open Response While riding one of the rides at the local amusement park, Zachary lost $5 from his pocket. **(Lesson 1)**

A. Write an integer to represent this situation. Explain.

B. Explain the meaning of zero in this situation.

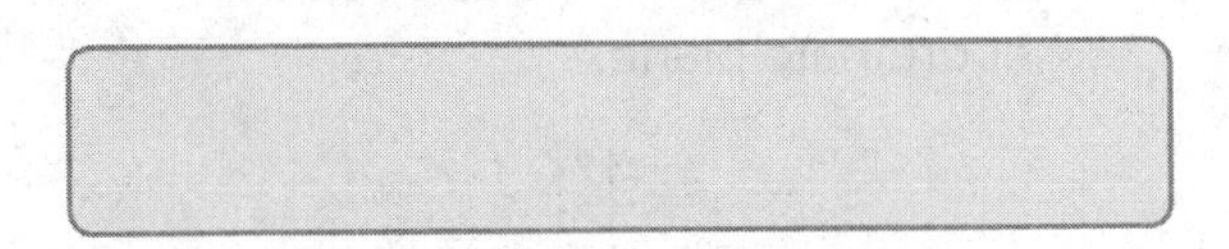

2. Grid Graph the set of integers {−6, −1, 0} on the number line. **(Lesson 1)**

−6 −5 −4 −3 −2 −1 0 1 2 3 4 5 6

3. Equation Editor Find −(−14). **(Lesson 2)**

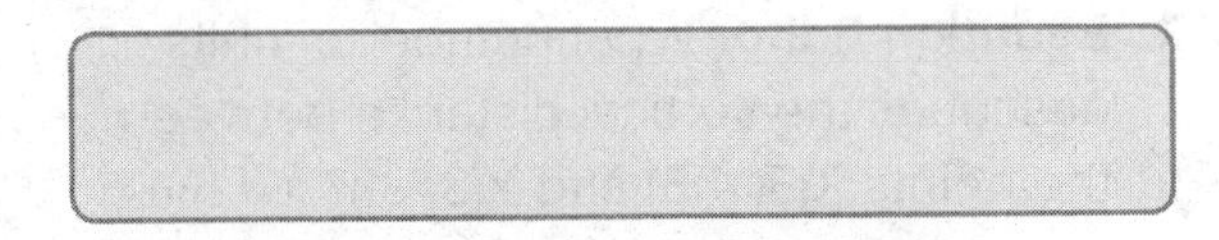

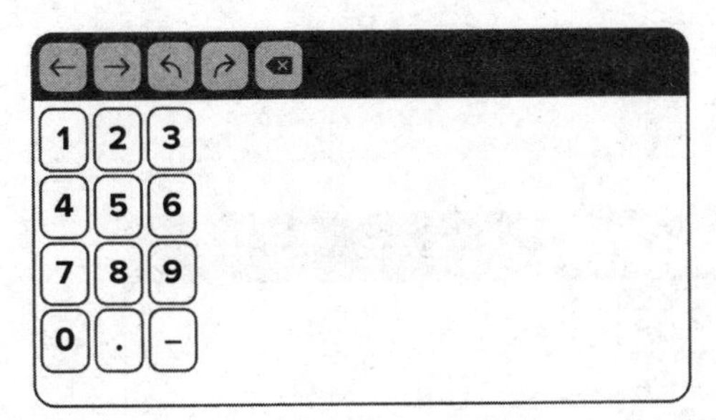

4. Table Item Indicate whether each inequality is correct or not correct. **(Lesson 3)**

	Correct	Not Correct
−7 < −9		
5 > −1		
−12 < −10		

5. Open Response The table show the boiling points, to the nearest degree Celsius, for six substances. Carbon dioxide boils at −79°C. Between which two substances is the boiling point of carbon dioxide? **(Lesson 3)**

Substance	Boiling Point (°C)
Ammonia	−36
Benzene	80.4
Acetylene	−84
Ethanol	79
Fluorine	−187
Water	100

6. Equation Editor Evaluate |−6.2|. **(Lesson 4)**

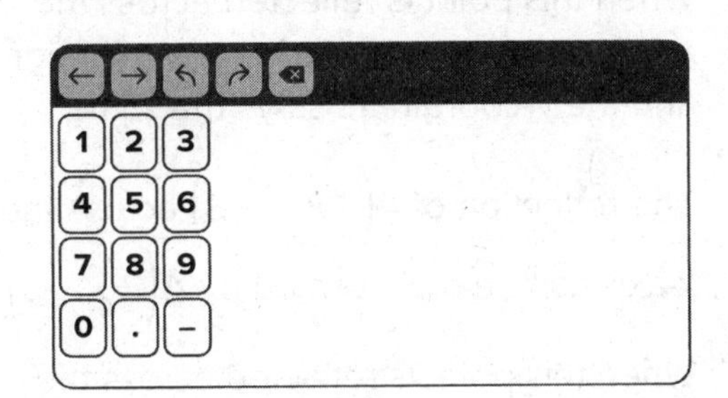

7. Open Response During the overnight hours, the temperature in Juneau fell from 0°F to −12°F. How many degrees did the temperature fall? **(Lesson 4)**

8. Multiple Choice Identify the quadrant in which the point $\left(\frac{2}{3}, -1\frac{1}{5}\right)$ is located. **(Lesson 5)**

Ⓐ Quadrant I

Ⓑ Quadrant II

Ⓒ Quadrant III

Ⓓ Quadrant IV

9. Table Item Indicate the axis on which each of these points lies. **(Lesson 5)**

	x-axis	y-axis
(−4, 0)		
(0, 9)		
(0, −6)		

10. Multiselect Consider the point $A\left(-2\frac{1}{4}, -3\right)$. Which of the following statements are true regarding the reflection of this point? Select all that apply. **(Lesson 6)**

☐ When this point is reflected across the *x*-axis, the *x*-coordinate reverses its sign and the *y*-coordinate stays the same.

☐ The reflection of $A\left(-2\frac{1}{4}, -3\right)$ across the *x*-axis can be represented by $A'\left(2\frac{1}{4}, -3\right)$.

☐ When this point is reflected across the *x*-axis, the *x*-coordinate stays the same and the *y*-coordinate reverses.

☐ The reflection of $A\left(-2\frac{1}{4}, -3\right)$ across the *y*-axis can be represented by $A'\left(2\frac{1}{4}, -3\right)$.

☐ When this point is reflected across the *y*-axis, the *x*-coordinate reverses its sign and the *y*-coordinate stays the same.

11. Grid Derrius draw a map of the community playground. He graphed the point $S\left(2\frac{1}{2}, -5\right)$ for the slide. The swings are located at S', a reflection across the *x*-axis. The restrooms are located at S'', a reflection across the *y*-axis. **(Lesson 6)**

A. Identify the ordered pair that describes the location of the restrooms.

B. Plot and label the point S'' on the coordinate plane.

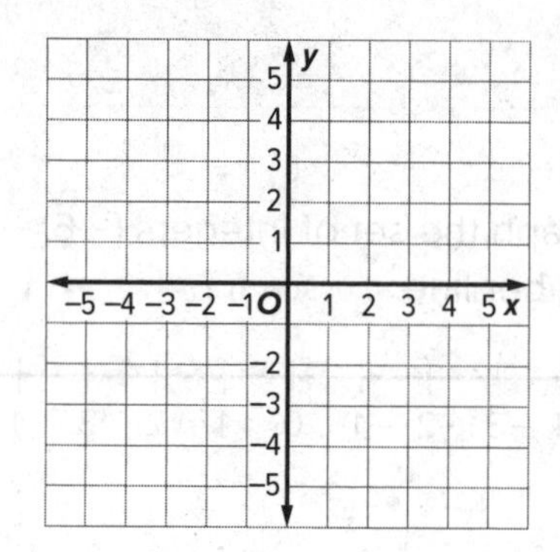

12. Equation Editor What number of units describes the vertical distance between the points $X(3, 4.5)$ and $Y(3, -1)$? **(Lesson 7)**

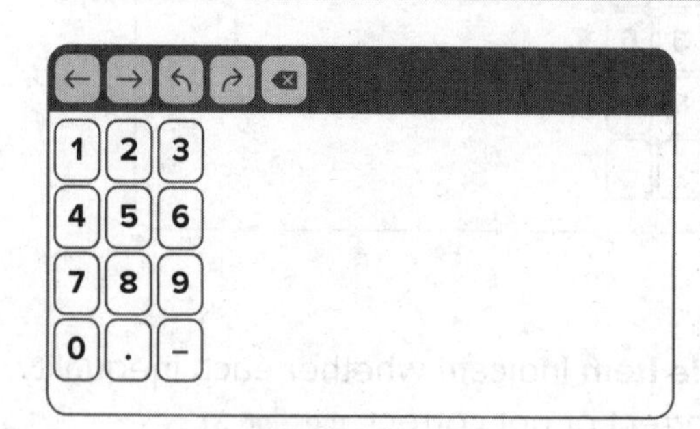

What Are Foldables and How Do I Create Them?

Foldables are three-dimensional graphic organizers that help you create study guides for each module in your book.

Step 1 Go to the back of your book to find the Foldable for the module you are currently studying. Follow the cutting and assembly instructions at the top of the page.

Step 2 Go to the Module Review at the end of the module you are currently studying. Match up the tabs and attach your Foldable to this page. Dotted tabs show where to place your Foldable. Striped tabs indicate where to tape the Foldable.

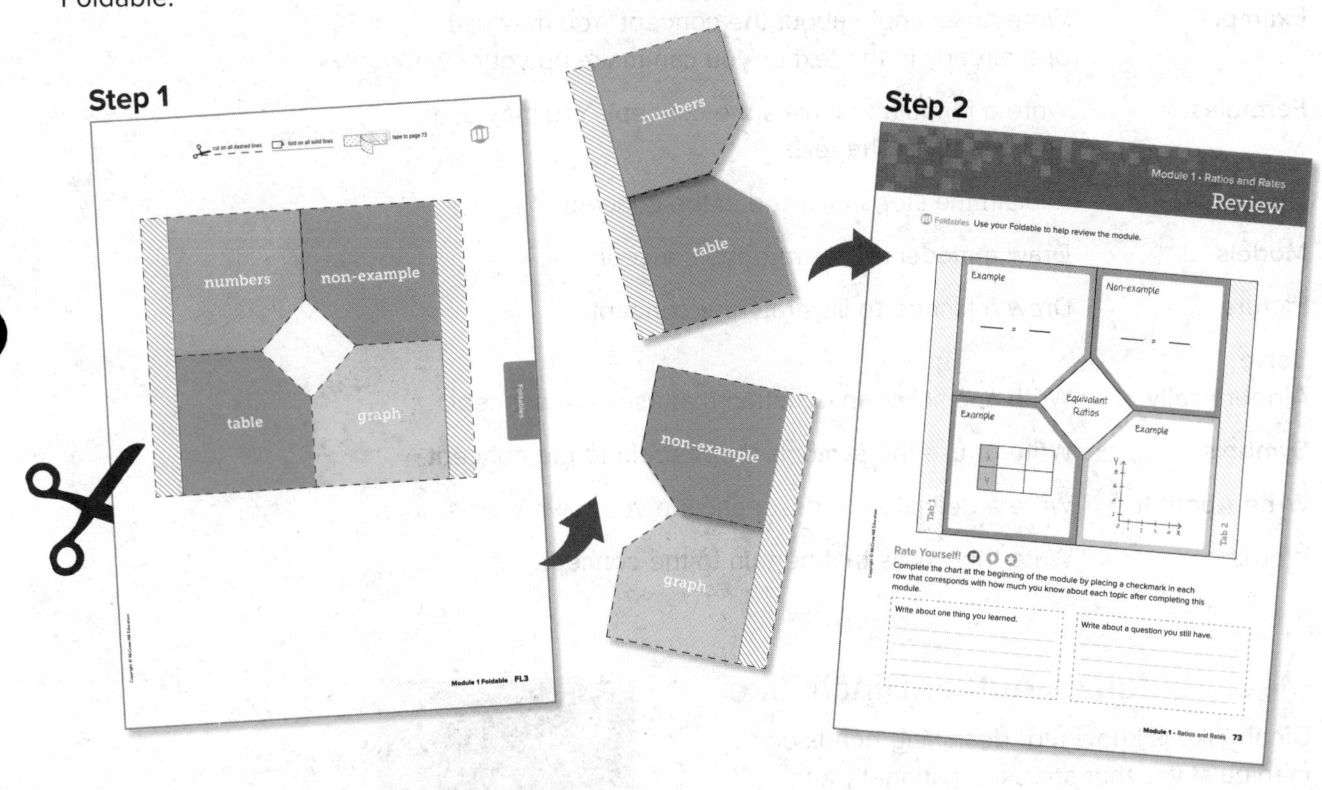

How Will I Know When to Use My Foldable?

You will be directed to work on your Foldable at the end of selected lessons. This lets you know that it is time to update it with concepts from that lesson. Once you've completed your Foldable, use it to study for the module test.

How Do I Complete My Foldable?

No two Foldables in your book will look alike. However, some will ask you to fill in similar information. Below are some of the instructions you'll see as you complete your Foldable. **HAVE FUN** learning math using Foldables!

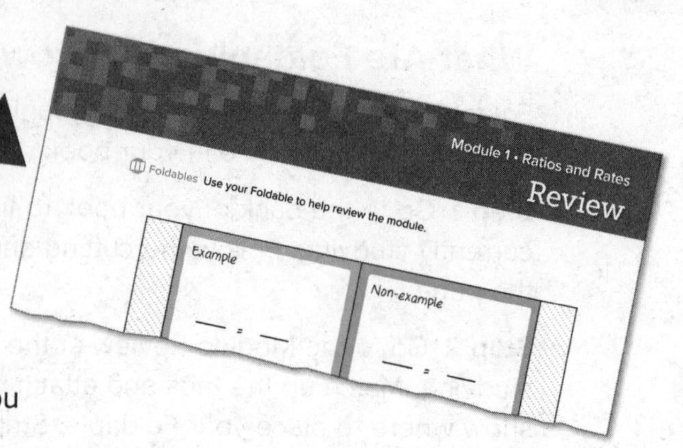

Instructions and What They Mean

Best Used to...	Complete the sentence explaining when the concept should be used.
Definition	Write a definition in your own words.
Description	Describe the concept using words.
Equation	Write an equation that uses the concept. You may use one already in the text or you can make up your own.
Example	Write an example about the concept. You may use one already in the text or you can make up your own.
Formulas	Write a formula that uses the concept. You may use one already in the text.
How do I ...?	Explain the steps involved in the concept.
Models	Draw a model to illustrate the concept.
Picture	Draw a picture to illustrate the concept.
Solve Algebraically	Write and solve an equation that uses the concept.
Symbols	Write or use the symbols that pertain to the concept.
Write About It	Write a definition or description in your own words.
Words	Write the words that pertain to the concept.

Meet Foldables Author Dinah Zike

Dinah Zike is known for designing hands-on manipulatives that are used nationally and internationally by teachers and parents. Dinah is an explosion of energy and ideas. Her excitement and joy for learning inspires everyone she touches.

cut on all dashed lines fold on all solid lines tape to page 73

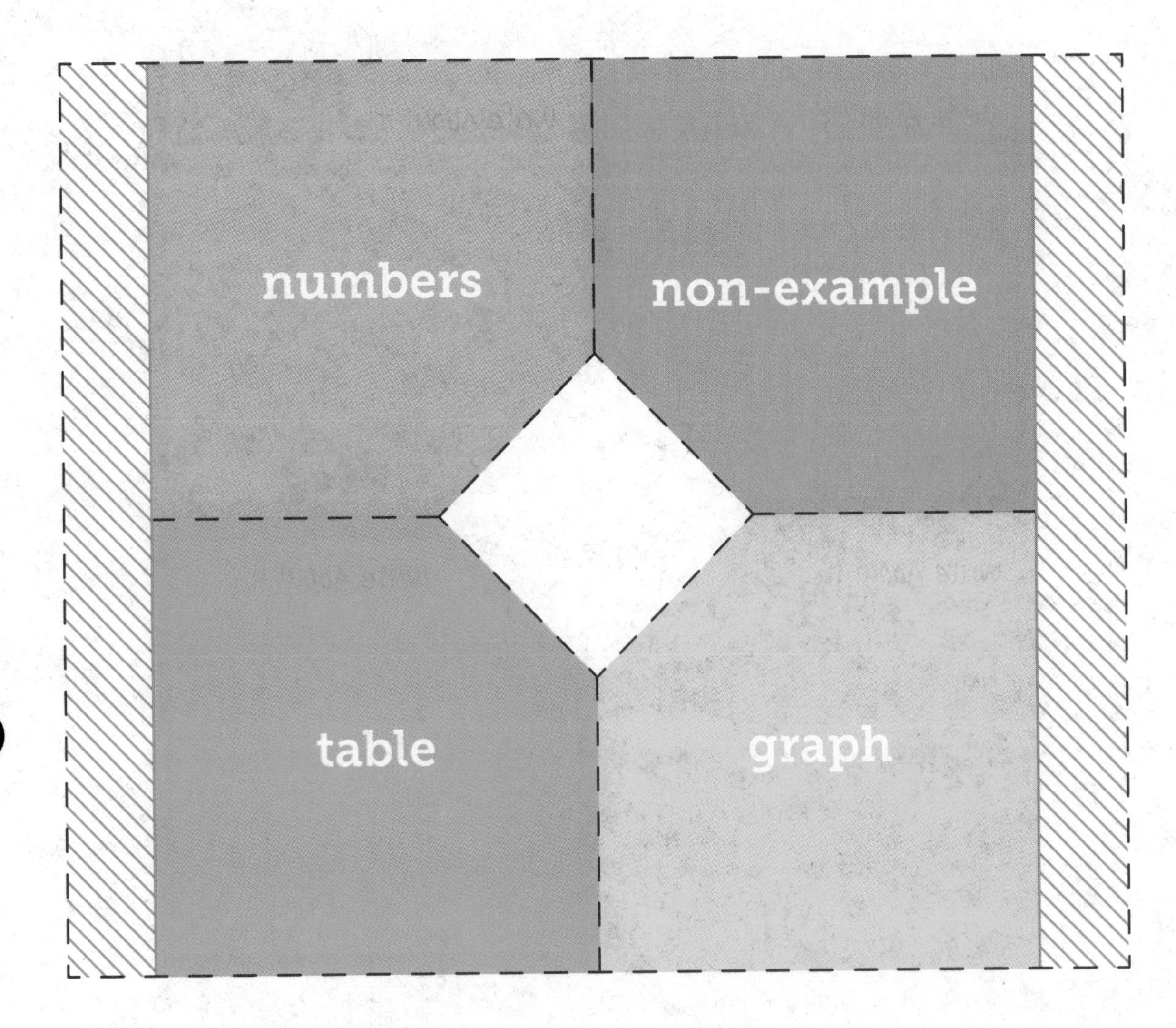

Foldables

cut on all dashed lines

fold on all solid lines

tape to page 73

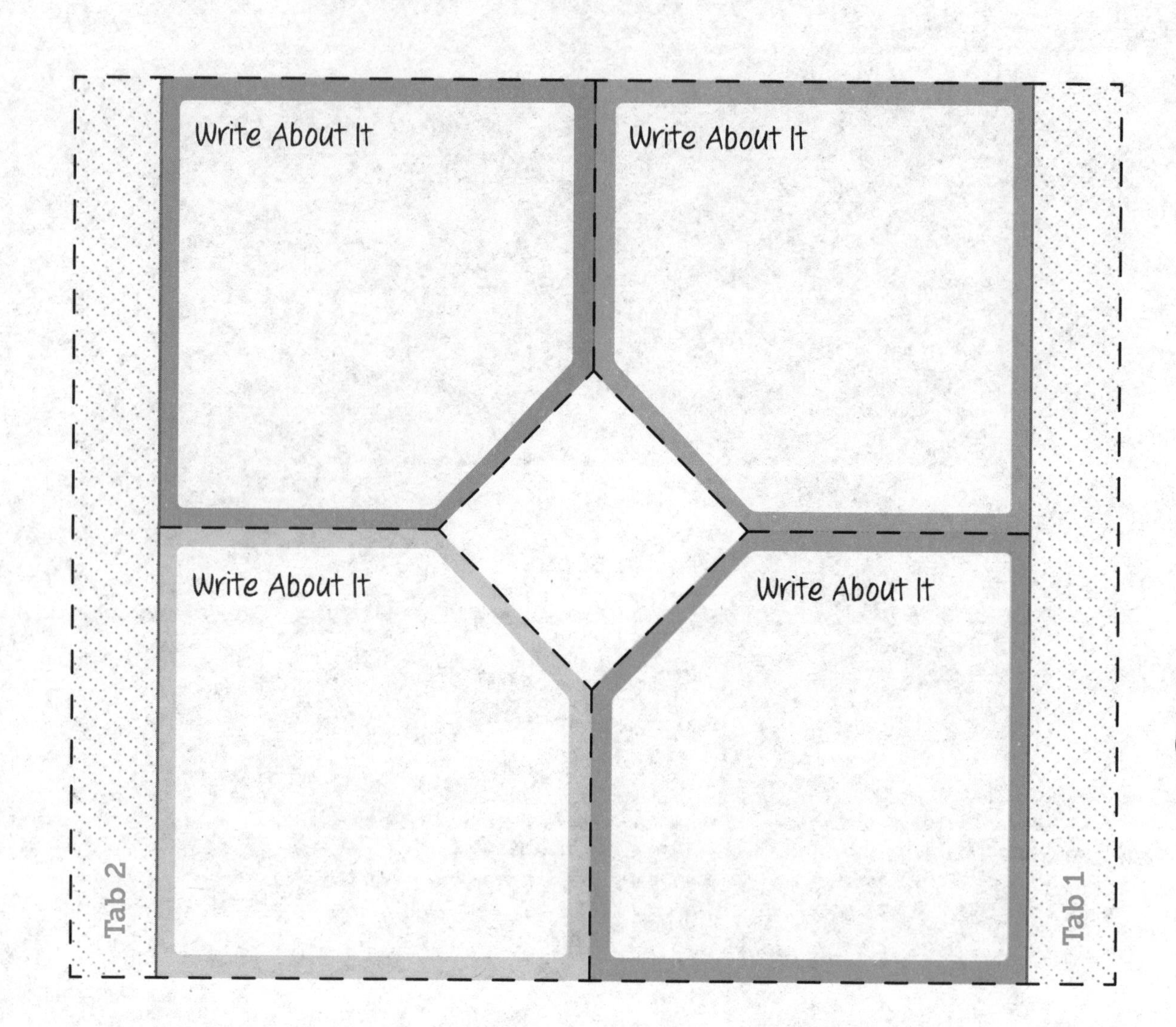

cut on all dashed lines fold on all solid lines tape to page 129

Fractions, Decimals, and Percents

percents and fractions

percents and decimals

percent of a number

Foldables

cut on all dashed lines

fold on all solid lines

tape to page 129

Write About it

Write About it

Write About it

cut on all dashed lines
fold on all solid lines
tape to page 187

Multiply and Divide Fractions

multiply	divide
Example	Example
fraction × whole number	whole number ÷ fraction
Example	Example
fraction × fraction	fraction ÷ fraction

Foldables

cut on all dashed lines fold on all solid lines tape to page 187

Tab 3

How do I divide a whole number by a fraction?

How do I multiply a fraction by a whole number?

Tab 2

How do I divide a fraction by a fraction?

How do I multiply a fraction by a fraction?

Tab 1

How do I divide a mixed number by a fraction?

How do I multiply a fraction by a mixed number?

cut on all dashed lines   fold on all solid lines   tape to page 255

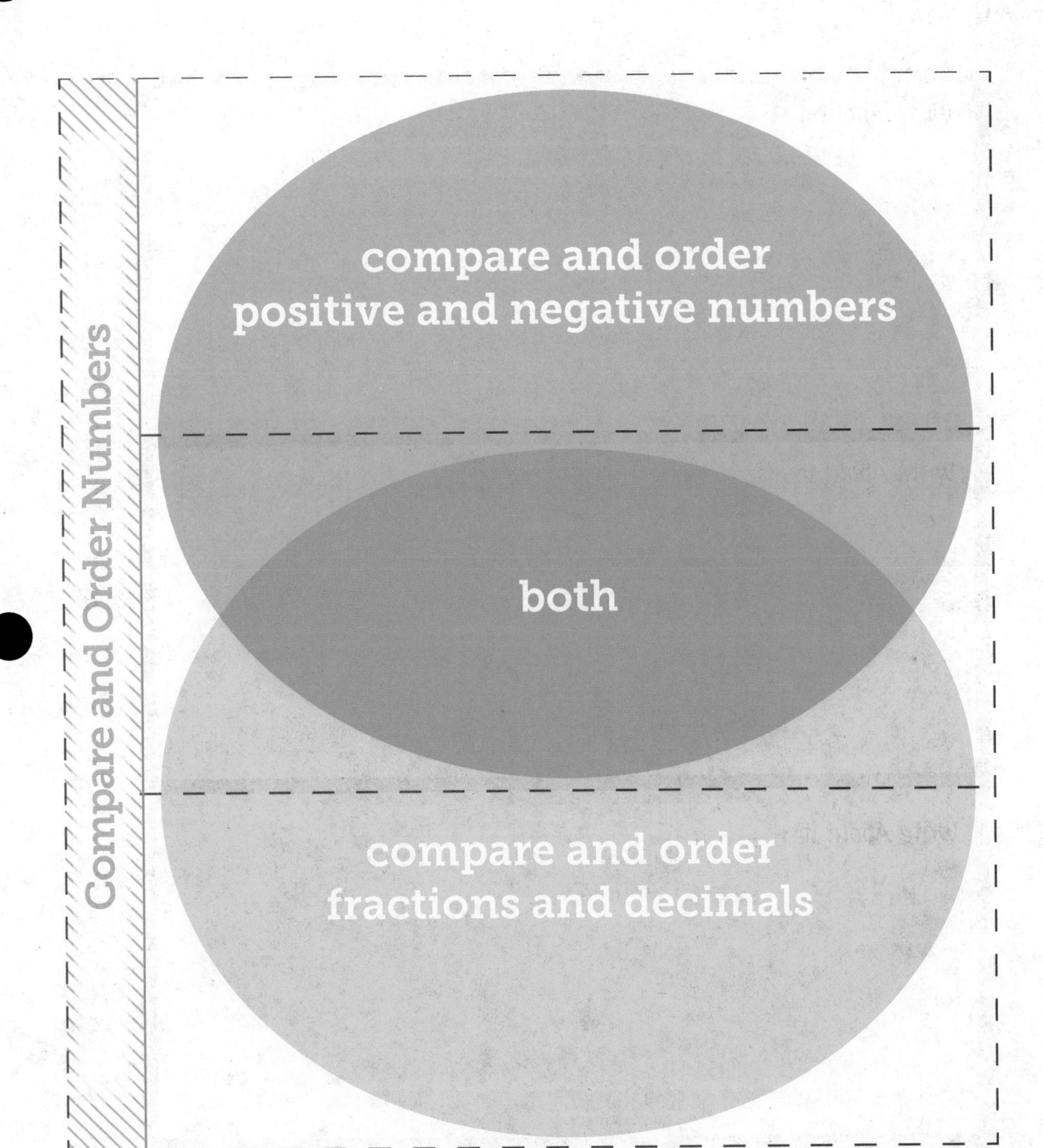

Foldables

cut on all dashed lines

fold on all solid lines

tape to page 255

Write About it

Write About it

Write About it

Glossary

The eGlossary contains words and definitions in the following 14 languages:

Arabic	**English**	**Hmong**	**Russian**	**Urdu**
Bengali	**French**	**Korean**	**Spanish**	**Vietnamese**
Brazilian Portuguese	**Haitian Creole**	**Mandarin**	**Tagalog**	

English	Español

A

absolute value (Lesson 4-2) The distance between a number and zero on a number line.

valor absoluto Distancia entre un número y cero en la recta numérica.

Addition Property of Equality (Lesson 6-8) If you add the same number to each side of an equation, the two sides remain equal.

propiedad de adición de la igualdad Si sumas el mismo número a ambos lados de una ecuación, los dos lados permanecen iguales.

algebra (Lesson 5-3) A mathematical language of symbols, including variables.

álgebra Lenguaje matemático que usa símbolos, incluyendo variables.

algebraic expression (Lesson 5-3) A combination of variables, numbers, and at least one operation.

expresión algebraica Combinación de variables, números y, por lo menos, una operación.

analyze (Lesson 10-1) To use observations to describe and compare data.

analizar Usar observaciones para describir y comparar datos.

area (Lesson 8-1) The measure of the interior surface of a two-dimensional figure.

área La medida de la superficie interior d una figura bidimensional.

Associative Property (Lesson 5-7) The way in which numbers are grouped does not change the sum or product.

propiedad asociativa La forma en que se agrupan tres números al sumarlos o multiplicarlos no altera su suma o producto.

average (Lesson 10-3) The sum of two or more quantities divided by the number of quantities; the mean.

promedio La suma de dos o más cantidades dividida entre el número de cantidades; la media.

B

base (Lesson 8-1) Any side of a parallelogram or any side of a triangle.

base Cualquier lado de un paralelogramo o cualquier lado de un triángulo.

base (Lesson 9-1) One of the two parallel congruent faces of a prism.

base Una de las dos caras paralelas congruentes de un prisma.

base (Lesson 5-1) In a power, the number used as a factor. In 10^3, the base is 10. That is, $10^3 = 10 \times 10 \times 10$.

base En una potencia, el número usado como factor. En 10^3, la base es 10. Es decir, $10^3 = 10 \times 10 \times 10$.

bases (Lesson 8-3) The bases of a trapezoid are the two parallel sides.

bases Las bases de un trapecio son los dos lados paralelos.

benchmark percent (Lesson 2-5) A common percent used when estimating part of a whole.

porcentaje de referencia Porcentaje común utilizado para estimar parte de un todo.

box plot (Lesson 10-4) A diagram that is constructed using five values.

diagrama de caja Diagrama que se construye usando cinco valores.

C

cluster (Lesson 10-7) Data that are grouped closely together.

agrupamiento Conjunto de datos que se agrupan.

coefficient (Lesson 5-3) The numerical factor of a term that contains a variable.

coeficiente El factor numérico de un término que contiene una variable.

common factor (Lesson 5-5) A number that is a factor of two or more numbers.

factor común Un número que es un factor de dos o más números.

Commutative Property (Lesson 5-7) The order in which numbers are added or multiplied does not change the sum or product.

propiedad commutativa La forma en que se suman o multiplican dos números no altera su suma o producto.

congruent (Lesson 8-2) Having the same measure.

congruente Ques tienen la misma medida.

congruent figures (Lesson 8-2) Figures that have the same size and same shape; corresponding sides and angles have equal measures.

figuras congruentes Figuras que tienen el mismo tamaño y la misma forma; los lados y los ángulos correspondientes con igual medida.

constant (Lesson 5-3) A term without a variable.

constante Un término sin una variable.

coordinate plane (Lesson 1-3) A plane in which a horizontal number line and a vertical number line intersect at their zero points.

plano de coordenadas Plano en que una recta numérica horizontal y una recta numérica vertical se intersecan en sus puntos cero.

cubic units (Lesson 9-1) Used to measure volume. Tells the number of cubes of a given size it will take to fill a three-dimensional figure.

unidades cúbicas Se usan para medir el volumen. Indican el número de cubos de cierto tamaño que se necesitan para llenar una figura tridimensional.

D

data (Lesson 10-1) Information, often numerical, which is gathered for statistical purposes.

datos Información, con frecuencia numérica, que se recoge con fines estadísticos.

defining the variable (Lesson 5-3) Choosing a variable and deciding what the variable represents.

definir la variable Elegir una variable y decidir lo que representa.

dependent variable **(Lesson 7-1)** The variable in a relation with a value that depends on the value of the independent variable.

variable dependiente La variable en una relación cuyo valor depende del valor de la variable independiente.

distribution **(Lesson 10-7)** The arrangement of data values.

distribución El arreglo de valores de datos.

Distributive Property **(Lesson 5-6)** To multiply a sum by a number, multiply each addend by the number outside the parentheses.

propiedad distributiva Para multiplicar una suma por un número, multiplica cada sumando por el número fuera de los paréntesis.

dividend **(Lesson 3-1)** The number that is divided in a division problem.

dividendo El número que se divide en un problema de división.

Division Property of Equality **(Lesson 6-4)** If you divide each side of an equation by the same nonzero number, the two sides remain equal.

propiedad de igualdad de la división Si divides ambos lados de una ecuación entre el mismo número no nulo, los lados permanecen iguales.

divisor **(Lesson 3-1)** The number used to divide another number in a division problem.

divisor El número utilizado para dividir otro número en un problema de división.

double number line **(Lesson 1-2)** A double number line consists of two number lines, in which the coordinated quantities are equivalent ratios.

línea doble Una línea numérica doble consta de dos líneas numéricas, en las cuales las cantidades coordinadas son proporciones equivalentes.

dot plot **(Lesson 10-2)** A diagram that shows the frequency of data on a number line. Also known as a line plot.

diagrama de puntos Diagrama que muestra la frecuencia de los datos sobre una recta numérica.

E

equals sign **(Lesson 6-1)** A symbol of equality, =.

signo de igualdad Símbolo que indica igualdad, =.

equation **(Lesson 6-1)** A mathematical sentence showing two expressions are equal. An equation contains an equals sign, =.

ecuación Enunciado matemático que muestra que dos expresiones son iguales. Una ecuación contiene el signo de igualdad, =.

equivalent expressions **(Lesson 5-7)** Expressions that have the same value, regardless of the values of the variable(s).

expresiones equivalentes Expresiones que poseen el mismo valor, sin importer los valores de la(s) variable(s).

equivalent ratios **(Lesson 1-2)** Ratios that express the same relationship between two quantities.

razones equivalentes Razones que expresan la misma relación entre dos cantidades.

evaluate **(Lesson 5-4)** To find the value of an algebraic expression by replacing variables with numbers.

evaluar Calcular el valor de una expresión algebraica sustituyendo las variables por número.

exponent **(Lesson 5-1)** In a power, the number that tells how many times the base is used as a factor. In 5^3, the exponent is 3. That is, $5^3 = 5 \times 5 \times 5$.

exponente En una potencia, el número que indica las veces que la base se usa como factor. En 5^3, el exponente es 3. Es decir, $5^3 = 5 \times 5 \times 5$.

F

face (Lesson 9-1) A flat surface of a prism or pyramid.

cara Una superficie plana de un prisma o pirámide.

factoring the expression (Lesson 5-6) The process of writing numeric or algebraic expressions as a product of their factors.

factorizar la expresión El proceso de escribir expresiones numéricas o algebraicas como el producto de sus factores.

first quartile (Lesson 10-4) The first quartile is the median of the data values less than the median.

primer cuartil El primer cuartil es la mediana de los valores menores que la mediana.

G

gap (Lesson 10-7) An empty space or interval in a set of data.

laguna Espacio o intervalo vacío en un conjunto de datos.

graph (Lesson 1-3) To place a dot on a number line, or on the coordinate plane at a point named by an ordered pair.

graficar Colocar una marca puntual en una línea numérica, o en el plano de coordenadas en el punto que corresponde a un par ordenado.

greatest common factor (GCF) (Lesson 5-5) The greatest of the common factors of two or more numbers.

máximo común divisor (MCD) El mayor de los factores comunes de dos o más números.

guess, check, and revise strategy (Lesson 6-1) A strategy used to solve a problem which involves narrowing in on the correct answer using educated guesses.

adivinar, comprobar y revisar la estrategia Una estrategia utilizada para resolver un problema que implica el estrechamiento en la respuesta correcta usando conjeturas educadas.

H

height (Lesson 8-1) The height of a parallelogram is the perpendicular distance between the base and its opposite side.

altura La altura de un paralelogramo es la distancia perpendicular entre la base y su lado opuesto.

height (Lesson 8-2) The height of a triangle is the perpendicular distance from the base to the opposite vertex.

altura La altura de un triángulo es la distancia perpendicular de la base al vértice opuesto.

height (Lesson 8-3) The height of a trapezoid is the perpendicular distance between the two bases.

altura La altura de un trapecio es la distancia perpendicular entre las dos bases.

histogram (Lesson 10-2) A type of bar graph used to display numerical data that have been organized into equal intervals.

histograma Tipo de gráfica de barras que se usa para exhibir datos que se han organizado en intervalos iguales.

I

Identity Properties **(Lesson 5-7)** Properties that state that the sum of any number and 0 equals the number and that the product of any number and 1 equals the number.

propiedades de identidad Propiedades que establecen que la suma de cualquier número y 0 es igual al número y que el producto de cualquier número y 1 es igual al número.

independent variable **(Lesson 7-1)** The variable in a relationship with a value that is subject to choice.

variable independiente Variable en una relación cuyo valor está sujeto a elección.

inequality **(Lesson 6-6)** A mathematical sentence indicating that two quantities are not equal.

desigualdad Enunciado matemático que indica que dos cantidades no son iguales.

integer **(Lesson 4-1)** Any number from the set {..., −4, −3, −2, −1, 0, 1, 2, 3, 4, ...} where ... means *continues without end.*

entero Cualquier número del conjunto {..., −4, −3, −2, −1, 0, 1, 2, 3, 4, ...} donde ... significa que *continúa sin fin.*

interquartile range (IQR) **(Lesson 10-4)** A measure of variation in a set of numerical data, the interquartile range is the distance between the first and third quartiles of the data set.

rango intercuartil (RIQ) El rango intercuartil, una medida de la variación en un conjunto de datos numéricos, es la distancia entre el primer y el tercer cuartil del conjunto de datos.

interval **(Lesson 10-2)** The difference between successive values on a scale.

intervalo La diferencia entre valores sucesivos de una escala.

inverse operations **(Lesson 6-2)** Operations which undo each other. For example, addition and subtraction are inverse operations.

operaciones inversas Operaciones que se anulan mutuamente. La adición y la sustracción son operaciones inversas.

Inverse Property of Multiplication **(Lesson 3-3)** A property that states that the product of a number and its multiplicative inverse is 1.

propiedad inversa de la multiplicación Una propiedad que indica que el producto de un número y su inverso multiplicativo es 1.

L

lateral face **(Lesson 9-4)** Any face that is not a base.

cara lateral Cualquier superficie plana que no sea la base.

least common multiple (LCM) **(Lesson 5-5)** The smallest whole number greater than 0 that is a common multiple of each of two or more numbers.

mínimo común múltiplo (mcm) El menor número entero, mayor que 0, múltiplo común de dos o más números.

like terms **(Lesson 5-3)** Terms that contain the same variable(s) to the same power.

términos semejantes Términos que contienen la misma variable o variables elevadas a la misma potencia.

M

mean **(Lesson 10-3)** The sum of the numbers in a set of data divided by the number of pieces of data.

media La suma de los números en un conjunto de datos dividida entre el número total de datos.

mean absolute deviation (MAD) **(Lesson 10-5)** A measure of variation in a set of numerical data, computed by adding the distances between each data value and the mean, then dividing by the number of data values.

desviación media absoluta (DMA) Una medida de variación en un conjunto de datos numéricos que se calcula sumando las distancias entre el valor de cada dato y la media, y luego dividiendo entre el número de valores.

measures of center **(Lesson 10-3)** Numbers that are used to describe the center of a set of data. These measures include the mean, median, and mode.

medidas del centro Numéros que se usan para describir el centro de un conjunto de datos. Estas medidas incluyen la media, la mediana y la moda.

measures of variation **(Lesson 10-4)** A measure used to describe the distribution of data.

medidas de variación Medida usada para describir la distribución de los datos.

median **(Lesson 10-3)** A measure of center in a set of numerical data. The median of a list of values is the value appearing at the center of a sorted version of the list, or the mean of the two central values, if the list contains an even number of values.

mediana Una medida del centro en un conjunto de datos numéricos. La mediana de una lista de valores es el valor que aparece en el centro de una versión ordenada de la lista, o la media de los dos valores centrales si la lista contiene un número par de valores.

mode **(Lesson 10-3)** The number(s) or item(s) that appear most often in a set of data.

moda Número(s) de un conjunto de datos que aparece(n) más frecuentemente.

Multiplication Property of Equality **(Lesson 6-5)** If you multiply each side of an equation by the same nonzero number, the two sides remain equal.

propiedad de multiplicación de la igualdad Si multiplicas ambos lados de una ecuación por el mismo número no nulo, lo lados permanecen iguales.

multiplicative inverses **(Lesson 3-3)** Any two numbers that have a product of 1.

inversos multiplicativos Cualquier dos números que tengan un producto de 1.

N

negative integer **(Lesson 4-1)** A number that is less than zero. It is written with a − sign.

entero negativo Número que es menor que cero y se escribe con el signo −.

net **(Lesson 9-2)** A two-dimensional figure that can be used to build a three-dimensional figure.

red Figura bidimensional que sirve para hacer una figura tridimensional.

numerical expression **(Lesson 5-2)** A combination of numbers and operations.

expresión numérica Una combinación de números y operaciones.

O

opposites **(Lesson 4-2)** Two integers are opposites if they are represented on the number line by points that are the same distance from zero, but on opposite sides of zero. The sum of two opposites is zero.

opuestos Dos enteros son opuestos si, en la recta numérica, están representados por puntos que equidistan de cero, pero en direcciones opuestas. La suma de dos opuestos es cero.

order of operations **(Lesson 5-2)** The rules that tell which operation to perform first when more than one operation is used.

1. Simplify the expressions inside grouping symbols.
2. Find the value of all powers.
3. Multiply and divide in order from left to right.
4. Add and subtract in order from left to right.

orden de las operaciones Reglas que establecen cuál operación debes realizar primero, cuando hay más de una operación involucrada.

1. Ejecuta todas las operaciones dentro de los símbolos de agrupamiento.
2. Evalúa todas las potencias.
3. Multiplica y divide en orden de izquierda a derecha.
4. Suma y resta en orden de izquierda a derecha.

ordered pair **(Lesson 1-3)** A pair of numbers used to locate a point on the coordinate plane. The ordered pair is written in the form (x-coordinate, y-coordinate).

par ordenado Par de números que se utiliza para ubicar un punto en un plano de coordenadas. Se escribe de la forma (coordenada x, coordenada y).

origin **(Lesson 1-3)** The point of intersection of the x-axis and y-axis on a coordinate plane.

origen Punto de intersección de los ejes axiales en un plano de coordenadas.

outlier **(Lesson 10-6)** A value that is much greater than or much less than the other values in a set of data.

valor atípico Dato que se encuentra muy separado de los otros valores en un conjunto de datos.

P

parallelogram **(Lesson 8-1)** A quadrilateral with opposite sides parallel and opposite sides congruent.

paralelogramo Cuadrilátero cuyos lados opuestos son paralelos y congruentes.

part-to-part ratio **(Lesson 1-1)** A ratio that compares one part of a group to another part of the same group.

proporción de parte a parte Una proporción que compara una parte de un grupo con otra parte del mismo grupo.

part-to-whole ratio **(Lesson 1-1)** A ratio that compares one part of a group to the whole group.

proporción de parte a total Una proporción que compara una parte de un grupo con todo el grupo.

peak **(Lesson 10-7)** The most frequently occurring value in a line plot.

pico El valor que ocurre con más frecuencia en un diagrama de puntos.

percent **(Lesson 2-1)** A ratio, or rate, that compares a number to 100.

por ciento Una relación, o tasa, que compara un número a 100.

positive integer **(Lesson 4-1)** A number that is greater than zero. It can be written with or without a $+$ sign.

entero positivo Número que es mayor que cero y se puede escribir con o sin el signo $+$.

powers **(Lesson 5-1)** A number expressed using an exponent. The power 3^2 is read *three to the second power,* or *three squared.*

potncias Números que se expresan usando exponentes. La potencia 3^2 se lee *tres a la segunda potencia* o *tres al cuadrado.*

prism **(Lesson 9-1)** A three-dimensional figure with at least three rectangular lateral faces and top and bottom faces parallel.

prisma Figura tridimensional que tiene por lo menos tres caras laterales rectangulares y caras paralelas superior e inferior.

pyramid (Lesson 9-4) A three-dimensional figure with at least three triangular sides that meet at a common vertex and only one base that is a polygon.

pirámide Una figura de tres dimensiones con que es en un un polígono y tres o mas caras triangulares que se encuentran en un vértice común.

Q

quadrants (Lesson 4-5) The four regions in a coordinate plane separated by the *x*-axis and *y*-axis.

cuadrantes Las cuatro regiones de un plano de coordenadas separadas por el eje *x* y el eje *y*.

quartiles (Lesson 10-4) Values that divide a data set into four equal parts.

cuartiles Valores que dividen un conjunto de datos en cuatro partes iguales.

quotient (Lesson 3-1) The result when one number is divided by another.

cociente El resultado cuando un número es dividido por otro.

R

range (Lesson 10-4) The difference between the greatest number and the least number in a set of data.

rango La diferencia entre el número mayor y el número menor en un conjunto de datos.

rate (Lesson 1-7) A special kind of ratio in which the units are different.

tasa Un tipo especial de relación en el que las unidades son diferentes.

ratio (Lesson 1-1) A comparison between two quantities, in which for every *a* units of one quantity, there are *b* units of another quantity.

razón Una comparación entre dos cantidades, en la que por cada *a* unidades de una cantidad, hay unidades *b* de otra cantidad.

ratio table (Lesson 1-2) A collection of equivalent ratios that are organized in a table.

table de razones Una colección de proporciones equivalentes que se organizan en una tabla.

rational number (Lesson 4-4) A number that can be written as a fraction.

número racional Número que se puede expresar como fracción.

reciprocals (Lesson 3-3) Any two numbers that have a product of 1. Since $\frac{5}{6} \times \frac{6}{5} = 1$, $\frac{5}{6}$ and $\frac{6}{5}$ are reciprocals.

recíproco Cualquier par de números cuyo producto es 1. Como $\frac{5}{6} \times \frac{6}{5} = 1$, $\frac{5}{6}$ y $\frac{6}{5}$ son recíprocos.

rectangular prism (Lesson 9-1) A prism that has rectangular bases.

prisma rectangular Una prisma que tiene bases rectangulares.

reflection (Lesson 4-6) The mirror image produced by flipping a figure over a line.

reflexión Transformación en la cual una figura se voltea sobre una recta. También se conoce como simetría de espejo.

regular polygon (Lesson 8-4) A polygon with all congruent sides and all congruent angles.

polígono regular Un polígono con todos los lados congruentes y todos los ángulos congruentes.

S

scaling **(Lesson 1-2)** The process of multiplying each quantity in a ratio by the same number to obtain equivalent ratios.

homotecia El proceso de multiplicar cada cantidad en una proporción por el mismo número para obtener relaciones equivalentes.

second quartile **(Lesson 10-4)** Another name for the median, or the center of a set of numerical data.

segundo cuartil Otro nombre para la mediana, o el centro de un conjunto de datos numéricos.

simplest form **(Lesson 5-4)** The status of an expression when it has no like terms and no parentheses.

forma más simple El estado de una expresión cuando no tiene términos iguales y no hay paréntesis.

slant height **(Lesson 9-4)** The height of each lateral face of a pyramid.

altura oblicua Altura de cada cara lateral de un pirámide.

solution **(Lesson 6-1)** The value of a variable that makes an equation true.

solución Valor de la variable de una ecuación que hace verdadera la ecuación.

solve **(Lesson 6-1)** To replace a variable with a value that results in a true sentence.

resolver Reemplazar una variable con un valor que resulte en un enunciado verdadero.

statistical question **(Lesson 10-1)** A question that anticipates and accounts for a variety of answers.

cuestión estadística Una pregunta que se anticipa y da cuenta de una variedad de respuestas.

statistics **(Lesson 10-1)** Collecting, organizing, and interpreting data.

estadística Recopilar, ordenar e interpretar datos.

Subtraction Property of Equality **(Lesson 6-2)** If you subtract the same number from each side of an equation, the two sides remain equal.

propiedad de sustracción de la igualdad Si sustraes el mismo número de ambos lados de una ecuación, los dos lados permanecen iguales.

surface area **(Lesson 9-2)** The sum of the areas of all the surfaces (faces) of a three-dimensional figure.

área de superficie La suma de las áreas de todas las superficies (caras) de una figura tridimensional.

survey **(Lesson 10-1)** A question or set of questions designed to collect data about a specific group of people, or population.

encuesta Pregunta o conjunto de preguntas diseñadas para recoger datos sobre un grupo específico de personas o población.

symmetric **(Lesson 10-7)** Data that are evenly distributed.

simétrica Datos que están distribuidos.

T

term **(Lesson 5-3)** Each part of an algebraic expression separated by a plus or minus sign.

término Cada parte de un expresión algebraica separada por un signo más o un signo menos.

third quartile **(Lesson 10-4)** The third quartile is the median of the data values greater than the median.

tercer cuartil El tercer cuartil es la mediana de los valores mayores que la mediana.

three-dimensional figure (Lesson 9-1) A figure with length, width, and height.

figura tridimensional Una figura que tiene largo, ancho y alto.

trapezoid (Lesson 8-3) A quadrilateral with one pair of parallel sides.

trapecio Cuadrilátero con un único par de lados paralelos.

triangular prism (Lesson 9-3) A prism that has triangular bases.

prisma triangular Prisma con bases triangulares.

U

unit price (Lesson 1-7) The cost per unit of an item.

precio unitario El costo por unidad de un artículo.

unit rate (Lesson 1-7) A rate in which the first quantity is compared to 1 unit of the second quantity.

tasa unitaria Una tasa en la que la primera cantidad se compara con 1 unidad de la segunda cantidad.

unit ratio (Lesson 1-6) A ratio in which the first quantity is compared to 1 unit of the second quantity.

razón unitaria Una relación en la que la primera cantidad se compara con 1 unidad de la segunda cantidad.

variable (Lesson 5-3) A symbol, usually a letter, used to represent a number.

variable Un símbolo, por lo general, una letra, que se usa para representar un número.

volume (Lesson 9-1) The amount of space inside a three-dimensional figure. Volume is measured in cubic units.

volumen Cantidad de espacio dentro de una figura tridimensional. El volumen se mide en unidades cúbicas.

x-axis (Lesson 1-3) The horizontal line of the two perpendicular number lines in a coordinate plane.

eje x La recta horizontal de las dos rectas numéricas perpendiculares en un plano de coordenadas.

x-coordinate (Lesson 1-3) The first number of an ordered pair. The x-coordinate corresponds to a number on the x-axis.

coordenada x El primer número de un par ordenado, el cual corresponde a un número en el eje x.

y-axis (Lesson 1-3) The vertical line of the two perpendicular number lines in a coordinate plane.

eje y La recta vertical de las dos rectas numéricas perpendiculares en un plano de coordenadas.

y-coordinate (Lesson 1-3) The second number of an ordered pair. The y-coordinate corresponds to a number on the y-axis.

coordenada y El segundo número de un par ordenado, el cual corresponde a un número en el eje y.

Index

N

O

Q

S

Lesson 1-1 Understand Ratios, Practice Pages 11–12

1. no; Sample answer: Suri's ratio is 6 : 4 and Martha's is 5 : 3. **3.** 6 cups **5.** 10 chocolate doughnuts **7.** 36 players **9.** 8 containers; Sample answer: She has 2 cups or 16 ounces of liquid starch. She will make 16 ÷ 4 or 4 batches of slime. Each batch makes 4 × 3 or 12 ounces, so she will make a total of 48 ounces of slime. If each container holds 6 ounces, she needs 48 ÷ 6 or 8 containers. **11.** 4 : 24; Sample answer: If 4 students bike to school, then 28 – 4 or 24 students do not bike to school. The ratio is 4 : 24. **13.** $\frac{3.14}{1}$ or 3.14

Lesson 1-2 Tables of Equivalent Ratios, Practice Pages 21–22

1. 30 snow cones **3.** 83 skips **5.** 98 minutes **7.** 25 pencils **9.** 20 biscuits **11.** no; Sample answer: If 5 goats and 5 chickens are added, there would be 26 goats and 40 chickens on the farm, with a goat-to-chicken ratio of 13 : 20. The ratio of goats to chickens was originally 3 : 5, which is not equivalent to 13 : 20. **13.** Sample answer: Seth's bouquet has 21 flowers with 15 roses. Keith's bouquet has 35 flowers with 25 roses. Are the ratios of roses to flowers the same? Yes, they both scale to 5 roses to 7 flowers.

Lesson 1-3 Graphs of Equivalent Ratios, Practice Pages 27–28

1. (1, 6), (2, 12), (3, 18), (4, 24); Sample answer: The points appear to be in a straight line. Each point is 1 unit to the right and 6 units up from the previous point. This confirms the relationship that for every package, there are 6 beach balls.

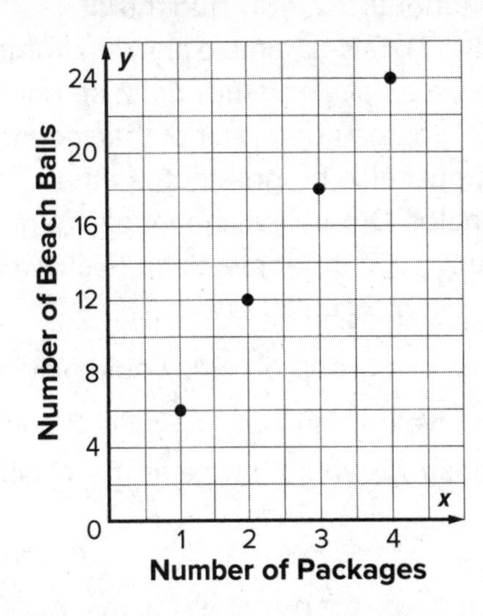

3. Sample answer: The ratio of photos to pages for Lexi's scrapbook is 4 : 1. The ratio of photos to pages for Audrey's scrapbook is 6 : 1. Audrey uses more photos per page than Lexi. **5.** dimes to dollars; Sample answer: The ratio of dimes to dollars is 10 : 1 and the ratio of quarters to dollars is 4 : 1. Since 10 is greater than 4, the ratio of dimes to dollars will have a steeper line. **7.** yes; Sample answer: A bracelet could have a length of 10.5 inches and 42 beads.

Lesson 1-4 Compare Ratio Relationships, Practice Pages 35–36

1. Brand B; Sample answer: When all three ratio relationships are graphed on the same graph, the graph for Brand B is the steepest. This means that Brand B has the greatest ratio of raisins to ounces of cereal. **3.** white bread **5.** Miguel **7.** Sample answer: Three packages of hot dogs cost $9.50. The relationship was displayed in words because it's easier and faster for people to understand while shopping.

Lesson 1-5 Solve Ratio Problems, Practice Pages 45–46

1. 640 students **3.** 15 baskets
5. 225 students **7.** 480 students
9. $1,015 **11.** false; Sample answer: For the ratios to be equivalent, they must be equivalent fractions. So, the numerator of the second fraction must also be greater than the denominator. Otherwise, the ratios are not equivalent. **13.** 8 people; Sample answer: Using equivalent ratios, $\frac{20}{140} = \frac{?}{540}$. So, 72 people in a group of 504, would play tennis. Using equivalent ratios, $\frac{1}{9} = \frac{?}{72}$. So, 8 people out of those 72 would have a tennis coach.

Lesson 1-6 Convert Customary Measurement Units, Practice Pages 55–56

1. 144 fluid ounces **3.** 12 cups **5.** $1\frac{1}{2}$ tons
7. 50 gallons **9.** 250 quarts **11.** $15.75
13. Sample answer: First, convert 20 miles to feet. There are 5,280 × 20 or 105,600 feet in 20 miles. Then convert one hour to seconds. There are 60 × 60 or 3,600 seconds in one hour. So, $\frac{105{,}600 \text{ ft}}{3{,}600 \text{ s}} \approx \frac{29.3 \text{ ft}}{1 \text{ s}}$ or about 29.3 feet per second. **15.** Sample answers: 7 cups; 56 fluid ounces

Lesson 1- 7 Understand Rates and Unit Rates, Practice Pages 63–64

1. 0.4 km per min **3.** 3 beats per second
5. 25 game tickets for $10 **7.** 6-pack of Student Tickets **9.** Party R Us; $0.25 less
11. Sample answer: 1 bagel for $0.50
13. 1 min; Sample answer: There are 60 minutes in 1 hour, so 1 mile per minute is equivalent to 60 miles per hour.

Lesson 1-8 Solve Rate Problems, Practice Pages 71–72

1. $9 **3.** $7.50 **5.** 390 donuts **7.** 36 minutes
9. yes; Sample answer: 2 hours = 120 minutes; Billie bikes at the rate of $\frac{45 \text{ min}}{9 \text{ mi}}$ or $\frac{5 \text{ min}}{1 \text{ mi}}$ and $\frac{5 \text{ min}}{1 \text{ mi}} = \frac{120 \text{ min}}{24 \text{ mi}}$. **11.** 48 mandarin oranges

Module 1 Review Pages 75–76

1. 18 **3.** B **5.** 65 miles per hour; 3 questions for each lesson **7.** no; Sample answer: Since the rates do not have the same unit rate, they are not equivalent. **9.** 25 students
11a. rate of speed downstream = 15 mph; rate of speed upstream = 10 mph; The rate of speed downstream was faster than the rate of speed upstream. **11b.** 5 miles per hour

Lesson 2-1 Understand Percents, Practice Pages 83–84

1. 60%
3.

5. 90%
7.

0% 50% 100%

9. yes; Sample answer: Each section of the model represents 20%. The 3 sections not shaded represent the percentage of students who did not vote for the tiger. So, 20% × 3 = 60% and 60% is greater than 50%. **11.** 20%; Sample answer: Each section represents 4%. Since 5 sections are shaded, 5 × 4% = 20%.
13. yes; Sample answer: To model 110%, use two bar diagrams each divided into 10 equal sections. Shade one bar diagram entirely to represent 100% and then shade the remaining 10% in the second bar diagram.

Lesson 2-2 Percents Greater Than 100% and Less Than 1%, Practice Pages 91–92

1. 136% **3.** 0.75%

5. 140%

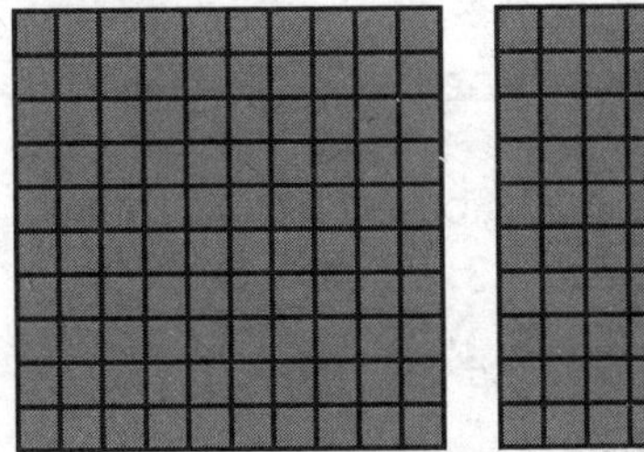

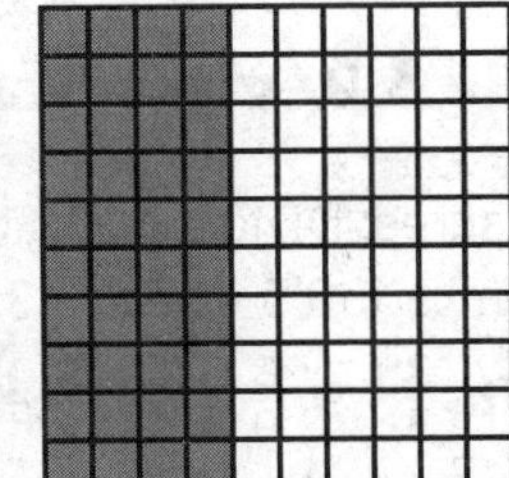

7. 0.0085 **9.** 30 mph **11.** Sample answer: The student modeled 2%, not 0.2%. To model 0.2%, only $\frac{1}{5}$ of one square should be shaded.

Lesson 2-3 Relate Fractions, Decimals, and Percents, Practice Pages 101–102

1. $\frac{9}{20}$, 0.45 **3.** $\frac{4}{5}$, 0.8 **5.** 175%, 1.75 **7.** 89%, $\frac{89}{100}$ **9.** 65%, $\frac{13}{20}$ **11.** $\frac{3}{10}$; 0.3 **13.** 0.85; $\frac{85}{100}$; $\frac{17}{20}$ **15.** $\frac{7}{10}$ **17.** no; Sample answer: 0.22 + 0.24 = 0.46 and 0.46 = 46%. Since 46% < 50%, chocolate milk and lemonade did not receive more than 50% of the votes. **19.** Sample answer: The percent will be less than 100% if the numerator is less than the denominator. The percent will equal 100% if the numerator and the denominator are equal. The percent will be greater than 100% if the numerator is greater than the denominator.

Lesson 2-4 Find the Percent of a Number, Practice Pages 111–112

1. 48 students **3.** 36 **5.** 8 **7.** 66 **9.** 0.525 **11.** 0.9 **13.** 43 students **15.** $103.08 **17.** Sample answer: 40% can be represented as 10% + 10% + 10% + 10%. 10% of 150 is 15. 15 + 15 + 15 + 15 = 60, So, 40% of 150 is 60.

Lesson 2-5 Estimate the Percent of a Number, Practice Pages 119–120

1. Sample answer: 30; 50% of 60 = 30 **3.** Sample answer: 80; 40% of 200 = 80 **5.** Sample answer: 20; 20% of 100 = 20 **7.** Sample answer: about $10; 25% of 40 = 10 **9.** Sample answer: about 225 people; 75% of 300 = 225 **11.** Sample answer: about 125 students; 25% of 500 = 125 **13.** about $55 **15.** about 14,250 people **17.** Sample answer: First, round 39% to 40% and $197 to $200. Next, find 10% of $200, which is $20. Last, multiply $20 by 4 to find 40% of 200, or $80.

Lesson 2-6 Find the Whole, Practice Pages 127–128

1. 25 members **3.** $25 **5.** 400 pictures **7.** 500 minutes **9.** 300 lunches; $1,050 **11.** no; Sample answer: A percent compares the part to the whole. In this case, the only known value is the part. To compare percents, the whole, the total number of sixth grade students and the total number of seventh grade students, must be known. **13.** Sample answer: James's soccer team won 68% of the games they played. If they won 17 games, how many did they play? 25 games

Module 2 Review Pages 129–130

1. B **3.** 110% **5.** 28%; $\frac{28}{100}$; $\frac{14}{50}$; $\frac{7}{25}$ **7.** 80 shots **9.** 27 students **11a.** 1,500 items **11b.** $16,425

Lesson 3-1 Divide Multi-Digit Whole Numbers, Practice Pages 141–142

1. 3,472 **3.** 36 **5.** 222.25 **7.** 28.125 **9.** 36.10625 **11.** 134 **13.** 24 bags **15.** 1,020 **17.** Sample answer: Check your answer by multiplying the quotient by the divisor. Compare this answer to the dividend. They should be equal.

Lesson 3-2 Compute With Multi-Digit Decimals, Practice Pages 153–154

1. 49.892 **3.** 80.027 **5.** 0.031 **7.** 2,042.125 **9.** 8.52 mi **11.** \$1.51 **13.** Sample answer: Since the decimal 0.95 is less than 1, the product of 5.5×0.95 must be less than the 5.5×1 or 5.5. **15.** Sample answer: If you add the whole numbers, the sum is 40. The sum of the decimals will be added to 40 which will make the sum greater than 40.

Lesson 3-3 Divide Whole Numbers by Fractions, Practice Pages 165–166

1. 2 **3.** $\frac{1}{8}$ **5.** $\frac{10}{7}$ **7.** 10 **9.** $11\frac{1}{5}$; Marie can make $11\frac{1}{5}$ scarves or 11 whole scarves. **11.** 27 **13.** $\frac{1}{3}$ cup **15.** yes; Sample answer: $20 \div \frac{1}{3} = \frac{20}{1} \times \frac{3}{1}$, or 60, is greater than 55, so Zach will have enough sandwich pieces. **17.** 4; Sample answer: The reciprocal of 4 is $\frac{1}{4}$, which is equal to 0.25, and $0.2 < 0.25 < 0.3$.

Lesson 3-4 Divide Fractions by Fractions, Practice Pages 275–276

1. 2 **3.** 6 **5.** $\frac{7}{8} \div \frac{1}{4} = 3\frac{1}{2}$; Chelsea can make 3 batches of icing. **7.** $2\frac{2}{5}$ **9.** 1 more bookmark **11.** yes; Sample answer: $\frac{9}{10} \div \frac{1}{3} = \frac{9}{10} \times \frac{3}{1} = \frac{27}{10} = 2\frac{7}{10}$. He only needs 2 flags. So, he has enough. **13.** Sample answer: $\frac{7}{8} \div \frac{7}{8} = \frac{7}{8} \times \frac{8}{7} = \frac{56}{56} = 1$

Lesson 3-5 Divide with Whole and Mixed Numbers, Practice Pages 285–286

1. $\frac{1}{2} \div 6 = \frac{1}{12}$; $\frac{1}{12}$ yd **3.** $\frac{7}{10}$ **5.** $\frac{7}{9}$ **7.** $2\frac{1}{4}$ **9.** 3 pairs **11.** $6\frac{1}{4}$ times greater **13.** Sample answer: A bag contains $22\frac{1}{2}$ cups of flour. A recipe for pancakes uses $1\frac{1}{4}$ cups of flour. How many batches of pancakes can be made with one bag of flour? 18 batches

15. less than; Sample answer: $\frac{9}{10} \div 3$ is divided into more parts than $\frac{9}{10} \div 2$. Since it is divided into more parts, each part represents a lesser amount. So, $\frac{9}{10} \div 2 > \frac{9}{10} \div 3$.

Module 3 Review Pages 289–290

1. 139.5 acres; Divide 8,370 by 60 to find that each farm is 139.5 acres. **3.** 0.032 **5.** D **7a.** 3 **7b.** $\frac{3}{5}$ **9.** $\frac{8}{9}$ **11a.** $\frac{3}{5} \div 6$ **11b.** $\frac{3}{5} \div 6 = \frac{3}{5} \times \frac{1}{6} = \frac{3}{30}$, or $\frac{1}{10}$ pound **13.** $2\frac{2}{3}$

Lesson 4-1 Represent Integers, Practice Pages 297–298

1. −2; The integer 0 represents no ounces gained or lost. **3.** −15; The integer 0 represents no money withdrawn or deposited. **5.** 3; The integer 0 represents average snowfall.

7.

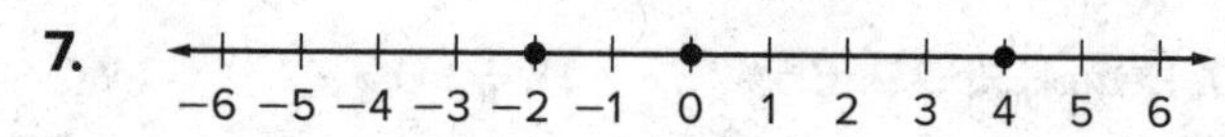

9.

11.

13.

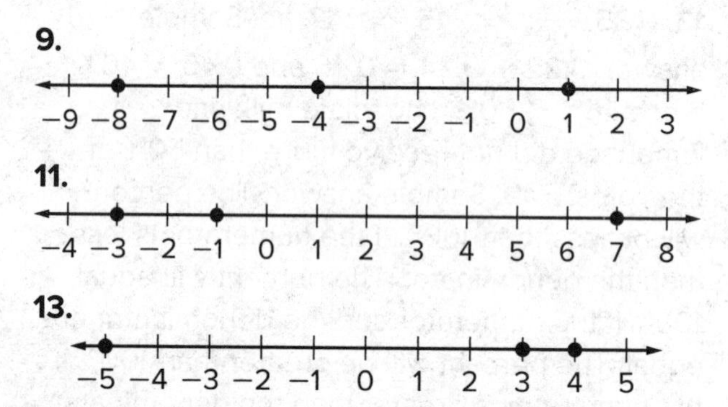

15. Beaker B; Sample answer: Beaker B is 2 units away from 0 on the number line, while Beaker A is 4 units from 0 on the number line. $4 > 2$.

−5 −4 −3 −2 −1 0 1 2 3

17. Sample answer: Use a number line. Graph 1 and −3 on a number line. Then count the units between each integer and zero. There is 1 unit from 0 and 1. There are 3 units from 0 and −3. So, 1 unit + 3 units = 4 units. **19.** Sample answer: A dog lost 6 pounds since his yearly checkup; −6.

Lesson 4-2 Opposites and Absolute Value, Practice Pages 203–204

1. 3 **3.** −6 **5.** −5; Sample answer: This is the opposite of the height of the hill. **7.** 11 **9.** −1 **11.** 100 **13.** 5 degrees **15.** Southern Moon; Sample answer: I found the absolute value of each minimum elevation and added the maximum elevation for each trail. The change in elevation for Southern Moon is 62 + 48, or 110, which is the least change of the three trails. **17.** false; Sample answer: Absolute value is a measure of distance and distance can never be negative. **19.** no; Sample answer: If x is a positive integer such as 1, then the result is −1. If x is a negative integer such as −1, then the result is 1.

Lesson 4-3 Compare and Order Integers, Practice Pages 213–214

1. $-4 < -1$; Since −4 is less than −1, John wins the game. **3.** ethane, helium, oxygen, carbon monoxide, argon, sulfur dioxide **5.** Sample answer: An elevation less than −5 feet is −10 feet. This means a distance of 10 feet from sea level, which is greater than a distance of 5 feet from sea level. **7.** Neil, Dawson, Felipe, Jesse **9.** Morocco and Argentina **11.** Sample answer: On Saturday the high temperature was −1°F. On Sunday the high temperature was −3°F; $-1 > -3$ **13.** −3, −2.5, −1, 0.66, 4, 5, 23

Lesson 4-4 Rational Numbers, Practice Pages 223–224

1.

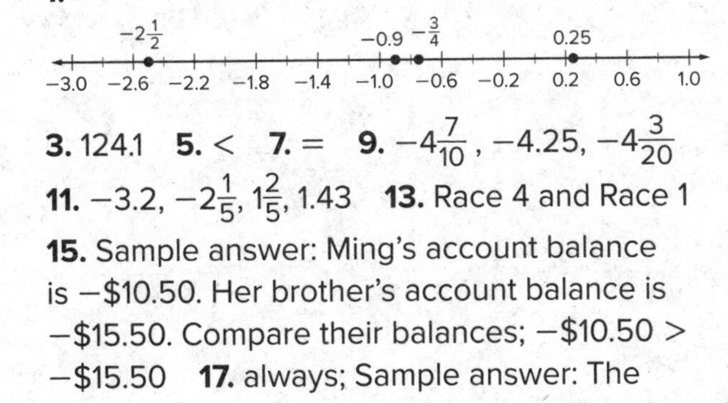

3. 124.1 **5.** $<$ **7.** $=$ **9.** $-4\frac{7}{10}$, −4.25, $-4\frac{3}{20}$ **11.** −3.2, $-2\frac{1}{5}$, $1\frac{2}{5}$, 1.43 **13.** Race 4 and Race 1 **15.** Sample answer: Ming's account balance is −\$10.50. Her brother's account balance is −\$15.50. Compare their balances; $-\$10.50 > -\$15.50$ **17.** always; Sample answer: The lesser the number, the closer it is to 0; therefore, it's opposite is also closer to 0. $x = -3$, $y = -2$

Lesson 4-5 The Coordinate Plane, Practice Pages 235–236

1. Quadrant III **3.** Quadrant I **5.** x-axis **7.** (−1.5, 1) **9.** (−1, −1.5) **11.** X

13.

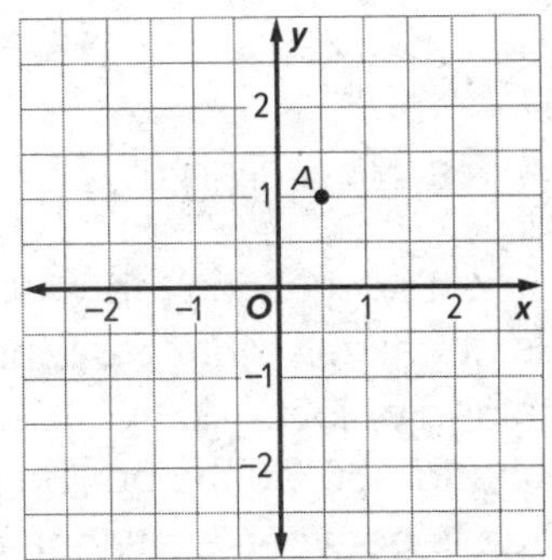

15. $\left(-\frac{1}{4}, -1\frac{11}{4}\right)$ **17.** m is a negative number; n is a positive number **19.** Sample answer: The student did not consider that b is positive, and therefore would be in either Quadrant I or II. The correct answer is Quadrant II.

Lesson 4-6 Graph Reflections of Points, Practice Pages 243–244

1. $\left(-2\frac{3}{4}, -1\right)$ **3.** $\left(-4, 2\frac{1}{2}\right)$ **5.** (−3.5, 3.5) **7.** y-axis

9.

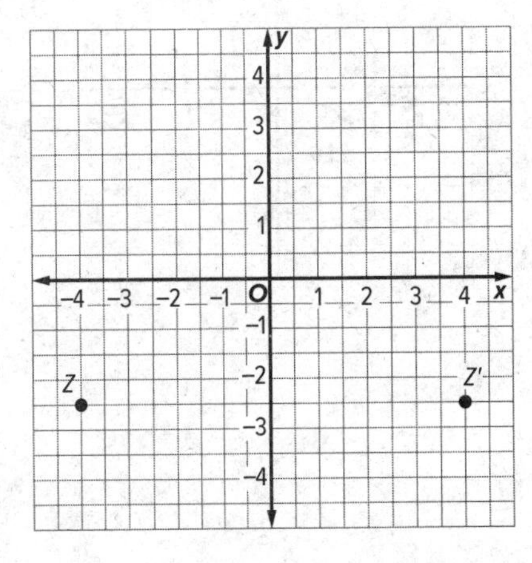

11. (4.5, −4.5) **13.** Sample answer: The student wrote the ordered pair for a reflection across the y-axis, not the x-axis. The correct ordered pair for point Y' is (1.5, 2). **15.** Sample answer: point $A(-1, -1)$; $A'(1, -1)$

Lesson 4-7 Absolute Value and Distance, Practice Pages 253–254

1. $\frac{1}{2}$ unit **3.** 3 units **5.** $1\frac{3}{4}$ units **7.** 3 units **9.** D **11.** Amber **13.** Sample answer: The student did not use the scale on the *y*-axis. The scale is 0.5 unit. So, the actual distance is 1.5 units. **15.** Sample answer: Distance cannot be negative. You have to find the absolute value of each coordinate.

Module 4 Review Pages 257–258

1a. –5; Because the situation represents a loss, the integer is negative. **1b.** Zero represents no money gained or lost. **3.** 14 **5.** –79 is between –84 and –36, so the boiling point of carbon dioxide is between acetylene and ammonia. **7.** 12°

9.

	x-axis	*y*-axis
(–4, 0)	x	
(0, 9)		x
(0, –6)		x

11a. $S''\left(-2\frac{1}{2}, 5\right)$

11b.

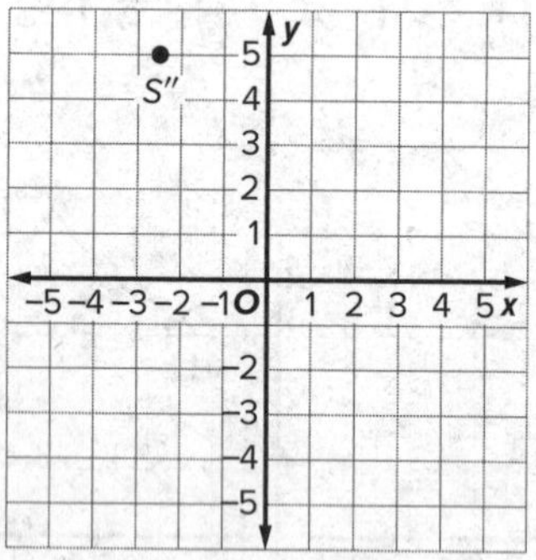